TEXT AND TECHNOLOGY

TEXT AND TECHNOLOGY

In Honour of John Sinclair

Edited by

MONA BAKER
GILL FRANCIS
ELENA TOGNINI-BONELLI

JOHN BENJAMINS PUBLISHING COMPANY
PHILADELPHIA/AMSTERDAM

1993

Library of Congress Cataloging-in-Publication Data

Text and technology : in honour of John Sinclair / edited by Mona Baker, Gill Francis, Elena Tognini-Bonelli.

p. cm.

Includes bibliographical references and index.

1. Discourse analysis. 2. Computational linguistics. 3. Linguistics--Methodology. 4. Technology. I. Baker, Mona. II. Gill, Francis. III. Tognini-Bonelli, Elena. IV. Sinclair, John (John M.)

P302.T356 1993

401'.41--dc20 93-17441

ISBN 90 272 2138 3 (Eur.)/1-55619-494-3 (US) (alk. paper) CIP

John Benjamins Publishing Co. · P.O. Box 75577 · 1070 AN Amsterdam · The Netherlands

John Benjamins North America · 821 Bethlehem Pike · Philadelphia, PA 19118 · USA

John McHardy Sinclair, born 14 June 1933
Professor of Modern English Language, University of Birmingham
Editor-in-Chief, Cobuild

Table of Contents

3. Text and Technology: Computational Tools

Foreword

John Sinclair is widely recognised as one of the most distinguished and innovative linguists of this century. Some of his achievements find expression in the research of a group of scholars who are currently working within the tradition he has established, in many cases directly under his guidance. The present volume includes contributions by several of these scholars.

There are three main themes in this collection: discourse, corpus-driven studies, and computational tools. The sections are arranged around these themes and ordered on historical and logical grounds. John's first major impact on British linguistics came as a result of his work on discourse analysis, and he later became known for his work on corpus linguistics. This predates his work on discourse but in a sense also follows naturally from it, providing, as it has, the means to explore naturally-occurring language on a much larger scale and yielding more reliable results. Corpus linguistics, in its turn, has found its own applications in other areas of language study, particularly in lexicography, and has necessitated the development of a wide range of software tools to allow more sophisticated analyses to be carried out on larger and larger corpora.

The introductory paper by Michael Stubbs, *British Traditions in Text Analysis: from Firth to Sinclair*, provides a comprehensive overview of British linguistics. It also outlines the contribution that John Sinclair has made to shaping the enduring and distinctive vision of language study which has come to be associated with the British tradition. Far from offering a mere chronology of people and influences, Stubbs provides an in-depth discussion of the principles which underlie the work carried out in British linguistics and examines the way in which Sinclair's work has "genuinely shifted the focus of the discipline".

Section 1, *Spoken and Written Discourse*, begins with a paper by Martin Warren, who explores the importance of *inexplicitness* as a feature of naturalness in conversation. Warren argues that context-dependent language is a product of a number of factors, namely the physical setting, the amount of shared knowledge between participants and the relevant linguistic context at a particular point in the discourse. Next, Hazadiah's paper introduces the notion of *topic* as a linguistic, organisational element in the routine construction of shared meaning. She proposes the Topic-Framework as a higher rank in discourse: a structural unit which organises discourse by successive acts of prospection at the level of the Exchange. Amy Tsui considers the lower ranks of discourse, namely 'move' and 'act', and investigates the relationship between acts in *multi-act moves*. Tsui offers a model for interpreting various types of moves and proposes a set of rules which allow us to identify the head act in each instance. Anna Mauranen discusses the functional role of *theme* in English and Finnish written discourse and shows that it establishes mutual relevance between consecutive sentences. The role of theme, however, is seen to vary with the prospections made by the previous sentence so that, following a strong prospection, its 'connecting' role is restricted. Susan Hunston addresses the question of *conflict* in academic discourse and explores the way in which this is presented in terms of a differential between knowledge claims. Different strategies are shown to be used for evaluating conflicting claims, thus influencing the discourse community in favour of one knowledge claim over another.

Section 2, *Corpus Studies: Theory and Practice*, begins with a paper by Gill Francis on the principles and practice of *corpus-driven grammatical description*, as it is currently being implemented at Cobuild. Francis argues that a grammar which emerges from the corpus will be essentially non-contrastive in orientation, and will be far more sensitive towards lexis and phraseology than traditional descriptive grammars. Bill Louw's paper takes up the lexical theme and explores an exciting new area of research, that of the *semantic prosody*, which he defines as "a consistent aura of meaning with which a form is imbued by its collocates". He argues that a departure in speech or writing from the expected profile of a semantic prosody may mark irony or the speaker's 'real' attitude; semantic prosodies, therefore, have a role to play in the composition and grading of suasive language and in 'radical stylistics'. Alan Partington follows this with another paper in which collocation is of central concern, this time from a perspective which involves the examination of diachronic and synchronic corpora. Partington looks at the function of *intensification* and attempts to explain its historical development in terms of delexicalisation and collocational shifts. Elena Tognini-Bonelli examines how, again through a process of delexicalisation, certain lexical items have acquired a structural role in discourse. She explores the functions of the words *actual* and *actually* and shows that they play a role in the way readers and hearers interpret stretches of discourse. Kirsten Malmkjær's paper, *Collocation, Translation, and Psycholinguistics*, represents a shift in topic to translation studies. Malmkjær addresses an important question which has been brought into sharp relief by corpus studies: what is the status of speakers' intuitions about meaning vis-à-vis corpus evidence which conflicts with those intuitions? She argues that translators' intuitions are more reliable when they are translating *out of* rather than *into* their mother tongues. Continuing with the theme of translation and corpora, Mona Baker concludes the section with a discussion of *the role of corpora in translation studies*. She suggests that the application of corpus techniques and methodology will bring about a major redefinition of the scope and goals of translation studies as an academic discipline.

Section 3, *Text and Technology: Computational Tools*, addresses the question of harnessing technology in the service of linguistic analysis. David Coniam describes a tool which parses text on the basis of the orthographic space between words: it attempts to determine what type of syntactic boundary is marked by each orthographic space it encounters. This tool forms part of a complementary suite of programs discussed in *The Automatic Analysis of Corpora* (Sinclair 1992). Jeremy Clear describes a 'collocate' program which uses two statistical methods for identifying significant collocations: Mutual Information and t-score. He shows that, by using appropriate measures of statistical significance, computational analysis can provide a satisfying and reliable account of the collocational behaviour of lexical items. Junsaku Nakamura's contribution continues the statistical theme and describes a method for classifying text types in large corpora on the basis of internal criteria. This method assumes that there is mutual dependency between types of texts and the linguistic categories used in them. Nakamura illustrates that, by using this statistical method, we can reliably account for three types of relationships: the relationships among linguistic categories, the relationships among texts, and the relationships between linguistic categories and texts. Geoff Barnbrook discusses the automatic analysis of dictionary definitions, with particular reference to parsing the explanations in Cobuild dictionaries. Barnbrook describes current work on developing a parser which identifies individual functional components in Cobuild explanations, for example superordinates and discriminators. This work has far-reaching implications for a variety of natural language processing applications, some of which are discussed and elaborated. Finally, Margaret Allan concludes the section, and the volume, with a discussion of hypermedia technology, describing the development of an audio-based hypermedia network for the study of discourse intonation by learners of English. The focus of this paper is somewhat different from the rest of the section in that it does not describe a tool for analysing language. Allan is more concerned with how the results of linguistic analysis can be made available to students of language via new technology. She argues that the new combinations of hypermedia and networking technologies could have a profound effect on the way we teach and learn languages.

This volume has been submitted to John Benjamins in camera-ready form. Graham Allport provided basic computational support and helped with the printing of the final manuscript. From John Benjamins, Mr. Cornelis Vaes gave us patient and competent guidance throughout. We are particularly grateful to Geoff Barnbrook, who spent many hours sorting out a variety of problems to do with layout and indexing. His help has been invaluable.

John Sinclair has taken it upon himself, throughout his career, to help his students and colleagues and to provide opportunities for them to realise their potential. The editors, and contributors, have all enjoyed his support at some stage in their working lives. This volume is a small token of our gratitude. It is affectionately and respectfully dedicated to him.

The Editors
March 1993

References

Sinclair, J. M. 1992. "The Automatic Analysis of Corpora." *Directions in Corpus Linguistics. Proceedings of Nobel Symposium 82, Stockholm 4-8 August 1991* ed. by J. Svartvik 379-97. Berlin and New York: Mouton de Gruyter.

British Traditions in Text Analysis
From Firth to Sinclair

Michael Stubbs
Universität Trier, Germany

The use of machines in linguistic analysis is now established. (Firth 1957:31).

It is my belief that a new understanding of the nature and structure of language will shortly be available as a result of the examination by computer of large collections of texts. (Sinclair 1991b:489).

This paper discusses a cluster of ideas which have been developed in British linguistics from the 1930s to the 1990s and which form an enduring and distinctive vision of language study. The basic philosophy has been very stable, but it has left great scope for individual initiatives, and it has recently been given a substantial new impetus by computer-assisted corpus linguistics. I will concentrate on the major contribution which John Sinclair has made to this tradition of language study. In addition, I will refer briefly to some of the historical origins of these ideas, especially in work by J. R. Firth (for example 1935, 1958), and I will point out how Firth's "contextual theory of meaning" has led to significant progress in grammatical and lexical theory and to significant grammars and dictionaries of English (for example Halliday 1985; Sinclair 1987a , 1990).

This paper, however, does not offer a chronology of people and influences. Rather, it concentrates on the principles underlying the work done in British linguistics. These principles demand studying the use of real language in written and spoken discourse, both across corpora of data, and with reference to the functions of language in social institutions. The major ideas have therefore clustered around the textual analysis of naturally occurring language, that is language as it is used without the intervention of the analyst, and also around the practical importance of language study in the real world.

Beyond the Firth-Halliday-Sinclair line of development,[1] there is also a broader British tradition of corpus linguistics, by Quirk,[2] Leech and others (see for example Quirk et al 1985, Garside et al 1987), and links to work elsewhere in Europe. In important ways however, this work differs from Sinclair's and I comment further on this below.

0. Some principles from Firth to Halliday and Sinclair

At the risk of over-simplifying the work of different linguists, I will try to sharpen the discussion by formulating as simply as possible some principles which have been central to much British linguistics, and which Sinclair has developed in detailed and concrete ways. These principles concern:

- the nature of linguistics: that it is essentially a social science and an applied science, with practical implications, especially in education;
- the nature of data for linguistics: that language should be studied in attested, authentic instances of use, not as intuitive, invented sentences; that language should be studied as whole texts, not as isolated sentences or text fragments; and that texts must be studied comparatively across text corpora;
- the essential subject of linguistics: that linguistics should study meaning; that form and meaning are inseparable; and that lexis and syntax are interdependent;
- the nature of linguistic behaviour: that language in use is a balance between routine and creation; and that language in use transmits the culture;
- the conceptual structure of the discipline: that Saussurian dualisms, especially langue/parole and syntagmatic/paradigmatic, are misconceived.

The principles themselves are as follows:

(i) Linguistics is essentially a social science and an applied science.
(ii) Language should be studied in actual, attested, authentic instances of use, not as intuitive, invented, isolated sentences.
(iii) The unit of study must be whole texts.
(iv) Texts and text types must be studied comparatively across text corpora.
(v) Linguistics is concerned with the study of meaning; form and meaning are inseparable.
(vi) There is no boundary between lexis and syntax; lexis and syntax are interdependent.
(vii) Much language use is routine.
(viii) Language in use transmits the culture.

(ix) Saussurian dualisms are misconceived.

1. Alternative traditions

Work by Halliday and Sinclair has spanned thirty years from the 1960s to the 1990s. This is precisely the period of the rise and heyday of Chomskyan linguistics, which legitimated a way of study for a whole generation of linguists. It must still be used as a point of reference, because it has formed the basic training for so many linguists and is still the source of so many assumptions in linguistics.[3]

A comparison of such Chomskyan assumptions with the neo-Firthian principles above shows how deep and systematic are the differences. Chomsky argues that linguistics is a branch of cognitive psychology (and ultimately of biology), that it can be based on intuitive data and isolated sentences, that corpus data is unrevealing, that the study of language in use is essentially uninteresting, and that linguistics can be based on a variant of the Saussurian langue/parole dualism. As an initial example, contrast these sharply opposed positions on the appropriate data for linguistics:

> *The critical problem for grammatical theory today is not a paucity of evidence but rather the inadequacy of present theories of language to account for masses of evidence that are hardly open to serious question.* (Chomsky 1965:19-20).

> *Starved of adequate data, linguistics languished - indeed it became almost totally introverted.* (Sinclair 1991a:1).

> *Like most facts of interest and importance ... information about the speaker-hearer's competence ... is neither presented for direct observation nor extractable from data by inductive procedures of any known sort.* (Chomsky 1965:18).

> *The ability to examine large text corpora in a systematic manner allows access to a quality of evidence that has not been available before.* (Sinclair 1991a:4).

The Chomskyan position has not significantly changed in over 25 years.[4] This paper is about work which provides substantial and well documented challenges to widely held assumptions about linguistic data and about form-meaning relations.

2. The practical applications of linguistics

Here then is the first of the principles mentioned in section 0 above:

Linguistics is essentially a social science and an applied science.

Strictly speaking, this formulation collapses two points into one: social scientific study need not have practical applications. However, it is easy to see why the two things have often gone together in language study.

Over the last fifty years much British linguistics has been deeply empirical, involved in practical problems of language and society, such as teaching English as a foreign language or mother tongue, speech pathology, teacher-training, and dictionary making. Moreover, British linguists have often been deeply involved in public debates and language planning around Standard English and the English curriculum in schools.[5] This is certainly true of the tradition developed by Firth, Halliday and Sinclair. They have made major descriptive and theoretical contributions to linguistics and have insisted on the integrity of linguistic methods. But they have never espoused the 'autonomous linguistics' of much recent American (and other British) work, in the sense of work divorced from all social relevance.[6] Firth describes his work as essentially sociological, and Halliday regards his work as ultimately applied and as being in particular a contribution to educational linguistics.

Sinclair has also done major work in educational linguistics: with reference to teaching English as both mother tongue and second language, and to teacher training. In the 1970s, he was influential in introducing linguistics into teacher training in Birmingham colleges and has developed innovative teaching materials. Along with colleagues at Birmingham University, especially David Lodge, he prepared material for English students, relating language and literature (see Sinclair 1971 for a general argument). He was Joint Director of the group which produced *Concept 7 to 9*: these were originally intended as classroom materials for children of Afro-Caribbean origin, but were then seen to be of more general teaching value in helping children to develop communicative and cognitive strategies (Sinclair 1973, Wight & Norris 1970). He was Academic Director of a project at the University of Malaya which prepared EFL materials based on discourse principles for people studying at university level in a foreign language (Morais 1980-81). And from work for the Cobuild dictionary (Sinclair 1987a) has come an EFL lexical syllabus (Willis & Willis 1988, Willis 1990).

In a series of battles with the English mother tongue teaching establishment, Sinclair is uncompromising about the sadly watered down and trivialised material about language which is often served to English teachers, and about the rejection of precise technical knowledge about language in their training. For example, he draws ironic analogies between the professional training of English teachers and of doctors (1982:21-22):

> Subjects like anatomy, physiology and all that are much too technical. Everyone knows medical students are not from the top of the academic pile, and they'll soon get confused by all those long Latin names and cutting people up. ... In reality, anyone knows when a chap's ill - you don't have to be a doctor to take his temperature - and if a few of us get together for a chat we'll surely work out what's wrong with him.

No-one would actually argue in this way with respect to medical students. But views of professionalism in teacher training are often very differently conceived. Sinclair argues here that it is patronising to think that teachers are incapable of an academically based training which is as demanding as that given to doctors. He has always emphasised that principled knowledge about language should underlie the teaching of language. He has made a significant contribution to the language awareness movement and is Chair of the journal *Language Awareness*.

Sinclair's work on linguistic description - of lexis, syntax and discourse - is inseparable from applied issues such as teacher education, the preparation of teaching materials and the writing of dictionaries. In the Foreword to the series *Describing English Language*, which he edits with Ron Carter, he expresses the aim of "exploit[ing] the relevance to teaching of an increased emphasis on the description of naturally occurring stretches of language" (1991a:xvii). In a different applied area, Sinclair was a member of the applied linguistics consortium responsible for designing Seaspeak (Strevens & Weeks 1985). This is the regularised sub-set of English designed for use over VHF radio in ship-to-ship and ship-to-shore maritime communications (analogous to the language of international air control). Because of its contexts of use (the vital importance of correctly understood messages, the frequently poor radio reception, its use by non-native speakers), the system demands use of a controlled vocabulary, syntax and discourse rules, and increased redundancy. Seaspeak is an important example of the planning of a restricted language (a central concept of Firth's).

3. Grammar and discourse, data and models

Sinclair is probably most widely known for his work on discourse, in particular on classroom discourse (Sinclair & Coulthard 1975, Sinclair & Brazil 1982). In fact, his work on lexis (Sinclair 1966b) and on the computer assisted analysis of a corpus of spoken language (Sinclair et al 1970) and other stylistic work (for example Sinclair 1965, 1966a, 1968) came first, and his computational work came to the fore again in work on lexis and grammar in the 1980s and 1990s.

A comparison of Sinclair's work in lexis, grammar and discourse in the 1970s with his later corpus work in the 1980s and 1990s is revealing. Some of the main progression is visible in the changing nature of the data:

1960s authentic audio-recorded spoken English as a small machine readable corpus for lexical and collocational studies; literary texts for stylistic analyses.
1970s invented sentences for a grammar of spoken English; authentic audio-recorded data, not machine readable, for discourse analysis.
1980s large machine readable corpora of authentic written and (less) spoken data (detailed references below), for various lexical and grammatical analysis.

His first grammar of English (Sinclair 1972) is based on Hallidayan scale-and-category theory, and analyses purely invented, intuitive, illustrative examples. It was significant at the time, however, as it represented a distinct shift away from the then dominant deep structural grammars, towards surface grammar, and therefore towards lexico-grammar.

The discourse model of exchange and transaction structure (Sinclair & Coulthard 1975) was constructed as the upward projection of this same Hallidayan rank-scale model, which had been designed for lower syntactic units. This was "a reasonable tactic for getting started" in a new area (Sinclair 1992a). But the model does not always fit the classroom data. (The appendix codes exchanges in ways ruled out by the descriptive system). And in a review of related work by Sinclair & Brazil (1982), I argued that the authors are unclear on the nature of the data: whether it is attested, or modified, or purely intuitive (Stubbs 1984).

In the classroom discourse studies, there is a clear commitment to authentic data in the form of transcriptions of audio-recordings made in real classrooms. This early work suffers from severe problems with the descriptive system: problems in grounding a pre-existing descriptive model in the recalcitrant data. However, this well known work on classroom discourse shows that spoken discourse is amenable to structural description, just as syntax is. The essential proposal is that exchanges in classroom discourse have a three-part structure of initiation-response-follow-up. This IRF structure can be seen intuitively in examples (Sinclair & Coulthard 1975:48) such as:

Teacher: Do you know what we mean by accent?
Pupil: It's the way you talk.
Teacher: The way you talk. This is a very broad comment.

It is evident from our experience of classrooms and also from much ethnographic educational research that teachers often ask such pseudo-questions, to which they already know the answer, and that they evaluate pupils' responses in line with their own expectations. Sinclair and Coulthard (1975) provide a detailed examination of such discourse.

One interpretative problem with this work was that it had two audiences, educational and linguistic, and was therefore read within very different intellectual contexts. Within education, it was naturally interpreted as a contribution to classroom ethnography and often criticised as being based on a narrow corpus, which represents only a narrow range of formal chalk-and-talk teaching style. The description of classroom discourse was further often interpreted by educationalists as a prescription of how teachers should or should not, depending on one's ideological position, talk to pupils. And all of this in the post-1968 debates over teacher control and pupil autonomy! From the beginning, however, this work was clearly presented by its authors as contributing to central linguistic problems. It was presented as "one stage in a continuing investigation of language functions and the organisation of linguistic units above the rank of the clause", and as an investigation of "the extent of linguistic patterning in long texts" (Sinclair & Coulthard 1975:1, 19).

A comparison of this 1970s work with more recent text analysis shows several clear developments, as the relations between data, existing linguistic models and new models are radically rethought. Rather than attempting to develop a pre-existing (rank scale) model of linguistic description by projecting it 'upwards' to larger discourse units, highly general and coherent models of prospective and retrospective organisation are outlined. From a beginning with relatively restricted data on classroom discourse, elegant hypotheses about spoken interaction in general are proposed. These more recent models relate observations about linguistic organisation at the levels of lexis, syntax and discourse. And rather than imposing such a model on the data, Sinclair looks for ways of deriving the theory from the data: the concept of 'data driven' description becomes central. One of the most frequent words in recent titles and articles by Sinclair is *evidence*. These points are developed further below.

4. Methodology: attested data and text corpora

At the centre of Sinclair's work are proposals about the empirical foundations of the discipline, which leads us to the second principle:

Language should be studied in actual, attested, authentic instances of use, not as intuitive, invented, isolated sentences.

This principle clearly contrasts the British neo-Firthian tradition with the American Chomskyan tradition, although there is other major American work, especially by C. C. Fries, H. Kučera and W. N. Francis, which made major contributions to corpus study, and other work (for example Labov 1975) which sharply attacks exclusive reliance on introspective data.

There are actually two points here. Much linguistics is based on invented sentences. In addition, often only a very small number of invented sentences are discussed. It is easy to forget or ignore how little data, either invented sentences or real texts, is actually analysed in the most influential literature in twentieth century linguistics. Saussure (1916) in the *Cours* analyses no textual data and few examples of any kind. Bloomfield (1933:22) rejects mentalism, and proposes turning to the "observation of normal speech". But the famous account of Jack, Jill and the apple, which immediately follows this proposal, and which is a key move in his argument about the analysis of meaning, is an entirely hypothetical story: he presents an invented and implausible anecdote as data. Chomsky (1957, 1965) analyses, respectively, only 28 and 24 invented sentences, some of them several times (De Beaugrande 1991:176 has counted them!). Other example sentences are quoted as either grammatical or deviant, but without analysis. Lyons (1977) in his massive two volume *Semantics* does not analyse a single authentic text or text fragment.

One of the ironies of speech act theory (influential for Sinclair & Coulthard 1975) is that it is based on "seeing what we can screw out of ordinary language" (Austin 1962:122). Yet, again, in the work by Austin and Searle which defines the field, all the data are invented. In Searle's proposals for an institutional theory of language (1969:50ff), there are no actual instances of language use in institutions.[7] The tradition continues in work by Grice and in further developments in relevance theory by Sperber and Wilson (1986). In all such work, there is the further problem with data that, traditionally, sentences were units of langue, and utterances units of parole. But utterances have here become a new genre, neither one thing, nor t'other. The analyst invents not only sentences, but utterances!

In so far as there is one single coherent thread throughout Firth's work, it is the contextual theory of meaning, and the principle that language cannot be studied as isolated sentences (Firth 1935:37):

> *... the complete meaning of a word is always contextual, and no study of meaning apart from a complete context can be taken seriously.*

Firth makes fun of the isolated sentences on which famous linguists have based their arguments: grammatically well-formed, but difficult or impossible to imagine in use, such as: 'The farmer kills the duckling'; 'Pussy is beautiful'; 'I have not seen your father's pen, but I have read the book of your uncle's gardener' (ibid:60-1). Yet even Firth, despite his sharp strictures quoted above (perhaps because of the rigour of his own demands?) analyses little text, sometimes merely presenting texts as though their features were obvious (see Firth 1964:118-34). Halliday provides detailed textual commentary and stylistic analysis of real texts, for example of Chinese-language material, of scientific and literary texts (in a famous analysis of Golding's *The Inheritors*, for example: Halliday 1971), and attested texts are used in his major grammar (Halliday 1985). But his arguments often rely nevertheless on invented sentences about aunts, dukes and teapots, or about Christopher Wren and a gazebo.

Especially with reference to data, it is important to check whether linguists do what they claim to do. For example, a major initiative in corpus linguistics, the Survey of English Usage, was begun at University College London in the 1960s. Substantial corpora of written and spoken data were collected, made available to scholars and published, in both book form and in computer readable forms, and used as evidence for two major grammars of English, most recently *A Comprehensive Grammar of the English Language* (Quirk et al 1985). However, the precise relation between the corpus data and the grammatical description is very vague. As Sinclair (1991a:100-1) points out, only occasional references are made in Quirk et al (1985) to the Survey corpus. In fact, almost all such references are used to provide frequency data: for example the frequency of different modals (ibid:136), of split infinitives in educated speech (498), and of types of noun phrase in different genres (1350-2). Only very occasionally is an example sentence explicitly taken from the corpus. The implication is that the example sentences are either invented, that is they are intuitive native speaker data, or are modified and idealised from the corpus in unspecified ways. This relation between corpus, example sentences and description is not discussed at all in the introduction to Quirk et al (1985), and the accountability to data of description and theory is therefore undefined.

I am emphasising here cases of linguists not following Principle 2. This emphasis is unfortunately necessary, because it is so easy to be blind to the very small amount of data on which contemporary linguistics is based, and this is fundamental to the whole intellectual organisation of the discipline: the theories which linguists develop, the types of corroboration they claim, the methods they use, and the ways in which students are trained. This negative set of comments does, however, usefully bring out two contrasts: between the rhetoric and the reality in much linguistics (where *data* often means invented data); and between Sinclair and other linguists, even within the relatively homogeneous British tradition of corpus linguistics.

Against a reliance on invented data, Sinclair (1991a:5) counters with the principle that examples are "not to be tampered with", and talks of "the absurd notion that invented examples can actually represent the language better than real ones". The mind plays tricks, and there is no justification for modifying data. Sinclair (1991a) provides a thorough critique of the use of intuitive data and gives precise examples of its limitations in lexical, grammatical and semantic studies. Intuitive judgements are particularly untrustworthy with respect to the frequency and distribution of different forms and meanings of words, and to the interaction of lexis, grammar and meaning. Furthermore, he points out (1987a:xv) the essential confusion between data and theory in much work:

> *... invented examples are really part of the explanations. They have no independent authority or reason for their existence, and they are constructed to refine the explanations and in many cases to clarify the explanation. ... Usage cannot be invented, it can only be recorded.*[8]

He also points out difficulties of perception due to training and warns against the imposition of pre-corpus beliefs on corpus data (1991a:29):

> *Linguistics usually operates with ... abstract categories. [... But ...] it is good policy to defer the use of them for as long as possible, to refrain from imposing analytical categories from the outside.*

The link between Sinclair's view of authentic data and his preference for surface grammar is here evident.

This discussion raises classic problems of the relation between language and text (Sinclair 1991a:102ff). The view that the abstract language system is realised in actual instances in concrete text is too simple. When they had only very limited corpus data, linguists studied actual instances, knowing that most were untypical. When the Chomskyan paradigm shift justified studying intuitive data, linguists ignored actual instances entirely, and studied invented instances, knowing that most could never occur (and were also untypical). A large corpus, searched with computer assistance, provides a way out of this dilemma. One of the main uses of a corpus is to identify what is central and typical in the language (Sinclair 1991a:17). These points raise questions of the representativeness of a corpus and as Sinclair (ibid:103) admits, it requires a theory of typicality which is yet to be fully worked out.

Authentic language occurs as language-in-text. Firth (1957:5) argues that "the actual text duly recorded" should be the main concern of the linguist. So the third and related principle is:

The unit of study must be whole texts.

The simple reason for this principle is that few linguistic features of a text are distributed evenly throughout (Sinclair 1991a:19). A clear example is that marked lexical and syntactic differences are often found between different parts of a scientific research article, say between abstract, introduction, main argument, and conclusion (Swales 1990).

Work by Halliday and Sinclair includes significant linguistic stylistic studies of literary texts, where the concern has to be with the interpretation of the whole work (for example 1965, 1996a, 1968). This principle has immediate consequences for the construction of corpora. It rejects the sampling methods necessarily used by the relatively small 'first generation' computer readable corpora (Brown, LOB and LUND), where 5,000 or 2,000 words text fragments are used. The Cobuild corpus (now the 'Bank of English' collection of corpora) therefore includes whole texts (Sinclair 1991a:19)[9], and this is in line with the fourth principle:

Texts and text types must be studied comparatively across text corpora.

The heterogeneity of language is a central assumption in Firthian linguistics (Firth 1935:67-8):

> *Unity is the last concept that should be applied to language. ... There is no such thing as 'une langue une' and there never has been.*

It is widely accepted that an important aspect of language variation is the varying frequency of lexical and grammatical features across different text types: in fact, register variation is usually defined as the systematic variation in such frequencies (for example Halliday 1991:33). Sinclair's analyses of literary style and of distinctive discourse structures in different discourse types, for example between teachers and pupils or doctors and patients, document such genre variations.

The view that language varies systematically across text types has implications for interpretation. It follows that texts are interpreted against a background of expectations: they are interpreted as much for what they omit as for what they express. As Firth (1964:183) puts it, in an uncharacteristically psychological formulation:

> *[The] patterns of language in the 'unconscious' background are just as important as the patterns of speech we make and hear.*

This view of language variation also has the methodological implication that text study must be comparative. As Sinclair (1965:76-7) points out, over 25 years ago, but when computer assisted techniques of corpus study were becoming feasible:

> *Any stretch of language has meaning only as a sample of an enormously large body of text; it represents the results of a complicated selection process, and each selection has meaning by virtue of all the other selections which might have been made, but have been rejected.*

This expresses the Hallidayan concept of the language potential from which selections are made as the potential is realised in a particular text. A currently fashionable phrasing of this point is that all texts make intertextual references. In what they include or omit, all texts make assumptions about their readers or listeners. Texts are shaped by prior texts, by repetitions or by being oriented to routines and conventions, and therefore all texts are inherently historical (Fairclough 1991).

Much work in the systemic tradition recognises that many texts are mixed: for example, there are many mixed modes between spoken and written language. In a major study, Biber (1988) shows that written and spoken genres overlap in their linguistic characteristics. With computer assistance, he studies similarities and differences across the twenty or so written and spoken genres in the 1.5 million words of the LOB and LUND corpora. He analyses some 60 lexical and syntactic features across the corpora and, using cluster analysis, studies co-occurrence patterns amongst these features. Nakamura (this volume) similarly uncovers distinctive patterns of distribution among semantic classes of verbs in different text types. Such analyses of expected patterns are beginning to document the 'unconscious background' of textual variability against which individual texts are interpreted.

There are various forces which can conspire to disguise the inherent variability of language, but the main one is an exclusive reliance on introspective data. Introspective data are consciously invented. Since they are selected to illustrate a theory, they are carefully considered and highly edited. But self-conscious editing and monitoring are characteristics of written rather than spoken language. The very fact that it is sentences which are invented biases the data towards written language, since sentences are units of written language, and only, at best, partially relevant to spoken discourse. The common Chomskyan claim to be studying language independent of its medium of transmission in speech or writing is therefore simply false.

Corpus studies show that language in use is characterised by spectacular regularities of pattern with endless variation (Sinclair 1991a:4). The aim must therefore be a theory of how the language system survives the fuzziness of the variable instances.

5. The essential content: form and meaning, lexis and grammar

At roughly the same time as Bloomfield (1933) was expressing the view that meaning was the 'weak point' of language study, Firth (1935) published a major article on 'the techniques of semantics' in which he emphasised that linguistics without meaning is sterile, thus establishing the fifth principle:

Linguistics is concerned with the study of meaning; form and meaning are inseparable.

Traditions deriving from Bloomfield and early Chomsky have always had extreme difficulties in combining rigorous distributional analysis of language forms with a theory of meaning. For example, Chomsky (1957:17) asserts that

> *grammar is autonomous and independent of meaning*

and later (ibid:102) only grudgingly admits that "the fact that correspondences between formal and semantic features exist ... cannot be ignored". Later Chomskyan positions, generative semantics (and its complete collapse), and more recent work inspired by speech act theory, take the debate in different directions, but do not solve the form-meaning problem. Contrast Chomsky's grudging admission with Sinclair's finding (1991a:6-7; 1991b:496):

> *... in all cases so far examined, each meaning can be associated with a distinct formal patterning. ... There is ultimately no distinction between form and meaning. ... [The] meaning affects the structure and this is ... the principal observation of corpus linguistics in the last decade.*

It is worth emphasising that corpus analysis was rejected in much American (Chomskyan) linguistics, on the argument that a corpus cannot give insight into meaning. Yet much British corpus linguistics has been concerned precisely with semantic analysis. The connecting thread has been Firth's notion of meaning as function in context, and more specifically, the concept of collocation, that is, syntagmatic relations between words as such, not between categories. As Firth (1957:11) puts it: "You shall know a word by the company it keeps. ... The habitual collocations [of words] are quite simply the mere word accompaniment". Collocations must be established in attested texts. They constitute one of the many levels to which meaning is 'dispersed' in Firth's concept.

These arguments are documented with detailed analyses, which are best illustrated under the sixth, related principle:

There is no boundary between lexis and syntax; lexis and syntax are interdependent.

The concept of lexico-grammar has long been proposed by Halliday. And a major claim in Sinclair's work in the 1980s and 1990s is that if a corpus is really used to construct the grammar, then the interdependence of lexis and syntax becomes obvious. But Sinclair's detailed lexico-syntactic studies take the argument considerably further than Halliday's position that "lexis is the most delicate syntax".

A central theme in neo-Firthian linguistics is the nature and importance of choices. Two related concepts can be illustrated most simply from phonology:

(i) Language is polysystemic (Firth 1968:24,43,200). Firth is here opposing the monosystemic principle of Meillet that a language is "un système où tout se tient". For example, there are different systems at different points in English phonological structure. All consonant phonemes except the velar nasal can occur word-initially; and all except /h/ can occur word-finally. The two structural slots, initial and final, permit different closed classes of paradigmatic choices.

(ii) Units are often co-selected. For example, English has both nasals and plosives at bilabial, alveolar and velar points of articulation. However, in words such as *limp, lint* and *link*, the word-final consonant pairs represent not two choices, but one. Due to assimilation to point of articulation, there are two bilabials, two alveolars or two velars, and no other combinations are possible. Firth talked here of prosodies across more than one unit.

Sinclair (1992a:14) gives a simple lexical example of co-selection of lexis and syntax. The word *lap* is more likely to occur in a prepositional phrase in adjunct position, than to occur as the subject or object of a clause. Detailed corpus-based work has now begun to show the ramifications of these concepts for grammatical description. In an elegant study using Cobuild data, Francis (1991) provides a more systematic demonstration of the phenomenon that all words have their own grammar. She takes a number of nouns from a specific frequency band of English (for example *accident, artist, darkness*). If syntax and lexis were independent systems, one would expect that a given selection of such words (lemmas) would be evenly distributed over different grammatical positions in the clause: subject, object, indirect object, adjunct, qualifier, and so on. Traditionally, grammarians have been primarily interested in syntactic structure, and lemmas have been used merely as slot fillers in such structures. The distribution of different lemmas in the same syntactic slots is, however, very uneven. For example, *context* and *darkness* are much more common in adjunct position than elsewhere, whereas, *impact* and *independence* are much more common in object position. Francis' article contains a careful discussion of the relevant distributional facts.

The explicitly pedagogical Cobuild grammar (Sinclair 1990) is a stage towards a thorough-going lexico-grammar. It associates structures with lexical items, and these are provided in the form of lists. It is clear that such lists are illustrative and incomplete, although they already provide information which is not available from introspection. Clear examples are the lists of verbs which either commonly or rarely occur in the passive (Sinclair 1990:407-8). An even simpler example (ibid:410) is the set of semantically related verbs (*adore, dislike, enjoy, hate, like, love, need, want*, etc) which occur in the structure:

all or *what* + PRONOUN + VERB + BE + NOUN

as in *What we want is Watney's*; *All you need is love*. Other papers provide analyses of various lemmas and give a large number of precise examples to further document these claims about polysystemic organisation and co-selection of lexis and syntax.

A further stage in the argument shows that different forms of a lemma pattern differently. In a paper on 'the nature of the evidence', Sinclair (1991a: 67ff) discusses the lemma *set* and in particular its uses in the phrasal verb *set in*. One of the forms, *set*, is much commoner than the others: *sets, setting*. In its verbal uses therefore, *set* is commonest in the past tense. *Set in* has a tendency to occur in end structures, and its subjects are generally unpleasant and largely abstractions, such as: *where the rot set in; disillusion had set in*. None of these points could be derived from intuition (see also Louw, this volume;

and Funden 1986).

For a small example, apparently innocent but far-reaching in its implications, Sinclair (1991b) shows the non-equivalence of singular and plural forms of nouns by documenting the different patterning of *eye* and *eyes:* "There is hardly any common environment" between the two word forms, and they "do not normally have the capacity to replace each other". The plural co-occurs with adjectives such as *blue, brown, covetous, manic.* The singular hardly ever refers to the anatomical organ, except when talking about injury or handicap. For both singular and plural, the phrasal and figurative meanings are not peripheral, but typical: singular and plural occur in different sets of fixed phrases (*all eyes will be on, rolling their eyes; turn a blind eye, keep an eye on*).

Such analyses show that the co-patterning of different forms and senses of lemmas and of syntax is much more detailed than is generally shown in grammars. New links are being proposed between lemmas, word forms and syntactic patterns. What corpus study shows is that lexis and syntax are totally interdependent. Not only different words, but different forms of a single lemma, have different grammatical distributions.

With reference to lexis, a further stage in the argument is that lexical items are selected more than one at a time. Using data from *The Times*, Sinclair (1992a) shows this co-selection and the consequent delexicalisation of what are normally thought of as lexical, content words. Examples include: *physical assault/damage/proximity*, where the meaning of the adjective duplicates a large part of the meaning of the noun; *scientific experiment/analysis/study*, where the adjective does little more than dignify the noun; and *full enquiry/account/circle*: a government is unlikely to announce a 'partial enquiry'!

Two important articles (Sinclair 1991a:81-98, Renouf & Sinclair 1991) analyse aspects of highly frequent grammatical words, which receive either a fragmented treatment or no treatment at all in traditional dictionaries or grammars. The word *of* is traditionally regarded as a preposition. But from concordance data, Sinclair (1991a) shows that *of*, and probably other very common words, are members of one-word classes. If we accept the traditional classification of preposition, this leads to the strange conclusion that the word *of* is the most common member of this class (more than double the next), yet is normally used in a structure which is not typical for the class. Prepositions normally form prepositional phrases which are used as clause adjuncts: *in the same week; behind the masks*. But by far the most common use of *of* is to elaborate a preceding noun or noun group: *the back of the van; a small bottle of brandy.*

Renouf and Sinclair (1991) discuss discontinuous pairings of grammatical words in collocations such as: *a(n) + ? + of; be + ? + to; too + ? + to.* They provide detailed statistics on the frequency of such collocations with intervening words. These collocational frameworks, as Renouf and Sinclair call them, are clearly not grammatical constituents. Nor are the triplets which they frame, for example *a couple of, a lot of, an indication of, an element of, be able to, be ready to, too easy to, too close to.* Such items are highly frequent, an integral part of the language, yet lie somewhere between word and group, and are missed both by current grammatical descriptions and also by conventional definitions of collocation.

These studies have far-reaching implications for assumptions which are traditionally made about form classes and grammatical constituents: "even major parts of speech are not as solidly founded as they might be" (Sinclair 1992a:14). Although a central topic of Chomskyan syntax has always been to motivate sentence constituents, note that the evidence for the form-classes which are assumed to be the ultimate constituents has been very thin indeed. Sinclair (1987b:114) talks of some of the main noun and adjective classifications crumbling under corpus evidence, and the Cobuild grammar (Sinclair 1990) gives many examples of the likely convergence of grammatical classes and lexical sets. The traditional 'parts of speech' system is under attack.

Central conclusions of such work on lexico-grammar, documented in detail from concordance data, are as follows:

(i) Any syntactic structure restricts the lexis that occurs in it; and conversely any lexical item can be specified in terms of the structures in which it occurs (see Francis, this volume). Note the difference from the Hallidayan position, quoted above. If lexis and syntax are co-selected, then lexis cannot merely be the "most delicate syntax".

(ii) Such restrictions are typically not absolute, but clear tendencies: grammar is inherently probabilistic.

(iii) Native speakers have no reliable intuitions about such statistical tendencies. Grammars based on intuitive data will imply more freedom of combination than is in fact possible (see below for a specific example). Grammar is 'corpus-driven' in the sense that the corpus tells us what the facts are.

(iv) Meaning is not constant across the inflected forms of a lemma. Traditional dictionaries do not show this.

(v) Every sense or meaning of a word has its own grammar: each meaning is associated with a distinct formal patterning. Form and meaning are inseparable.

(vi) Words are systematically co-selected: the normal use of language is to select more than one word at a time.
(vii) Since paradigmatic choices are not made independently of position in syntagmatic chain, the relation between paradigmatic and syntagmatic has to be rethought.
(viii) Traditional word-classes and syntactic units also have to be rethought.

A contrast between Sinclair's work and other British corpus linguistics emphasises some of these points. Consider a simple example from Quirk et al (1985:1393) on the extraposition of the clausal object in sentences such as: *You must find it exciting working here; I made it my objective to settle the matter; I owe it to you that the jury acquitted me*. Quirk et al imply by the omission of any discussion of lexis that such extraposition is possible for any verb. But corpus data shows (Francis, personal communication) that two verb lemmas *find* and *make* account for the vast majority of such structures. Such relations between lexis and syntax should find a place in the grammar.

It is easy to underestimate the radical implications of these points for many widely held assumptions of grammatical description. The principle of co-selection means that the linear (syntagmatic) nature of linguistic relations, and indeed the whole distinction between syntagmatic and paradigmatic, must be rethought. Further, if the relation between lexis and syntax is rethought, then the whole relation between dictionaries and grammars must be rethought.

Possibly the most negative effect of the introspective methodology in much recent linguistics is that 'descriptivist' became a term of criticism or simply abuse (Sampson 1980:146ff). The recording and description of languages, on the basis of cumulative documentary evidence, slowed to a trickle. Introspection was often seen not merely as supplementary evidence, but was even given an authority denied to observation, despite the fact that one linguist's introspections are private and not open to criticism by others, and that such private introspections are particularly dubious when the linguist has a vested interest in collecting evidence to support a given theory.

As a descriptivist in the positive sense, who produces not only studies of the details but also comprehensive dictionaries and grammars, Sinclair stands in the broader European tradition.

6. The nature of linguistic behaviour

A view of language study as a social science, with the consequent observation of authentic language use in texts, leads to principle seven:

Much language use is routine.

In a famous passage, Firth (1935) discusses the often routine and stereotyped forms of everyday conversation, and his proposals here are at the basis of much later work on discourse analysis (ibid:66,70-1):

> *Conversation is much more of a roughly prescribed ritual than most people think. Once someone speaks to you, you are in a relatively determined context and you are not free just to say what you please. We are born individuals. But to satisfy our needs we have to become social persons ... it is [in] the study of conversation ... that we shall find the key to a better understanding of what language really is and how it works.*

Firth's discussion provides a clear statement of the tension between individual and social, freedom and constraint, creation and routine,[10] and all this thirty years before work in speech act theory and conversational analysis by Austin, Searle, Sacks, et al.[11] Compare the quote from Firth with these statements:

> *... each turn in a discourse [is] a predictive classification of all possible next turns - and even though literally anything can happen next it is not difficult to supply a comprehensive guide to the next utterance.* (Sinclair 1975).

> *... each utterance provides a framework within which the next utterance is placed.* (Sinclair 1992b:83).

There are several clear developments in Sinclair's work on discourse between the 1970s and the 1990s:
(i) from a description of one rather narrow discourse type (teacher-dominated classroom discourse) to hypotheses about spoken and written text in general;
(ii) from a rather elaborate rank-scale model of exchange structure to simple and elegant hypotheses about linear prospective and retrospective text structure;
(iii) and from work on spoken interaction to simple and powerful hypotheses about the different organisation of spoken and written language.

At the centre of these general hypotheses is the concept of prospection in spoken interaction. Sinclair (1992b) regards the creation and maintenance of prospections as the defining criterion of exchanges. He considers whether a two-part adjacency pair such as 'question-answer' is the basic exchange structure, or whether the three-part IRF structure, identified for classroom discourse, is basic. He concludes that the three-part structure, as in

I Did you wake up late today?
R Yeah, pretty late.
F Oh dear.

is always the option, and that in many types of discourse it is virtually obligatory.

Moves in spoken interaction prospect ahead: speakers cannot remember exactly what has been said, so cannot refer back with accuracy. In contrast, written text is still present, readers can look back at it, and written language may therefore be retrospectively structured. Sinclair (1992a:10-11) proposes a that:

> *each sentence ... contains a single act of reference which encapsulates the whole of the previous text. ... [The] previous states of the text ... are present in the current sentence in so far as they are needed. ... In many cases [written text] also prospects forward.*

This is a very simple and powerful hypothesis, which is intuitively understandable, but which requires to be documented in detail from corpus data.

Sinclair (1980) and Sinclair and Brazil (1982) also demonstrate some of the relations between the interpretations of initiations and their grammatical structure. They show that the following syntactic forms are progressively less ambiguous in their functional interpretation: declarative, interrogative (positive polar), interrogative (negative polar), interrogative (tag), wh-interrogative, imperative. Those earlier on the list can be interpreted as those later, but not vice-versa. For example, it is normal to interpret a declarative as a command, but impossible to interpret an imperative as a statement. To take a classic Austin example, a declarative sentence such as *There is a bull in the field* is notoriously ambiguous. It might be used as a statement, warning, or request for clarification: indefinitely many interpretations are possible. But sentences in interrogative form, *Could you put the bull in the field*, or imperative form, *Put the bull in the field*, are much less ambiguous, and perhaps not ambiguous at all.

Superficially, the studies on discourse and on lexis seem far apart, but many of the same principles hold in both areas. In particular, the concept of fixed routines and the concept of linear syntagmatic structure and prospection hold for both discourse and lexis. Thus Sinclair (1991a) gives a central place to fixed expressions, clichés, recurring collocations, idioms, and so on (see especially Sinclair 1991a:109ff on the idiom principle and open-choice principle in lexis).

One of Firth's major insights is that a balance is required between variation and routine, individual and social. Hence principle eight:

Language in use transmits the culture.

Firth also points out clearly the importance of studying the transmission of the culture (though he does not quite make the connection clear between everyday routine and cultural transmission). For his "contextual and sociological technique" of semantics, he proposed (1935:40) "research into the distribution of sociologically important words"; one example concerns words such as *work* and *leisure*. Firth, as was often the case, was here making a programmatic proposal for which he never carried out the actual textual analysis. But, to be fair, he was proposing a study of social change and ideology and demanding techniques of data collection and analysis, which became feasible only some fifty years later with computational methods and the access to large corpora.

This is a theme developed in detail by Halliday (for example 1978:111) in his discussion of semantic habits and coding orientations. It is not developed so explicitly in Sinclair's own work. Sinclair's study of classroom discourse models some of the sociolinguistic knowledge that children learn as they become socialised into patterns of classroom behaviour, and this theme is developed in the work of others. Firth's position is clearly sociological, largely because of his pessimism about our ability to study private psychological states (he unfortunately creates another dualism here). Halliday relates his work clearly to sociological theory, especially Bernstein's. However, despite its origins in pedagogical concerns, Sinclair's work is strangely detached from any psychological or social theory, and his corpus work is formulated with no discussion of how the linguistic analysis could support "a linguistically centred social analysis" (Firth 1968:177). The evidence from "corpus, concordance and collocation" (Sinclair 1991a) is not used to explain "class, codes and control" (Bernstein 1990).

It is, however, not difficult to see how Sinclair's work could be developed in such directions. For example, the detailed observations about the co-selection of lexical items are ripe for input into psycholinguistic theory on language learning. Another line of development is demonstrated in detail in several publications by Alexander (for example 1992), on the basis of comparative analyses of English and German lexis. Alexander sees lexicalisation as a way of fixing, and hence transmitting, experience. The study of fixed expressions, idioms, clichés and recurrent wordings can therefore be given a cultural interpretation through a study of how the culture is expressed in lexical patterns. Keywords, in the sense of fixed expressions, are both linguistic and cultural units, where the interplay of language and ideology can be studied. Baker (1992), with explicit reference to the Cobuild project, demonstrates how work on fixed expressions is relevant to translation theory. Observations about the regular distributions of syntactic choices in text are similarly ripe for integration into textual analyses of language and ideology (see Stubbs & Gerbig 1993).

8. Beyond Saussurian dualisms

A general interpretation of the British work from Firth to Sinclair is that it is trying to find a position which overcomes Saussurian dualisms, a position which is expressed in principle nine:

Saussurian dualisms are misconceived.

Firth (1957:2n) denounces such dualisms as "a quite unnecessary nuisance" and explicitly describes his position as monistic (1935:53, 1957:2). Similarly, both Halliday (1978:38) and Sinclair (1991a:103) reject the competence-performance distinction, and I have already noted above the Firthian rewriting of the syntagmatic-paradigmatic opposition.

There are many aspects to the Saussurian paradox (Labov 1972:185ff). In the autonomous linguistics tradition from Saussure to Chomsky, langue/competence is conceived as systematic and as the only true object of study, although it is abstract and therefore unobservable. Parole/performance is considered unsystematic and idiosyncratic, and therefore, at best, of only peripheral interest. But in addition, parole is, by definition, observable only in passing fragments: as a whole, it is also unobservable. Mainstream twentieth century linguistics has therefore defined itself with reference to a dualism, both halves of which are unobservable. Much of the significance of computer-assisted study of large corpora lies in possibilities for breaking out of this impasse, since concordancers and other software allow millions of words of data to be searched for patterns which could not be observed unaided.

On the relation between the instance and the system, Halliday's (1991, 1992) metaphor of the weather and the climate is particularly helpful. The weather and the climate are the same phenomenon but regarded from different time depths. If we are thinking of the next few hours, then we are thinking of the weather: and this perspective determines what kinds of actions we might take, for instance going to the beach or taking an umbrella. If we are thinking of the next decade or the next century, then we are thinking of the climate: and this perspective also determines what kinds of actions we might take, for example legislating against industrial processes which are destroying the ozone layer. If the climate changes, then obviously the weather changes. But conversely, each day's weather affects the climate, however infinitesimally, either maintaining the status quo or helping to tip the balance towards climatic change. Instance and system, probability and categoriality, micro and macro, are two sides of the same coin, relative to the observer's position. Halliday is fond of this way of thinking. Thus syntax and lexis are not two phenomena, but one: the same thing seen by different observers. And every instance in a text perturbs the overall probabilities of the system, if only to an infinitesimal extent: the system is therefore inherently probabilistic (Halliday 1992).

It was Saussure who said that it is the viewpoint which creates the object of study for linguistics. Halliday also points out that it is the position of the observer which creates the illusion of different objects of study: thus abolishing the dualism which Saussure's viewpoint created. Sinclair's work shows in a precise and concrete way how a large corpus and an associated technology create a viewpoint which can lead to innovations in linguistic description and theory.[12]

8. Looking up

Looking Up is the punning title of a recent book by Sinclair (1987b): we look things up in dictionaries; and linguistics is looking up - thanks to new methods.

The essential vision underlying corpus linguistics is that computer-assisted analysis of language gives access to data which were previously unobservable but which can now profoundly change our understanding of language :

> *Analysis of extended naturally-occurring texts, spoken and written, and, in particular, computer processing of texts have revealed quite unsuspected patterns of language ... The big difference has been the availability of data ... [The] major novelty was the recording of completely new evidence about how the language is used. ... [The] contrast exposed between the impressions of language detail noted by people, and the evidence compiled objectively from texts is huge and systematic ... The language looks rather different when you look at a lot of it at once.* (Sinclair 1991a: xvii, 1, 2, 4, 100).

In this vision of the subject, a corpus is not merely a tool of linguistic analysis, but an important concept in linguistic theory.

This paper has emphasised, as Sinclair does, the relation between data, description and theory in linguistics. Labov (1975) has analysed the problems with intuitive data in recent linguistics, and has pointed out that every major paradigm shift in linguistics has been led by a shift in the concept of the basic data for the subject. Examples are: differences between words across languages and over time as data for historical comparative linguistics; synchronic rather than diachronic data for Saussurian linguistics; unwritten Amerindian languages for Bloomfieldian structuralism; the native speaker's intuitions for Chomskyan linguistics; and phonological variables across a speech community for Labovian sociolinguistics. The implication of Sinclair's work is that the data now available from computer readable corpora will lead to major changes in linguistic theory. This is one sense of the concept of 'data-driven linguistics'.

Sinclair has not (yet?) produced a comprehensive statement of his theoretical position, and arguably, he is interested less in theory for its own sake, and more in what practical consequences a theory has. However, his book (1991a) expresses some of the vision, based on textual evidence of a kind not previously accessible: a vision of "the emergence of a new view of language, and the technology associated with it" (ibid:1).

Sinclair's work provides a new way of thinking about some of the classic and recalcitrant questions that linguistics finds new ways of formulating, but never solves, concerning the relations between data and theory, examples and generalisations, rules and tendencies, knowledge and use, and between linguistics and other disciplines. In this concluding section I will try to identify some of these broader implications of Sinclair's work and some points for development.

8.1 Data-driven linguistics

Sinclair's discussions of the relation between evidence and theory strike at the heart of many Popperian assumptions in recent linguistics. He explicitly tackles questions of data and theory with reference to the inadequate speculative evidence (a contradiction in terms!) on which much linguistics is based. See for example his comments on "the cult of the counter-example" (1991a:99). The rhetoric is confident, with talk of a move towards "scientific rigour" (1992a:12). But the concept of corpus-driven linguistics must confront the classic problem that there is no such thing as pure induction. As Firth frequently warns, in this case quoting Goethe:

> *The highest truth would be to grasp that everything factual is already theoretical. ... There are no brute facts. ... The notion of a mere fact is the product of the abstractive method.* (Firth 1957:1, 29).

Sinclair (1992a) admits that the linguist always approaches data with hypotheses and hunches, however vague. And given his insistence on wholly automatic computational analysis (1992c), it would be useful, 30 years on, and with technological possibilities unforeseen by Chomsky, to have an explicit statement on Chomsky's original criticisms of corpus work.

8.2 Probabilistic grammar

The claim is that computer-assisted work will revolutionise not only dictionaries but grammars. At the time of writing (1993), a new lexico-grammar of English is being prepared in Birmingham. The test of many of the descriptive proposals will be whether this grammar leads to lasting revisions in the description of English and in the theory of (lexico)grammar.

One question is whether this grammar will be a full-blown probabilistic grammar. Halliday (1991, 1992; Halliday & James 1993) makes precise, but limited, proposals on the form of probabilistic choices in a systemic grammar. But the production of a comprehensive probabilistic grammar is uncharted territory.

Corpus work depends on the interpretation of frequency and distributional data. Statistics provides ways of summarising complex numerical data, so that inferences can be drawn from them, showing which variation is likely to be significant or random, and which data sets are homogeneous or heterogeneous. As Sinclair (1991a:3) admits, "the numerical and statistical side has hardly begun". And it may well be that the new kinds of language data now available will lead to the development of new statistical parameters, which might look very different indeed from the parameters known from the statistical tradition of Zipf (1935).

8.3 Descriptive and pedagogical grammars

The connections between an innovative descriptive and a pedagogical grammar - another classic problem - also require to be spelled out. Different descriptions for different purposes would be a Hallidayan formulation (though this formulation explains little).

The link between commerce and academia is also important, and a topic of wide significance in contemporary academic life. Without substantial commercial funding, the work would never have taken place. Organisational and intellectual questions are not independent. Dr Johnson notwithstanding, writing dictionaries and grammars, and studying hundreds of millions of words, is not a one-person affair. As Sampson (1987a&b) points out, people who invent their own data can work alone, but people who work with probabilistic models of corpora tend to work in groups. However, commercial links then place pressures on the work. In the production of comprehensive dictionaries and grammars, the nub of the matter is the relation between descriptive statement and normative authority. Such work is, after all, of considerable political significance in a country obsessed with teaching grammar and Standard English in schools.

8.4 Linguistic variability

In the work on lexis and grammar, long texts are used as evidence of the language as a whole. The theory emphasises variability, but the necessary fiction of 'the English language' has to be maintained in preparing dictionaries and grammars. Differences within individual long texts and across text types are therefore averaged away. There is large scope both for studies of genre variation (Biber 1988; Nakamura in this volume), and also for analyses of the internal organisation of whole texts (Phillips 1989).

8.5 The public nature of science

An important implication of technology is that even desk-top PCs can give access to linguistic patterns in significant corpora (up to a few million words). Some projects need large resources of machines and people, but individual researchers with modest technology can now check claims on real data. There is no longer any excuse for exclusive reliance on intuitions - the ultimately private data, and the ultimate protection against counter-example and alternative theory.

8.6 Linguistic facts

Then there are questions about the subject matter of linguistics itself. Linguists who work on intuitive sentence data and linguists who work on attested corpus data are regularly either puzzled or dismissive of each others' work. Are such groups studying the same subject from different points of view? Or are they simply studying different phenomena (as Fasold 1992 argues)? One group is studying isolated sentences as abstract objects about which speakers have intuitions (about entailment, ambiguity and so on). The other group is studying utterances in context, events in time, which can be recorded, and which are amenable to distributional and statistical description. It is well known that sentences and utterances do not map easily on to each other, and it is arguable that language in use consists of a mixture of sentence-like and non-sentential elements. The position is confused by sentence grammars which are based in unspecified ways on corpus data (Quirk et al 1985).

8.7 The problem of order

The crux of the matter is that linguistics is part of the attempt to find explanations for human behaviour and to formulate a theory of social action. A cognitive, intentionalist theory captures some aspects of language behaviour: language is individual, intentional and creative. But this is only half the picture. Language is also social, partly unintentional and routine, based on social conventions which are not open to introspection. There are many linguistic patterns - revealed in collocations, and in co-selections of lexis and grammar - which are due to habit. Intentional behaviour has many unintended consequences (Giddens 1984). Adapting Sinclair (1965), quoted above:

> *Any instance of intentional, individual language behaviour has meaning only against the background of an enormously large body of conventional, routine social action.*

This is the classic problem of how order is possible in the social world, and our theories must be balanced between the individual and the social.

It is a measure of the importance of Sinclair's work that it provides many new ways of thinking about these recurrent and recalcitrant problems at the basis of the discipline.

9. Descriptions and theories, methods and applications

John Sinclair is one of the few linguists who have genuinely shifted the focus of the discipline. He has made contributions to educational linguistics, to stylistics, to discourse analysis, to lexicography, and to syntax and semantics. He denies the suggestion that there is a "Birmingham School ... in the sense of a group of scholars working in a co-ordinated manner, increasing the dimensions of a shared position" (1992b:83). He has, however, been a major organisational force in British linguistics. For over twenty-five years he has held a Chair in Modern English Language at the University of Birmingham, where he has always attracted large numbers of research students and other co-workers. Much of the work has been possible due to his obtaining funding for research projects which could only be completed by many people working together,[13] and there is no doubt that he has had a profound influence on many people who have worked with him. His work has resulted in major developments in:

(i) applied linguistics: especially in educational linguistics;
(ii) linguistic methodology: especially in the analysis of large corpora;
(iii) linguistic description: in major dictionaries and grammars of English;
(iv) linguistic theory: not only in the extension of linguistic concepts from syntax to lexis and discourse, but also in work which has clear implications for theories of grammar and meaning, and therefore for core mainstream linguistics.

It is perhaps particularly important to emphasise Sinclair's contribution to linguistic description. He has, quite simply, discovered a large number of new facts about English: not something which can be said of many linguists. As Firth (1957:32) said, in his blunt Yorkshire way: "The business of linguistics is to describe languages". The detailed descriptive work now being done in Birmingham will bring far-ranging revisions to many received ideas in syntax and semantics.

Any one of the above contributions - intellectual and organisational - would give John Sinclair a firm place in contemporary linguistics. Together, they make him one of its major figures.

Author's Address:
Universität Trier, FBII-Anglistik, Postfach 3825, D-5500 Trier, Germany.

Notes

1. Firth (1890-1960) held the first Chair of General Linguistics at a British university, from 1944 at the School of Oriental and African Studies in London. Halliday (1925-) was a pupil of Firth's. From 1965 Halliday held a Chair of General Linguistics at University College London, and from 1976 a Chair of Linguistics at the University of Sydney. Sinclair (1933-) taught at Edinburgh, where he worked with Halliday, and since 1965 has held the Chair of Modern English Language at the University of Birmingham. Major axes are the Universities of Edinburgh, London, Birmingham and Sydney. Firth did not leave a tidy and consistent body of writings, and the relation of his writings to more recent work is not always clear. But the third generation of linguists is now working out some of the implications of his ideas. Robins' obituary of Firth (1961) is a useful brief account, and De Beaugrande (1991) has useful chapters on Firth and Halliday.
2. But see section 4 for a critique of the way in which Quirk uses the corpus.
3. I first got to know John Sinclair in 1973-74, when I was employed on his SSRC-funded

project at the University of Birmingham, the purpose of which was to analyse spoken discourse in trade union negotiations and doctor-patient consultations. This was the follow-up to the classroom discourse project (Sinclair & Coulthard 1975). As an undergraduate, I had been trained in Chomskyan linguistics at Cambridge, but then spent three years as a postgraduate in Edinburgh, where people seemed tolerantly amused that I hadn't read any Halliday (and where I floated indecisively between the departments of linguistics, education and computing). Halliday was later my PhD external examiner: he spent most of the viva trying patiently to explain to me why the competence - performance distinction, which I had taken for granted in my thesis, was not necessary. At the time, perhaps inevitably in a PhD viva, I did not grasp the point at all. Perhaps I worked with John at an early enough age to have some hope of getting free of an early Chomskyan training.

4. There have of course been changes of direction within Chomskyan linguistics. Major events include the rise (and collapse) of generative semantics, and the development of work in universal grammar. But in the underlying approach to language, very little has changed. See for example the introduction to a recent major textbook on government and binding theory (Haegeman 1991): the assumptions about the data and the explanatory tasks for linguistics have not changed at all from Chomsky 1965.

5. Sinclair submitted important evidence to the Bullock Report (1975) on English teaching. And at the time of writing (1993), the Kingman and Cox Reports and their effects on English teaching in British schools are being widely debated in the press. Several British linguists, notably Quirk, are involved in the public debate over Standard English. Stubbs (1989) gives an account of these reports and of the political and educational debate.

6. Bloomfield (1933) makes irritated references to the ways in which linguistics could help in education, both in mother tongue and foreign language teaching, if only educationalists were willing to listen. He also helped to prepare language teaching materials for the American military. But the line of thought seems to have gone from theory to practice: behaviourism is the way to view things, therefore one should use drills and repetition in language teaching. In famous statements, Chomsky actually denies the educational relevance of linguistic ideas.

7. Stubbs (1992) discusses an alternative concept of 'institutional linguistics' which might be developed from proposals by Firth.

8. Mey & Talbot (1988), in their highly critical review of Sperber & Wilson (1986), make the same point as Sinclair (1987a:15) that in relevance theory, "examples are not ... chosen to vouchsafe for the outside reality ... but to make a point of theory".

9. The Brown corpus, established at Brown University, consists of one million words of written American English, published in 1961, and sampled as text fragments of 2,000 words each. The LOB (Lancaster-Oslo-Bergen) corpus is the British equivalent. The London-Lund corpus consists of about 500,000 words of spoken British English. The Bank of English is the title of the current archive of texts held at Cobuild, Birmingham. This consists of several corpora totalling over 200 million words of running text.

10. Again a contrast is inevitable with the Chomskyan position, where the emphasis is on creativity, and where routine can be conceived only negatively as mere habit formation (with behaviourist connotations). Firth sees the need to achieve a balance, in theories of human behaviour, between freedom and constraint.

11. I cannot in this paper discuss details of people and influences, but a study remains to be

written of the circulation of ideas on 'real language', speech acts, conversation, etc. As one specific example: Sinclair (1992b:80) regards work by C. C. Fries (1952) as a "forgotten landmark" which made "major advances in description"; and Sacks (1992:95, 189) regards Fries' work as "extremely close" to what he was trying to do, "the only grammar of English which [is] an attempt to handle actual conversation itself". Sinclair makes frequent references to Sacks' work on adjacency pairs. A related study would examine changing concepts of evidence and of 'linguistic facts': See Labov 1975.

12. Stubbs (1992) discusses in more detail such work which attempts to overcome Saussurian dualisms, especially with reference to J. R. Firth, T. Hill and work by the British sociologist A. Giddens on agency and structure.
13. Sinclair is currently involved in establishing a policy for corpus-building in Europe (the Network of European Reference Corpora), together with a Council of Europe project on bilingual lexicography and a Eurotra project on semantic analysis.

Acknowledgements

For much helpful criticism on drafts of this article I am very grateful to Joanna Channell, Gabi Keck and Andrea Gerbig, and to the editors (who also provided much other material). Richard Alexander has discussed many of the issues in this article with me. I am of course responsible for cases where I declined this good advice and went my own interpretative way.

References

Note: This list contains only a selection of John Sinclair's publications.

Alexander, R. 1992. "Fixed Expressions, Phraseology and Language Teaching: a Sociosemiotic Perspective". *Zeitschrift für Anglistik und Amerikanistik* 40, 3. 237-49.

Austin, J. L . 1962. *How to Do Things with Words.* Oxford: Clarendon Press.

Baker, M. 1992. *In Other Words.* London: Routledge.

Bernstein, B. B. 1990. *Class, Codes and Control.* London: Routledge.

Biber, D. 1988. *Variation across Speech and Writing.* Cambridge: Cambridge University Press.

Bloomfield, L. 1933. *Language.* Chicago: Chicago University Press.

Chomsky, N. 1957. *Syntactic Structures.* The Hague: Mouton.

________. 1965. *Aspects of the Theory of Syntax.* Cambridge. Mass: MIT Press.

De Beaugrande, R. 1991. *Linguistic Theory: the Discourse of Fundamental Works.* London: Longman.

Fairclough, N. 1991. *Discourse and Social Change.* Cambridge: Polity.

Fasold, R. 1992. "Linguistics and Grammatics". *Thirty Years of Linguistic Evolution* ed. by M. Pütz , 161-76. Amsterdam: Benjamins.

Firth, J. R. 1935. "The Technique of Semantics". *Transactions of the Philological Society.* 36-72.

________. 1957a. "A Synopsis of Linguistic Theory, 1930-1955". *Studies in Linguistic Analysis* (Special Volume of the Philological Society). 1-32.

_________. 1957b. *Papers in Linguistics.* London: Oxford University Press.

_________. 1964. *Tongues of Men and Speech.* ed. by P. Strevens. London: Oxford University Press. [Reprint of 1937, 1930.]

_________. 1968. *Selected Papers of J R Firth 1952-1959.* ed. by F. R. Palmer. London: Longman.

Francis, G. 1991. "Nominal Groups and Clause Structure". *Word* 42,2. 145-56.

Fries, C. C. 1952. *The Structure of English.* New York: Harcourt, Brace & Co.

Funden, E. R. 1986. "Text in Technologie, Technologie in Text". *Wissenschaftliche und Technische Zeitschrift* 14, 6. 19-33.

Giddens, A. 1984. *The Constitution of Society.* Oxford: Polity.

Haegeman, L. 1991. *Introduction to Government and Binding Theory.* Oxford: Blackwell.

Halliday, M. A. K. 1971. "Linguistic Function and Literary Style". *Literary Style* ed. by S. Chatman, 330-68. London: Oxford University Press.

_______________. 1978. *Language as Social Semiotic.* London: Arnold.

_______________. 1985. *An Introduction to Functional Grammar.* London: Arnold.

_______________. 1991. "Corpus Studies and Probabilistic Grammar". *English Corpus Linguistics* ed. by K. Aijmer & B. Altenberg, 30-43. London: Longman.

_______________. 1992. "Language as System and Language as Instance: the Corpus as a Theoretical Construct". *Directions in Corpus Linguistics* ed. by J. Svartvik, 61-77. Berlin: Mouton de Gruyter.

_______________ & Z. James. 1993. "A Quantitative Study of Polarity and Primary Tense in the English Finite Clause". *Techniques of Description* ed. by J. M. Sinclair, M. Hoey & G. Fox, 32-66. London & New York: Routledge.

Hill, T. 1958. "Institutional Linguistics". *Orbis* 7. 441-55.

Labov, W. 1972. *Sociolinguistic Patterns.* Philadelphia: University of Pennsylvania Press.

_________. 1975. *What is a Linguistic Fact.* Lisse: Peter de Ridder.

Lyons, J. 1977. *Semantics.* (2 volumes). London: Cambridge University Press.

Mey, J. & Talbot, M. 1988. "Computation and the Soul". *Journal of Pragmatics* 12. 743-89.

Morais, E., ed. 1980-81. *Skills for Learning.* (4 volumes). Walton-on-Thames: Nelson/ University of Malaya Press.

Phillips, M. 1989. *Lexical Structure of Text.* Discourse Analysis Monograph 12. University of Birmingham: English Language Research.

Quirk, R., S. Greenbaum, G. Leech & J. Svartvik. 1985. *A Comprehensive Grammar of the English Language.* London: Longman.

Renouf, A. & J. McH. Sinclair. 1991. "Collocational Frameworks in English". *English Corpus Linguistics* ed. by K. Aijmer & B. Altenberg, 128-44. London: Longman.

Robins, R. H. 1961. "John Rupert Firth". *Language* 37, 2. 191-200.

Sacks, H. 1992. *Lectures on Conversation.* (Volume 1). ed. by G. Jefferson. Oxford: Blackwell.

Sampson, G. 1980. *Schools of Linguistics.* London: Hutchinson.

_________. 1987a. "Probabilistic Models of Analysis". *The Computational Analysis of English* ed. by R. Garside, G. Leech & G. Sampson, 16-29. London: Longman.

_________. 1987b. "The Grammatical Database and Parsing Scheme". *The Computational Analysis of English* ed. by R. Garside, G. Leech & G. Sampson, 82-96. London: Longman.

Saussure, F. de. 1916. *Cours de Linguistique Générale*. ed. by C. Bally et al. Lausanne: Payot.

Searle, J. 1969. *Speech Acts*. London: Cambridge University Press.

Sinclair, J. McH. 1965. "When is a Poem like a Sunset?" *A Review of English Literature* 6, 2. 76-91.

___________. 1966a. "Taking a Poem to Pieces". *Essays on Style and Language* ed. by R. Fowler, 68-81. London: Routledge & Kegan Paul.

___________. 1966b. "Beginning the Study of Lexis". *In Memory of J. R. Firth* ed. by C. E. Bazell et al, 410-30. London: Longman.

___________. 1968. "A Technique of Stylistic Description". *Language and Style* 1. 215-42.

___________. 1971. "The Integration of Language and Literature in the Curriculum". *Educational Review* 23, 3. 220-34.

___________. 1972. *A Course in Spoken English: Grammar*. London: Oxford University Press.

___________. 1973. "English for Effect". *Commonwealth Education Liaison Committee Newsletter* 3, 11. 5-7.

___________. 1975. "Discourse in Relation to Language Structure and Semiotics". Paper read to Burg Wartenstein Symposium.

___________. 1980. "Discourse in Relation to Language Structure and Semiotics". *Studies in English Linguistics for Randolph Quirk* ed. by S. Greenbaum, G. Leech & J. Svatrvik,110-24. London: Longman.

___________. 1982. "Linguistics and the Teacher". *Linguistics and the Teacher* ed. by R. Carter, 16-30. London: Routledge & Kegan Paul.

___________, ed. 1987a. *Collins Cobuild English Language Dictionary*. London: Collins.

___________, ed. 1987b. *Looking Up*. London: Collins.

___________, ed. 1990. *Collins Cobuild English Grammar*. London: Collins.

___________. 1991a. *Corpus, Concordance, Collocation*. Oxford: Oxford University Press.

___________. 1991b. "Shared Knowledge". *Georgetown University Round Table on Languages and Linguistics 1991*, 489-500. Washington D.C.: Georgetown University Press.

___________. 1992a "Trust the Text". *Advances in Systemic Linguistics* ed. by M. Davies & L. Ravelli, 5-19. London: Pinter.

___________. 1992b. "Priorities in Discourse Analysis". *Advances in Spoken Discourse Analysis* ed. by M. Coulthard, 79-88. London: Routledge.

___________. 1992c. "The Automatic Analysis of Corpora". *Directions in Corpus Linguistics* ed. by J. Svartvik, 379-97. Berlin: Mouton de Gruyter.

___________, S. Jones & R. Daley. 1970. *English Lexical Studies*. University of Birmingham, for Office for Scientific & Technical Information.

___________ & R. M. Coulthard. 1975. *Towards an Analysis of Discourse*. London: Oxford University Press.

___________ & D. Brazil. 1982. *Teacher Talk*. Oxford: Oxford University Press.

Sperber, D. & D. Wilson. 1986. *Relevance: Communication and Cognition*. Oxford: Blackwell.

Strevens, P. & F. Weeks. 1985. "The Creation of a Regularized Subset of English for

Mandatory Use in Maritime Communications: SEASPEAK". *Language Planning Newsletter* 11, 2.

Stubbs, M. 1984. "Review of Sinclair & Brazil". *Applied Linguistics* 5, 1. 71-4.

________. 1989. "The State of English in the English state: Reflections on the Cox Report". *Language and Education* 7, 1: 1-25.

________. 1992. "Institutional Linguistics: Language and Institutions, Linguistics and Sociology". *Thirty Years of Linguistic Evolution* ed. by M. Pütz , 189-211. Amsterdam: Benjamins.

________ & A. Gerbig. Forthcoming. "Human and Inhuman Geography: on the Computer-assisted Analysis of Long Texts". *Corpus-Driven Linguistics* (provisional title) ed. by M. Hoey & G. Fox. London: Collins.

Swales, J. 1990. *Genre Analysis*. Cambridge: Cambridge University Press.

Wight, J. & R. Norris. 1970. *Teaching English to West Indian Children*. Schools Council Working Paper 29. London: Methuen.

Willis, D. 1990. *The Lexical Syllabus*. London: Collins.

Willis, J. & D. Willis. 1988. *Collins Cobuild English Course*. London: Collins.

Zipf, G. K. 1935. *The Psychobiology of Language*. Boston: Houghton Mifflin.

1. Spoken and Written Discourse

Inexplicitness
A Feature of Naturalness in Conversation

Martin Warren
University of Luleå, Sweden

Abstract
This paper explores John Sinclair's assertion that it is characteristic of a competent discourse to be inexplicit. The case is made that the use of inexplicit language is a feature of naturalness in conversation. The level of explicitness used by participants in a conversation depends on overlapping factors such as the physical setting and shared knowledge. It is argued here that, in conversations, utterances are produced at a lower level of explicitness than in other discourse types. The implications of this for language teaching and for developments in artificial intelligence are then briefly discussed.[1]

0. Introduction

Sinclair (personal communication) claims that inexplicitness is a characteristic of competent discourse. 'Inexplicit' here is used of language that is context-dependent, where context is recognised to be an integral part of the meaning conveyed in discourse. I maintain that we can associate different levels of explicitness with different discourse types. McCarthy (1991) gives examples of several discourse types and claims that it is possible to predict the level of explicitness one might find in them. A natural conversation is heavily context-dependent, and therefore inexplicit, while "a broadcast lecture on the radio may be quite 'freestanding' in that everything is explicit, self-contained and highly structured" (ibid:149).

This paper discusses the important role played by context and argues that context-dependent language is a feature of naturalness in conversation. Conversely, failure to employ context in conversation produces a discourse which is unnatural. Naturalness refers to the well-formedness of utterances in text, as produced by native speakers in normal situations (Sinclair 1988:11).

1. The notion of context

At this point, it might be helpful to clarify what the term 'context' is meant to cover. In this study, context is understood to cover the two widely accepted uses of the term, as documented by the *Collins Cobuild English Language Dictionary* (Sinclair 1987):

> 1 The **context** of something consists of the ideas, situation, events or information that relate to it and make it possible to understand it fully.
> 2 The context of a word, sentence, or text consists of the words, sentences or text that come before and after it and help to make its meaning clear.

It should be noted that such a general definition of context would probably be unacceptable to a number of scholars. For example, Halliday & Hasan (1976:31-3) choose to use 'context' to refer to the accompanying text and 'context of situation' to describe the field, tenor and mode of a particular discourse. Thus when discussing reference in texts, they distinguish between exophoric (textual) and endophoric (situational) references. In this study, however, it is felt that to distinguish between textual and situational contexts, even if we assume that a reliable distinction is always possible, is unnecessary when the emphasis is being placed on the importance of context in general, rather than on the role of different types or elements of context.

Others working on the analysis of conversation have been inclined to place to one side the kinds of classification offered above and use 'context' in a very wide and sometimes rather vague sense. For example, Wardhaugh (1985:101) argues that context includes the linguistic environment as well as all the various aspects of the non-linguistic environment, including previous conversations between participants in a discourse, any relevant aspects of their life histories, and whatever rules of behaviour and assumptions about the world they subscribe to. In a similar vein, Schiffrin (1987:5) claims that language always occurs in a context and is context sensitive, and, "even more strongly, that language *reflects* those contexts because it helps to constitute them." Schiffrin's interpretation of context might well be summarised as life minus language.

I would like to argue that both Schiffrin's and Wardhaugh's notions of context are too wide and vague to be usable in defining a technical concept. Context in the broad sense, as defined by Wardhaugh, must also be questioned on the grounds that the premise of there being such a thing as a firm external context, to which all the participants in a conversation relate, is too simplistic. Language makes rather than reflects the context, so that an utterance such as *Mellor's gone* creates a putative context; if it means nothing to the hearer s/he will often strive to construct a plausible interpretation. External context is not fixed: shared knowledge is constantly being negotiated and is not some kind of unsubstantiated library to which conversational participants have full and equal access. We need to recognise just how dynamic is the role of the ongoing negotiation of context in conversation.

Those who try to separate textual and situational contexts run the risk of ignoring or understating the interplay between the two, while those who advocate the kind of broad definition described above seem to imply that shared knowledge of the world is fixed at any particular moment and is identical for all the participants in a conversation who then 'reflect' it in their language. The reality is, however, very different: there is *sharing* rather than *shared* knowledge between conversational participants. Shared knowledge is thus subject to negotiation in conversation and it is in this way that discourse makes changes in the real world.

I will now go on to look at inexplicitness and how it is manifested in discourse. I will consider its linguistic realisations as well as the way these are influenced by the physical setting and the degree of shared knowledge between participants.

1.1 Context and conversational coherence

In a study which involved non-native speakers, Warren (1985:36-38) showed that when the participants in a discourse are denied access to elements of the context as a language resource, the level of explicitness becomes greater than it would otherwise be. Context is not only reflected in a conversation but it is also fully incorporated into it and plays a role at least as important as that of the spoken word and non-verbal behaviour.

Context then is not simply a colourful backdrop against which participants in a conversation are able to produce a more meaningful string of utterances. Rather, it provides speakers with an essential resource without which conversations would have to operate at an unacceptable level of explicitness. If they were entirely independent of context, all everyday conversations would have to start from scratch in order to be coherent. However, while context-free and hence highly explicit language may pose serious problems for our effectiveness in communicating with each other, the opposite is not necessarily the case. One does not have to look very far to find examples in which the need for conversation to help convey meaning is frequently rendered redundant by speakers opting for total reliance on context and non-verbal behaviour in order to communicate meaning and intention. An awareness of context is therefore essential if one is to appreciate the overall coherence of a conversation. Hobbs (1982:223) observes that when we come across naturally occurring discourse which appears to be incoherent "we always have the suspicion that if we only knew enough we could see the hidden coherence." This hidden coherence is usually a product of context-dependent language.

Grice (1975:45) discusses the underlying co-operative nature of conversation; an important component of this co-operation, it is argued here, involves the speakers fully utilising the context. This, in turn, has consequences for the level of explicitness produced in the ensuing discourse. Grice's maxims of Quantity and Manner (1975:45-6) add weight to the argument that inexplicitness is a feature of naturalness in conversation. The maxim of Quantity relates to speakers making their contributions no more or less informative than is required; that of Manner requires the avoidance of obscurity and ambiguity and stipulates the need for brevity and orderliness. Participants in a conversation who ignore the context in which they are operating can be expected to produce a discourse which exceeds the level of information required for the purposes of the exchange. Also such a conversation will not be as brief as one that utilises all the implicit resources available to the participants. Similarly, a conversation in which the level of explicitness required is underestimated by one or more speakers will promote ambiguity and obscurity. Context is therefore an invaluable resource to a speaker, and failure to utilise it fully will result in the conversation displaying an unnatural level of explicitness or failing to reach an adequate level of intelligibility.

2. Linguistic realisations of inexplicitness

If, as has been suggested here, the context can determine a speaker's choice of words, then the manipulation of the context should alter the level of explicitness employed as speakers adapt their utterances to suit the new context. In order to test this claim, an activity was devised to explore whether fluctuation in the level of explicitness in the discourse was a product of changes in the context. The particular activity chosen to illustrate the importance of context was based on a popular game known as 'Spot the Difference'. This is a game played by two participants, with two pictures which are identical except for certain details. The object of the game is for each participant to find these differences without looking at his or her partner's picture. In Warren (1985:182), I suggested that by devising a highly artificial situation one must expect to obtain highly artificial language, displaying a level of explicitness far in excess of that normally expected. For the purpose of this study, this hypothesis was put to the test by recording participants playing two versions of the game: one where each of the participants had access to both pictures, the other where they had access to only one picture each. Extracts from the recordings are presented below.

Extract 1: Participants with access to both pictures. C: man; J: woman

J: no flower
C: what
J: looks like a flower - there's no flower there
|
C: oh yea
um - right that's one difference - flower - er - let's have a look - er - oh there's a difference the there's a lamp there on that one
J: yea
C: right that's two - um - oh yes yes there's a boy missing from the coach - on that one
J: only one chimney pot - there's two there
C: oh yea - um - oh you're good at this - ah I can see that's different as well - the - whatdoyoucallit
J: signal
C: signal yea - um
J: oh two wagons - only one wagon
|
C: how many's that
oh yea (laughs) yes
|
J: that's six (coughs)
C: um - you're getting more than me er aren't you - er - oh alright yes there's a a tube thing coming down there isn't there
|
oh mmm seven
J: and there two things there and there's only one there
C: one there yea right so how many's that
|
J: eight

Extract 2: Participants with access to only one picture each

C: right - okay (chairs moving) (coughs) er
J: right you start
C: is it on oh yes - um - right I've got picture A er goodness me I
|
J: mmm

C: goodness me I don't know where to start really there's a - have you got a bus
J: yes - bus got (laughs)
C: er
J: has it got two
|
C: I can see three people on the bus
J: yea coloured in black
|
C: you've got three have you - one getting off the back
J: yea got that
|
C: is that right - um
J: yea
C: um
J: has he got a hat on
C: yea he's wearing a hat yea
J: what about the traffic lights is one coloured in and two not
C: er the top one - is coloured yes
J: yea
C: um
J: what about the woman with the umbrella has it got - six spokes
C: hang on the woman with the umbrella - I can't see a woman with an umbrella
J: right in the middle
C: in the middle - is that an umbrella
J: it's sort of er not - it's
|
C: it's a bag isn't it
J: no she's putting the umbrella up
C: is she
J: yea
C: oh no I don't think I've got that
J: oh alright that's

The two extracts above illustrate clearly the ways in which utterances in a discourse can be influenced by the context in which they are produced. In Extract 1, the participants have equal access to both pictures and this, as one would expect, has consequences for the discourse. For example, the participants are able to assist each other in the production of the discourse, as when C asks J for help in naming the *whatdoyoucallit.* Non-verbal communication is obviously being employed to supplement utterances such as *there's a tube thing coming down there.* A striking feature throughout Extract 1 is the extensive, and on one occasion, exclusive use of pro-forms and the general noun *thing*, as when J says *there two things there and there's only one there.* It would seem that when speakers are in a position to avoid the use of highly accurate, heavily descriptive language they do just that. By avoiding such explicitness and relying on shared knowledge and the physical setting they are actually enhancing the likelihood of a successful outcome. The discourse consists, for the most part, of a series of statements with an occasional question concerning the number of differences found. The absence of questions relating to the content of the pictures again reveals a context in which the speakers have equal access to the salient information. When speakers have shared knowledge concerning the topic under discussion, there is far less need for establishing common ground. Indeed, it is suggested by Sinclair et al (1982) that the choice of a question is in itself a move towards explicitness. This assertion would appear to be validated by the extracts under consideration.

Extract 2 differs from Extract 1 in a number of important ways. It begins, interestingly, with the participants selecting who is to begin, which is in itself a departure from convention. It is not the norm for speakers to explicitly select who should initiate the discourse since naturally occurring conversations are, of course, spontaneous. Speaker C is also quick to comment on the difficulty of the task in hand when he says *goodness me I don't know where to start.* Both of these observations also draw attention to the fact that the participants are declaring their unwillingness to accept responsibility for the discourse. Furthermore, the externally imposed constraints make it necessary for them to ask a lot of questions in a very hit-and-miss fashion. In terms of achieving the stated goal, namely to find all the differences between the two pictures, Extract 2 gives ample evidence of the speakers having to produce many more utterances to complete the activity. Moreover, the language employed in Extract 2 is different from the language in Extract 1 in that, without access to both pictures, the level of explicitness increases dramatically as the participants are required to produce much more accurate, heavily descriptive language. The use of pro-forms is far less frequent and utterances such as *there two things there and there's only one there* are simply no longer possible. Instead, utterances of the type *what about the woman with the umbrella has it got six spokes* are in evidence throughout; the speaker first has to indicate clearly the part of the picture to be scrutinised and then ask specific questions about the details contained in it.

The kind of differences described above confirm some of Ure's work on lexical density (1971), where she describes the proportional occurrence of lexis in different text types. Ure shows that the language we use for everyday practical purposes usually has a low lexical density due to its highly context-dependent nature. This is another aspect of inexplicitness that would be well worth exploring in a future study.

3. Contextual components contributing to the level of inexplicitness

Three main components have been identified as comprising the context in which speakers produce their utterances, and therefore as contributing to the level of inexplicitness in discourse: physical setting, shared knowledge and linguistic context. It will be seen, however, that there is a great deal of overlap among the three components and that it is not possible to isolate each one clearly in order to measure its contribution to determining the level of inexplicitness in a given encounter.

3.1 The physical setting and its institutionalisation

It is generally accepted that the physical setting plays a key role in determining how inexplicit a particular discourse is likely to be. For example, Ventola (1983:243), in her examination of the language found in service encounters, recognises the uniqueness of every discourse. She further offers evidence to suggest that the physical setting enables the researcher to assign discourses to a specific register. Once a register has been established, she claims, it is possible to predict the ways in which the participants might be expected to express themselves.

The way in which the physical setting is built into the semantics and cohesion of the language is best illustrated by looking at data. The conversation quoted below was recorded surreptitiously in a newsagent/grocery store where a customer is purchasing various goods. The physical setting, therefore, can be objectively defined and offers a familiar setting for explaining the level of explicitness produced by the participants.

Extract 3: A: customer B: shop assistant

B: that's 32 37 67 77 85 love - thank you
A: two 17p stamps
(pause)
B: that's 85 and 34 then isn't (pause) 85 34 is 9 one nineteen love thank you - ta thank you 19 20 30 50 that's your two pounds thank you shall I put them in a bag for you
A: no no no
B: thank you OK
A: ta
B: thanks a lot

The above discourse begins with the shop assistant counting out the customer's change. Clearly, in this particular setting, with the customer able to serve himself and then simply present the goods and money to the shop assistant, his intentions are sufficiently explicit to require no spoken interaction at all. He purchases six items and yet only one is mentioned in the discourse, namely the stamps. The other items are simply referred to collectively as *them*. The stamps are kept behind the counter in the shop and so they have to be specifically mentioned by the customer. However, it is interesting to note that the stamps are not explicitly requested (e.g. *Can I have two 17p stamps?*) because the setting is a shop and the intention of the customer is taken for granted.

This service encounter provides examples of inexplicit utterances which are heavily dependent on the physical setting, but the level of inexplicitness is not such that the register and setting are unrecognisable from the transcript. However, it is possible to provide examples of conversations in which the level of inexplicitness is such that it is quite impossible to ascertain from the transcript what type of transaction took place as the conversation was being held. In a conversation of this kind, dependence on the physical setting may mean that nothing need actually be said about the goals of the encounter, or at least nothing related to the immediate activity taking place. An example of this type of discourse is given below.

Extract 4: F: customer B: shop assistant

F: morning M____
B: hello hello
F: nice morning
B: yes it is
F: no church this morning Mabel (laughs)
B: no - the vicar will want to
      | |
F: that's right   he's just gone by the vicar has
B: has he
F: he's got to go round
B: he'll be looking for me I'll bet - ah I bet he was (laughs) I haven't been there for a long while
F: no that's right - never mind (laughs)

Extract 4 is a complete record of a conversation which took place in the same shop as Extract 3. During the course of this conversation, a purely business transaction was completed by the participants, and yet this is not made explicit in the conversation. It is clear that the speakers are not strangers to one another and the event taking place, namely the purchasing of a newspaper, is quite probably a daily occurrence that has been enacted by the speakers countless times. Consequently, the customer simply picks up the newspaper and hands the money to the assistant, who hands back the change. Instead of the conversation following similar lines to those in Extract 3, this one ignores what is implicit in the encounter and proceeds to a social plane.

The ongoing building and consolidation of social relationships, so often dismissed as 'small talk', can relegate service encounters to the point where they have no linguistic realisation at all. Ventola (1987:48-50) is not unaware of the effects of conversation on service encounter talk, viewing them as an interesting area for research into "how differences in social distance influence the linguistic realisations in service encounters." Unfortunately, however, Ventola excludes all such examples from her study, presumably because they fail to conform to the service encounter genre.

A distinction should be made between conversational goals (i.e. changes in the real world) and elements such as the maintenance, reinforcement and reification of social relationships, which are also important reasons for the conversation being as it is. Berger and Luckman (1979) state that social relations have to be continuously created or they do not exist, and this particular conversation provides a very good example. The more familiar the physical setting and/or the more intimate the participants, the less reliant the discourse need be on the use of explicit language. The greater the shared knowledge between speakers with respect to the topic, the less they have to put explicitly into a conversation to express themselves and in turn be fully understood. The importance of shared knowledge and the ways in which it is linked to the physical setting will now be discussed.

3.2 Shared knowledge

Sinclair (1991: 496-9) provides a comprehensive definition of shared knowledge and identifies four relevant types: subliminal mastery of phraseology, received information about grammar and lexis, strict linguistic inference (including textual inference), and aspects of culture which are not signalled anywhere in the text but which have to be known. When expressed in these terms, it is difficult to imagine a discourse which is entirely free of shared knowledge. Indeed, Sinclair's definition could well fit the general use of the term 'context' as outlined in section 1. It should be noted, however, that there are other factors which aid the interpretation of conversations, including the pure inductive and deductive skills of participants. Even if context, in the broadest sense, were deleted, it would still in theory be possible to interpret an inexplicit utterance at some tolerable level. Let us examine some data which highlights the way in which shared knowledge influences the level of inexplicitness in discourse.

Extract 5 is a record of a conversation which took place between two colleagues in a car park outside their place of work. It is particularly interesting because the participants discuss the workings of a camera tripod and the procedure for using silica crystals as a dehumidifying agent, and yet neither the tripod nor the silica are explicitly named. The tripod is actually held in the hands of one of the participants during the conversation. The physical setting thus dispenses with the necessity for explicitly naming the item. Moreover, shared knowledge between the participants, about both the tripod and the silica, means that the speaker who initiates the conversation is able to dispense with a lengthy preamble.

Extract 5: M: man P: man

M: here P____ (pause) you need to put it in the oven though to turn it blue again
|
P: oh thank you thanks
P: absolutely magic - I was just there's wondering there's no there's nothing to that you just pull that and then they come
M: yes yes
P: I'm I'm just yes I know and there's no there's no quirks at all
|
M: yea yea
M: no it works perfectly but there are many knobs and
|
P: (inaudible)
M: whatsits you know
|
P: yea - no that does that one and this does - kind of thing - the I get that one
M: ah no that one er
P: I think I'm alright actually - if it I mean if it if there's no
|
M: I can't remember
P: quirks there's there's no problem
M: no no there's nothing wrong with it but er once you've got it set
|
P: (inaudible)
M: up I'd check what's what

P: yea right
|
M: this is er - I think that's for going
|
P: that's that's this one isn't it
M: oh yea cranking it up higher yea
P: oh thanks ever so much do you want some money for this
|
M: it's a while since I used that
M: no it's a while since I've used that
P: and and what do you do you you just leave it
M: you put it spread it out on an oven tray in the
oven on a low oven and it goes and it'll go that colour - very very
| | | |
P: right yea I see and then
M: blue you know real royal blue then and you have to quickly then put it in whatever
|
P: right
M: you're going to put it in because
|
P: like an air tight container
M: or I used I was using these large tupperware boxes
P: yea and then put your things things in does it have a lid on the top of it
M: yea I put a then I put a piece of paper over it to keep it away from whatever was in so the little balls don't get inside my camera

In Extract 5, the initial topic, silica, is introduced by M. P succeeds in changing the topic by using his expression of thanks as a frame and then introducing the tripod as the new topic in the conversation. The physical presence of the tripod obviously removes the need to explicitly name it. A combination of pro-form and non-verbal behaviour is all that is required in this context. As mentioned above, this can be seen as an example of how the physical setting can influence inexplicitness. P now returns to the topic of the silica by expressing his thanks once again and then asking M if payment was expected for the silica. Again, the level of explicitness employed by P is very low; it is perfectly adequate for M, though, because of the extent of shared knowledge between the two participants: only the silica is being given to P (the tripod is on loan) and so a pro-form is all that is necessary to identify the topic now under discussion. We have a case here where physical setting and shared knowledge clearly overlap to contribute to the inexplicitness of the discourse.

Extract 5 is a particularly good example of what Sinclair calls strict linguistic inference (1991:497). Sinclair claims that, in language in use, when something is obvious it is optional and, paradoxically, "the more obvious it is, the less likely it is to occur, and the more marked for meaning its occurrence." This is clear in the above example where neither the tripod nor the fact that it is on loan are explicitly mentioned. The first is obvious because of being physically present; the second is obvious because it is part of shared knowledge. Here, as in the earlier examples, the speakers can be seen to be operating at a level of explicitness that fully incorporates context into the conversation, enabling them to employ the most effective and efficient means available to communicate with one another.

3.3 Linguistic context and its influence on inexplicitness

The linguistic context consists of the accompanying text. In the case of conversation, the accompanying text is the utterances which surround a particular point in the discourse. Unlike the case made for physical setting and shared knowledge, it is more difficult to argue that a specific level of inexplicitness is directly attributable to the linguistic context. There is often a certain degree of overlap between general shared knowledge and more specific linguistic context so that it is often not possible to draw a clear line between them. A preceding utterance, for example, may be deemed to fall into either or both the linguistic context or the shared knowledge components of context. The occurrence of ellipsis as a realisation of inexplicitness is a good example of this indeterminacy of categorisation. Ricento (1987:751-2) points out that it can be viewed as a product of prior discourse and/or physical setting. Similarly, the presence of deixis in spoken discourse can be seen as falling within the realm of semantics or pragmatics, according to whether it is viewed as a linguistic element referring to the preceding text or a shared knowledge element referring to the world outside (Levinson, 1985:55).

4. Interpreting inexplicitness in conversation

As has been stated earlier, one of the factors distinguishing one discourse type from another is the level of explicitness displayed. All conversations are not going to have the same level of explicitness, but a feature of naturalness in conversation is the presence of a significantly lower level than that to be found in other discourse types.

In an experiment which involved describing a series of pictures, Bernstein (1972; 1975) found that working-class children employed levels of inexplicitness which were different from those of middle-class children of the same age. They failed to appreciate what was expected of them in the unfamiliar setting of a research experiment. They assumed, quite understandably, that the researcher with the series of pictures was familiar with their content and so behaved as if this was shared knowledge. For the purposes of the experiment, however, they appeared 'restricted' because the level of explicitness employed in their answers was inappropriate for the context in which they found themselves. Middle-class children handled similar tasks very differently, using much less context-bound language.

Bernstein's concept of 'restricted code' (1972; 1975) is sometimes attacked for apparently suggesting that context-dependent language is inherently inferior to less context-bound language. Analysis of conversational data could indeed lead one to the conclusion that, because it is inexplicit, it is somehow a poorer means of effective communication compared to other more explicit types of discourse. However, the suggestion that context-bound language is inferior to context-free language in all contexts misses the central thrust of Bernstein's research.

Bernstein's comments centred on the question of language appropriacy. As far as this is concerned, one could just as easily argue from the opposite point of view. If one accepts that context-bound and context-free language attending to the same point are equivalent in terms of the communication load that they carry, and given that context-bound language is briefer, one could say that it is lower in redundancy and therefore more efficient. The case being made here is the central importance of context and the relative unnaturalness of discourses which attempt to ignore it or deny participants in a discourse full access to it. At this point in the discussion it is perhaps appropriate to look at how participants in conversation interpret inexplicitness.

Brown and Yule (1983:58-67) suggest that participants in a discourse have two principles for disentangling the various contextual features: the principle of local interpretation and the principle of analogy. The principle of local interpretation "instructs the hearer not to construct a context any larger than he needs to arrive at an interpretation" (ibid:59). The principle of analogy is based on hearers assuming that "everything will remain as it was before unless they are given specific notice that some aspect has changed" (ibid:65). In other words, a combination of relevant past experience and the principle of local interpretation will lead hearers to interpret a string of utterances appropriately.

It is one thing, however, to argue that inexplicitness is a feature of naturalness in conversation, and quite another to suggest that all speakers are equally adept at employing an appropriate level of explicitness in every discourse. Brown (1989) examines the language used by poor academic achievers in British schools and concludes that there are skills to be learned and developed if speakers are to communicate adequately with each other.

5. Conclusion

The importance of inexplicit language as a feature of naturalness in conversation has been described and exemplified. It has been argued that the context in which the participants in a conversation find themselves forms the basis of the level of explicitness at which they communicate. This feature of conversation enables the speakers to produce a discourse which relies on the mutual appreciation of context. This mutual appreciation is a prerequisite for a successful outcome and underlines the co-operative nature of conversation. The level of explicitness displayed by participants in a conversation is influenced by a number of factors. These include the physical setting, the amount of shared knowledge that exists between participants and the linguistic context relevant to a particular point in the discourse. It is suggested here that participants in a conversation produce utterances at the lowest level of explicitness required to reach a successful outcome in a particular context. Exceptions occur if something is not fully clear in the terms of the transaction, or if an unusual or unexpected event takes place. In all other cases, to raise the level of explicitness markedly is to behave oddly and therefore point to something unusual in the social relationship.

The above observations have implications for areas other than conversational analysis. Firstly, in the design of language teaching materials and activities, the importance of context is mostly ignored, leading to the creation of unnaturally explicit drills and dialogues in the classroom, dialogues which masquerade as authentic conversation. Consequently, as learners of the target language are always taught to be explicit, they are at a serious disadvantage in this regard. It is not part of the brief of this article to encourage or discourage the implementation of particular methodologies in the language classroom. However, it is perhaps appropriate to emphasise that while most simulated communicative activities do not in fact reproduce the high level of inexplicitness found in conversations outside the classroom (Warren, 1985:36-8), such activities do create opportunities for students to speak the target language and offer oral practice, especially in the formulation of questions, in a less prescribed format than those often to be found in the English language classroom.

Nevertheless, it should be recognised by language teachers that activities aimed at maximising talk by artificially manipulating the context inevitably result in levels of explicitness far in excess of those found in natural conversation.

Secondly, scholars involved in developing artificial intelligence systems should not ignore Sinclair's observation that it is characteristic of a competent discourse to be inexplicit. It is clearly quite a difficult task to design and write computer software that is able to operate at the levels of explicitness appropriate for naturally occurring conversation. A more realistic and achievable goal might be to accept certain limits imposed by computers and to develop a less context-dependent 'computerspeak'. Computers could be programmed to handle language at certain levels of inexplicitness and 'computerspeakers' trained to conduct discourse at levels of explicitness which would be deemed unnatural in everyday conversation. There is a need for much more research and co-operation between the various disciplines in order to determine just what these levels of explicitness could be. The language produced by many communicative language teaching activities provides a possible insight into the likely forms that the discourse between man and machine might take. Indeed, such activities could well form the basis for training the computer operators of the future.

Notes

1. This paper is based on a section of my PhD thesis (forthcoming) which attempts to describe the features of naturalness in conversation. The main source of inspiration for this work has come from John Sinclair, and I am greatly indebted to him for his ideas, insight and support. Thanks are also due to Antoinette Renouf who provided detailed comments on an earlier version of this paper.

Authors's address:
University of Luleå, Luleå. S-951 87 Luleå. SWEDEN.

References

Berger, P. L. & T. Luckman 1979. *The Social Construction of Reality*. London: Penguin Books.

Bernstein, B. 1972. "Social Class, Language and Socialisation". *Language and Social Context* ed. by P. P. Giglioli, 157-78. London: Penguin Books.

__________. 1975. *Class Codes and Control 3: towards a Theory of Educational Transmissions*. London: Routledge and Kegan Paul.

Brown, G. 1989. "Making Sense: the Interaction of Linguistic Expression and Contextual Information". *Applied Linguistics* 10,1. 97-108.

_________ & G. Yule. 1983. *Discourse Analysis*. Cambridge: Cambridge University Press.

Grice, H. P. 1975. "Logic and Conversation". *Syntax and Semantics III: Speech Acts* ed. by P. Cole & J. L. Morgan, 44-58. New York: Academic Press.

Halliday, M. A. K. and R. Hasan. 1976. *Cohesion in English*. London: Longman.

Halliday, M. A. K. 1978. *Language as a Social Semiotic: the Social Interpretation of Language and Meaning*. London: Edward Arnold.

Hobbs, J.R. 1982. "Towards an Understanding of Coherence in Discourse". *Strategies for Natural Language Processing* ed. by W. G. Lehnert & M.H. Ringle, 223-44. Hillsdale, N.J.: M.H. Lawrence Erlbaum Associates.

McCarthy, M. 1991. *Discourse Analysis for Language Teachers*. Cambridge: Cambridge University Press.

Ricento, T. 1987. "Clausal Ellipsis in Multi-party Conversation in English". *Journal of Pragmatics* 11. 751-75.

Schiffrin, D. 1987. *Discourse Markers*. Cambridge: Cambridge University Press.

Sinclair, J. M., ed. 1987. *Collins Cobuild English Language Dictionary*. London: Collins.

_____________. 1988. "Naturalness in Language" in *Naturalness in Language* ed. by M. McCarthy. English Language Research Journal (University of Birmingham) 2. 11-20.

_____________. 1991. "Shared Knowledge". *Georgetown University Round Table on Languages and Linguistics 1991. Linguistics and Language Pedagogy: State of the Art* ed. by James E. Alatis, 489-500. Washington D.C.: Georgetown University Press.

_____________ & D. C. Brazil. 1982. *Teacher Talk*. Oxford: Oxford University Press.

Stubbs, M. 1983. *Discourse Analysis: the Sociolinguistic Analysis of Natural Language*. Oxford: Basil Blackwell.

Ure, J. N. 1971. "Lexical Density and Register Differentiation". *Applications of Linguistics: Selected Papers of the 2nd International Congress of Applied Linguistics, Cambridge 1969* ed. by G. E. Perren & J. L. M. Trim. Cambridge: Cambridge University Press.

Ventola, E. 1983. "Contrasting Schematic Structures in Service Encounters". *Applied Linguistics* 4.3. 242-9.

_________. 1987. *The Structure of Social Interaction: a Systemic Approach to the Semiotics of Service Encounters*. London: Frances Pinter.

Wardhaugh, R. 1985. *How Conversation Works*. Oxford: Basil Blackwell.

Warren, M. J. 1985. "Discourse Analysis and English Language Teaching: a Contrastive Study of Discourse-based and Communicative Activities". Unpublished M.A. thesis, University of Birmingham.

Topic as a Dynamic Element in Spoken Discourse

Hazadiah M. D.
ITM, Shah Alam, Malaysia

Abstract
This paper examines the status of topic in the management of naturally occurring conversation. From a semantic point of view, topic is identified as a complex of meanings created and developed over several utterances. It is argued here that it is not useful to locate topic merely as a property of the sentence; rather we should go beyond the sentence and view topic as a discourse rank of a higher order. The importance of sequential organisation in joining exchanges of different types is highlighted in the unfolding of coherent discourse. Topic is seen as the element that holds these exchanges together. A linguistic unit, namely the Topic-Framework, is proposed to carry the patterns of topic.

0. Introduction

Topic is viewed in this paper as a structural unit operating at the level of the exchange, that is as an element which organises discourse into linguistically and socially relevant exchanges by successive acts of prospection. Different types of prospection are accomplished in the unfolding of topic.

The term 'Topic Framework' (hereafter TF) is borrowed from Brown & Yule (1983) where it is taken to mean "the contextual framework within which the topic is constituted" (ibid:75). In this paper, it is used to define a rank in discourse which lies between Exchange and Transaction in the Sinclair & Coulthard model (1975). Each exchange is seen as an element of the structure of a TF, and a TF consists of one or more exchanges. This view assumes the description of structure as a taxonomic hierarchy (Halliday 1961).

The descriptive apparatus proposed in this paper enables us to see how the Topic Framework initially relies for its development on the consensus of the participants in the conversation; this consensus is then structured by successive acts of prospection at exchange level. This process makes relevant certain aspects of discourse that contribute to linguistic well-formedness and social appropriatness, such as the fact that a topic needs to be introduced, developed, and maintained. By recognising the topical structure of conversation we will be able to:

(i) demonstrate that spoken discourse is inexplicit, because topic is not shared initially but develops as the conversation unfolds,

(ii) argue for the prospective functions of topic,

(iii) record and account for topic shifts in terms of relevance within the proposed structure of the Topic-Framework.

I would like to suggest that participants have perceptions of topic and non-topic per se; a perception of topic assumes linguistic realisation once it is established as an exchange. A single utterance made at the beginning of the conversation may carry an offer of a topic but it does not state the topic as such. This inexplicitness allows for the fact that the 'offer' can be interpreted within a broad range of possible topics; after the initial offer, the way the topic develops will depend on the utterances that follow it.

0.1 The notion of topic

One of the main characteristics of topic in conversation is that it is an entity that is not known initially, but is created in the process of negotiation; spoken discourse unfolds in real time and none of the participants has total control over it or knows exactly where the conversation is going to lead to. This specific characteristic of topic affects for example the degree of explicitness of the text. It also affects the development of discourse as each contribution, at the level of the exchange, prospects the following one. Topic can thus be seen as a progressive move from a vague category towards explicitness realised by progressive stages of prospection.

0.1.1 Inexplicitness as an outcome of topic creation and development

Warren (this volume) treats inexplicitness as a feature of naturalness in conversation and discusses the essential role of context in the negotiation of meaning in any discourse. Indeed, seen from the perspective of *topic*, inexplicitness is often the natural result of 'topic creation' and 'topic development' (Hazadiah 1992a) within a process of negotiation.

A view of topic as not initially shared by participants but rather as becoming gradually shared as conversation progresses is different from that of conversation analysts, for example Schegloff & Sacks (1973), who suggest that speakers place topics so that they 'arise naturally' from other topics.

Schegloff & Sacks adopt a retrospective approach to topic and give the impression that topic is a known entity to participants, who are able to manipulate it and place it at the appropriate point in the discourse so that it can occur *naturally*. Conversation analysts therefore imply that topic is shared early in the conversation and tend to take its stability and explicitness for granted. This leads them to ignore the dynamic element which underlies topic development.

In my description, what Schegloff & Sacks see as delay tactics is taken to be the simple manifestation of the fact that topic is often not shared in detail until a late stage of development. Consequently, it is not likely to be stated explicitly by any participant, but rather offered or suggested as a potential topic in a tentative and inexplicit fashion. Another plausible explanation for this inexplicitness could be that the articulation of topic in an informal conversation can be a face threatening act. Risking an early statement of topic may be seen as an imposition on fellow participants, who may or may not accept one's topic. Wilson (1989), for instance, argues that in a speech event the maintenance of equality of speaker rights results in less explicit forms of topic initiation. In this case the existence of explicit forms would characterise symmetrical distribution of power and consequently of speaker rights.

There are of course situations where the articulation of topic may be less face threatening and where topic announcements are therefore often explicit. Classroom talk, to a certain extent, belongs to this type of discourse, and so do chaired meetings with well defined agenda, for instance a staff-meeting or the annual budget meeting of an organisation, where topic stands a good chance of being expressed and agreed as part of the business of the encounter.

0.1.2 Prospection

The initial, often implicit, offer of a topic, carried by one exchange, can be made by any participant in the conversation at any time. Such an offer is 'prospective' because it preclassifies the next set of utterances in terms of expecting support from the next exchange to develop any aspect of the budding topic. Each new contribution moves the state of the topic forward and prospects the following exchange in various ways. An utterance that does not fulfil the presupposition of support built in the previous utterance (this is the case of a challenge) either aborts a topic, or checks it, or steers it in a new direction.

By way of illustration, let us consider a short exchange taken from Tsui's corpus (1986:366).[1] Following Sinclair & Coulthard (1975) we can explain the discourse in terms of exchanges which, as mentioned in the introductory section, are the units of the Topic Framework:

Example 1 : (Data A/p.43) (Side A: 188-200)

Exchange		Moves
Exc 1	1 B	I'm going to see Mike Harding next Saturday
	2 C (a)	Oh you're not musically inclined then (pause)
Exc 2	(b) 3 B	Where's he on ? The Dominion
Exc 3	4 C 5 B	Not the Rainbow No
Exc 4	6 C 7 B 8 C	Are you sure you don't want a cigarette ? No, I couldn't your last but one well the last one actually - that would be my last one

In Exchange 1 a potential topic about *Mike Harding* is introduced. I see this as a budding topic since any aspect of it has the potential to be developed. The exchange now prospects support from the exchange that will follow it. The prospection is fulfilled by exchange 2 which selects the specific element *Mike Harding* from the preceding utterances. Other potential topics are also prospected: a topic on some other activity involving *next Saturday* for instance. We know that this exchange carries the topic prospected by the preceding exchange because it maintains some characteristic features of it. It also adds new information to the topic by moving it a step further and focusing on the place of performance (*where is he on?*). Exchange 3 (*Not the Rainbow - No*) highlights a presupposition which is evaluative in nature but still lies within the orientation of the preceding exchanges in that it is still focusing on the place of performance. By contrast, in exchange 4 that follows (*are you sure you don't want a cigarette? ...*), there is a change of subject: what has gone before is not taken any more as part of shared knowledge. The topic has changed to a new focus.

The specific role of prospection as a feature of structure is discussed in more detail in section 4.3.1 below. For the moment, it is sufficient to note that the development of topic can be influenced by each exchange made in the construction of discourse; as a prospected potential topic is fulfilled, others are discarded or set aside. With each movement, topic becomes more explicit and better defined. Topic should therefore be viewed as a dynamic category in discourse.

1. A dynamic model of discourse

This discussion has so far brought up several issues relating to the status of topic, one of which is its prospective function. Section 3.2 below deals with the way in which different types of prospection are accomplished. There is, however, a problem that needs to be attended to first. Topic is perceived and negotiated 'prospectively' by participants in discourse but it is viewed 'retrospectively' by analysts. Topic can be considered as a product in spoken language only when, through recordings and transcriptions, we view and review a piece of text as if it had the characteristics of a written text. But spoken discourse unfolds in real time and it must, therefore, be described as a continuous process and in prospective terms. Participants set out with different perceptions of what they want to talk about; a single topic is then identified while others are dropped in the process of negotiation. If the description of discourse is to carry credibility, it must be capable of assigning values to utterances *as they occur*, without the benefit of knowing what might happen next. I therefore maintain that a dynamic model of discourse will allow us to reconcile retrospection and prospection, or topic seen as a product and topic seen as a process.

1.1 Planes of discourse

The dynamic model I am proposing as the basis for a description of topic is derived from Sinclair (1985, 1991, 1992). Sinclair sees discourse as a "continuous movement from one state of affairs, or posture, to another" (1985:15) and language in use as a two-fold process, a "continuous negotiation between participants" on the one hand, and a "developing record of experience" on the other (ibid). These two aspects are realised simultaneously on two 'planes of discourse' which Sinclair identifies as the 'interactive plane' and the 'autonomous plane'.

In the framework of a dynamic model, the actual activity of constructing discourse is carried out on the interactive plane. This is where participants use a variety of tactics to realise their communicative intent of sharing experience, running checks on one another and informing each other of how their discourse is developing. Some of the tactics they employ are prediction, discourse labelling, cross-referencing and self referencing (Sinclair 1981:81). The interactive plane facilitates the construction of shared experience between two participants, the 'I' and the 'you' of the discourse. As the process is prospective, the interactive plane describes and captures the creation of text and the negotiation of topic. Part of this process consists of the taking up of options which are not yet in use, the offer of a potential topic.

The options that are taken up on the interactive plane, which I regard as interactive elements, are transferred to the autonomous plane as actualised elements, that is options that have already been taken up, where they are organised and maintained as text structure. Once the transfer is made, the focus shifts to the syntactic and lexical coding of text content. By building on what has gone before, readjusting, and consolidating the old material with the new, the autonomous plane presents itself as product-oriented.

Sinclair (personal communication) suggests that even though the autonomous plane is product-oriented, it should be seen as an updating process; the items in the 'accumulated significance' of experience on the autonomous plane can be retrieved later on. The autonomous plane is, therefore, responsible for the organisation of the text, whereby the gradual sharing of relevant experience on the interactive plane (which is realised by recalling previous words and phrases) is continuously reworked in the new context. This process is what Sinclair calls the "internalisation of experience, from the world outside to the inner space of language" (1985:3).

On the interactive plane, topic is an interactive construct whose development beyond the first step of introduction depends on the collaborative efforts of the participants in a discourse. At this level each exchange introduces a topic potential, supports it or doesn't support it. A record of this process is kept on the autonomous plane for constant updating. The exchange here has the role of changing the profile of the overall topic.

In Sinclair's model every utterance is capable of being described on both planes. The two planes can be distinguished in an explicit performative such as *I promise you that I'll be good from now on.* We can see the interactive performance of an action explicitly in the first clause, i.e. promising, while the second clause articulates the nature of the promise. A reporting structure therefore transfers attention from the interactive to the autonomous plane within an utterance. By doing so, the report does not attend to the interactive potential of the reported clause. Similar to report, in terms of its discourse function, is another phenomenon which Sinclair (1981) names "plane change".

1.2 Plane change

Each successive move in spoken discourse, or sentence in written discourse (Sinclair 1993:6-31) is taken either as fulfilling a prospection set up by a preceding unit or as replacing its predecessor as the point of interaction. Each can make a further prospection. This progression is normal; the old state of interaction, having been replaced by the new, is now stored on the autonomous plane as an actualised element. In a linear and uncomplicated text, this is the routine way of constructing discourse. Plane change, as a discourse device, achieves similar results, only in a more explicit way; thus, for participants, it is a way of labelling the preceding move or sentence, refusing to attend to its interactive potential, and transferring it to the autonomous plane. Plane change is often used in discourse as the principal technique of evasion. Consider the following example:

Example 2 (from field notes) : A : I can't be doing everything you know.
B : What do you mean by that ?
A : Well, what do you think ?

Note how each of the moves in turn refuses to attend to the full interactive potential of the previous one. Each new utterance dismisses the potential development of a topic and merely consigns the preceding one to the autonomous plane. The discourse does not get anywhere since Grice's (1975) co-operative principle is violated. In the TF, plane change is often used as a device to challenge the prospected topic in some way. This is dealt with in more detail in section 4.2.1 below.

2. Topic-Framework

A fundamental feature of conversation is that topic has to be introduced, and once introduced it has to be supported and developed and then concluded. This, however, may not always be the case because, given the range of speaker options in discourse, anything can happen. A speaker may choose to support a topic, ignore it or simply challenge it (as illustrated in example 2 above). All options available to a participant in spoken discourse affect the structure of talk. It is possible, however, to claim that participants are oriented towards dealing with discourse by developing a series of topics; the supportive and co-operative behaviour of conversationalists argues a common understanding of the normal progress of talk, that is its topical structure.

Previous research on naturally occurring data (Hazadiah 1991) supports the assumption that there is a regular sequential organisation which joins different exchanges to form a body of coherent discourse. These exchanges have different roles in the structure while maintaining a consistency of subject matter; the Topic Framework is the element that holds these exchanges together.

2.1 Topic-Framework as a rank in discourse

The orderly sequential combination of units is a fundamental feature of language at all levels of structure, and yet linguistic enquiries have tended, until recently, to stop short at the level of the sentence in written discourse. In spoken discourse, very few attempts have been made to examine the internal structures of the ranks above the exchange.

In order to establish topic as a structural unit in discourse I have proposed the Topic-Framework, as a discourse rank, to carry the patterns of topic. This means that topic can now be given an organisational function: that the elements of the structure can be assigned and those that are obligatory or optional in the structure can be ascertained.

By proposing the notion of a TF as a rank in discourse we have to extend the model for discourse analysis originally proposed by Sinclair & Coulthard (1975). This model was used to describe teacher-pupil interaction in terms of a hierarchical rank scale, consisting of Acts, Moves, Exchanges, Transactions and Interactions. In my proposed model, a TF fits in above the Exchange and below the Transaction, though a minimal Transaction can consist of a single TF. The highest unit is the Interaction which consists of several Transactions. Each Transaction consists of one or more TFs, each of which in turn consists of one or more Exchanges. Each Exchange consists of a combination of moves; and each move consists of acts which form the lowest unit in the rank scale (Tsui, this volume).

The Transaction may have an organisational role by carrying what we might call a 'macro-topic'. This is responsible for groupings of Topic-Frameworks that may suggest a more general topic covering a number of related topics. It is beyond the scope of this paper, however, to pursue the organisational details of macro-topics and Transactions. What concerns us here is topic at the intermediate level. Hence, the Exchange is recognised as a unit which captures a participant's conscious awareness of topic, and topic, in turn, is seen as a linguistic, organisational element in the routine construction of shared meaning.

2.2 The structure of the Topic-Framework

A Topic Framework may consist of a minimum of one exchange in a 'simple structure', going up to several structured exchanges in a 'complex structure'.

A simple structure consists of one introductory exchange which is neither prospected by the exchange that precedes it nor is in itself prospective, that is it does not expect to be supported or developed. I call this 'speech-in-action' exchange or SIA-exchange.[2] This category is often used for topic interruption during an ongoing talk, or as a transaction boundary marking TFs. Exchanges 6, 7 and 8 in Example 1 (beginning with *are you sure you don't want a cigarette?*) represent a typical SIA-exchange, assuming that the participants did not choose to talk about cigarettes in order to generate several exchanges on the subject; if the latter had been the case, the structure of the TF would have become a complex one.

A complex structure has as its minimal, obligatory elements the 'Introductory Exchange' (IE), the 'Supportive Exchange' (SE) and the 'Pre-closing Exchange' (PE). Other elements are optional, and include for example the 'Maintenance Exchange' (ME), the 'Focus Exchange' (FE) and the 'Appraisal Exchange' (AE).

Example 1 illustrates how a TF is accomplished in three exchanges. We can now rename the exchanges as follows:

Exc 1 & 2 (a)	Introductory Exchange (IE), prospecting support
Exc 2 (b)	Supportive Exchange (SE), fulfilling the prospection
Exc 4 & 5	Pre-closure Exchange (PE), proposing to close
Exc 6, 7 & 8	SIA Exchange, marking a new TF

The discourse value of each exchange is identified according to its position and its categorical function within the structure: the relevance of each exchange in the structure is interpreted in terms of how it relates to the preceding exchange and what structural expectation it sets up for the next. Thus each exchange invites the next one simultaneously to share its current pool of meaning and to update in order to be an integral part of a TF. Let us now go on to consider in more detail the specific functions of the different types of exchange outlined above.

The Introductory Exchange can serve several functions in the TF:

(i) to carry an offer of a topic potential from which other aspects of topic can develop; this is the case in Exchanges 1 and 2 (a) where *I'm going to see Mike Harding next Saturday - Oh you're not musically inclined then* have this function;

(ii) to change the direction of a current TF to a new TF; and

(iii) to indicate the close of a prior TF when there has been an explicit closing behaviour (a Pre-closing Exchange for instance), or to merely terminate the structure when there is none. This exchange also sets up a prospection that the next exchange, which is the Supportive Exchange, will develop the budding topic.

The Supportive Exchange is uniquely prospected by the Introductory Exchange. Further instances of Supportive Exchanges are optional. This exchange supports the preceding IE by maintaining the same message. Hence a SE is interpreted with reference to the previous exchange, with the clear implication that the latter is part of its shared knowledge. An SE can either specify, elaborate or clarify the budding TF, still acting within the framework established by its predecessor, though. Exchanges 2 (b) and 3 (*where's he on? - The Dominion*) can be seen as a Supportive Exchange indicating agreement amongst participants on the topic introduced in the preceding exchange.

The Focus Exchange brings about a shift towards a different aspect of the same TF; it opens up a new orientation while still maintaining connections with preceding exchanges and retaining some salient features of the TF. This exchange characteristically extends the TF. As mentioned before, the Focus Exchange is optional.

The Pre-closing Exchange has a function which is almost parallel to that of the Introductory Exchange but inverted: as the IE announces a new topic, so the PE announces a proposal to close of the old topic. Exchanges 4 and 5 (*Not the Rainbow - No*) represent the PE exchange proposing to close the TF. In structural terms, an actual closure can never be precisely predicted. We only know that a TF has closed when an IE succeeds it or when a conversation comes to a close. In example 1 above, the SIA Exchange (*Are you sure you don't want a cigarette? ...*) follows the PE Exchange to signify that the TF has come to a close.

Other types of exchanges are found in a complex TF which incorporates a narrative extension, that is which engages in narrating a story. These are the Maintenance Exchange and the Appraisal Exchange; they are not, however, dealt with in this paper.[3]

3. Relevance in the Topic-Framework

The model I am proposing incorporates Sperber & Wilson's (1986) principles of relevance into a linguistic framework. On the basis of their work, I maintain that relevance in the Topic-Framework must ultimately be seen as a property of discourse, whereby the position of the exchange in a given structure determines its relevance in terms of what it does for that structure. I will first present my understanding of relevance as a semantic notion, and then as a property of structure.

3.1 Relevance as a semantic notion

The exchange, as an element of TF, carries a proposition or a set of propositions in a semantic network. Following Beekman & Callow (1974:272-3) a proposition is defined as a grouping of concepts where one of the concepts is central, or nuclear. A proposition is a unit of communication: it affirms, denies, questions, or commands. It is also a minimal semantic unit consisting of a concept or a combination of concepts and communicating an 'event' or 'relation'.

The exchange, defined here in semantic terms, is taken as the upper limit of semantic units and syntax. This confirms Stubbs' view of it as an 'information unit' definable in terms of increasing predictability and semantic completion as manifested "in increasingly elliptic syntax as the exchange progresses" (1983:109).

3.2 The notion of aboutness

Philips (1989) uses the term 'aboutness' as a statistical notion accounting for the inter-collocation of collocates. He maintains that groups of words with shared environments can be built up, and the most pervasive groups can be selected as those most likely to express the aboutness of a text. In this paper, though, 'aboutness' is used as a semantic notion.

Discourse is about something. An Introductory Exchange, for instance, makes relevant a whole range of propositions which the discourse may be about; as talk progresses the formulation of 'being about' is the way in which individual knowledge is organised into shared knowledge.

In discussing aboutness, the underlying assumption, following Sperber & Wilson (1986:164), is that a communicator has made manifest his/her intention to communicate, that is to make clear to the addressee some sets of assumptions {I}. The addressee has to identify these assumptions. The Principle of Relevance identifies one member of {I} which is the 'presumption of relevance'. An important conclusion from Sperber & Wilson's presumption of relevance is that an assumption is not just a member of {I}, it is *about* {I}, and as such it can be confirmed or falsified by the contents of {I}.

Sinclair (personal communication) argues that the end of a unit delivers a certain proposition (or set of propositions) which is established temporarily as part of shared knowledge or accumulated significance of the participants. The state of the exchange as a unit (structural element) is an understanding of the significance of this proposition; this in turn contributes to the aboutness constraints. The key concept here is 'understanding', for it suggests that establishing relevance is a matter of negotiation. In a TF, the proposition carried by an exchange is relevant if it is interpreted as sharing the aboutness of the previous exchange. By identifying the TF as a rank in discourse we can observe the participants' orientation to aboutness: it is the participants' awareness of a TF that is conducive to their formulation of an aboutness relation.

Example 3: Data D1.2

Exchange		**Moves**
	J:	Dave was saying how well Mike's getting on with Melon
IE	S:	Yes he said that
	J:	Incredible now that they're actually married with a baby
	S:	He was saying he was going to claim in the end that it is his baby he obviously knew what she =
SE 1	J: (a)	= a::h just like Dynasty
	(b)	Brilliant wouldn't it =
PE	D: (c)	= em
FE 1	(d)	But what does she see in him ?
SE 2	S:	Sorry ?
BE 3	D:	But what does she see in him ?
SE 4	J:	What does he see in her ?
SE 5	S:	Ya exactly
SE 6	J:	The bloke she married ?
	S:	Ya
	J: (e)	He dribbles doesn't he ? Perhaps that's the attraction ((laughter))
FE 2	(f)	Do you remember him coming round to our rooms
	S:	Oh dear I couldn't believe it

The notion of aboutness can perhaps be exemplified if we work on the assumption that at any one time there is only one TF that is being dealt with in an exchange. This topic earns its status in the TF in two ways: (i) the opening exchange in the TF carries a potential topic and prospects support and development from the following exchange, and (ii) the next exchange establishes the message of the preceding exchange as part of its shared knowledge. A certain consistency of presupposition is thus demanded of an exchange and this gives it a coherence that precludes much diversity of subject matter. Example 3 illustrates the argument. In this complex TF, three friends talk about a couple whose courtship they have been following quite closely with some scepticism. In this example the IE opens up the channel of communication by activating a range of propositions about Mike's relationship with Melon - their compatibility with each other, their marriage and their parenthood, all of which are evaluated against the participants' presupposition that the relationship is fragile. The prospection that this exchange sets up is that one of its activated propositions will either be clarified, expanded or specified by the next exchange.

SE 1 fulfils the prospection by focusing on the status of the baby, thereby clarifying doubts about its legitimacy; the PE (*em*) in the structure offers an assessment of the information and the possibility of closing at this juncture. This would become an actual close if one of the participants took the opportunity to introduce a new TF, and another participant acknowledged this claim; in this case though we see a change in orientation achieved by changing the immediate subject without changing the overall topic. We see here that a change of subject does not necessarily require a change of TF, though in a new TF there is, of course, a change of subject matter. The exchange FE 1 (*but what does she see in him?*) marks a shift in the participants' interest from a general to a specific concern. The presupposition is still that 'she sees something in him', so we are still within the same TF.

Looking at all the exchanges, we find that the propositions they carry maintain a consistency of focus which can be considered as instantiating the higher level notion of aboutness underlying the first exchange. We may verbalise it as 'The actual state of Mike & Melon's relationship'. Each subsequent exchange expands our knowledge about the nature of their relationship, the baby and the possible reason for Melon's attraction for Mike. Each exchange interprets what has gone before and builds it in as part of its shared knowledge; having done that, the exchange can now prospect ahead. We can say that the notion of aboutness determines the relevance of an exchange in terms of its relationship to previous and subsequent exchanges.

It is interesting to consider how, after several exchanges, the TF introduced in example 3 comes to a close. In example 4 below, the IE no longer fulfils the aboutness criterion. What has gone before is not interpreted as shared knowledge. In this case, while not relevant any more to the existing TF, the new IE (*You didn't see my gorgeous bathroom did you? ...*) marks a new

TF in terms of the overall structure of discourse. The presumption of relevance changes with a change of subject. This also illustrates the fact that an exchange can still be felt as relevant at discourse level even if it is marked as a digression. It may not make a relevant contribution to the existing TF but the expressive behaviour is relevant in that its linguistic manifestation encodes a desire to change the subject. As Sperber & Wilson (1986:121) put it, "relevance may be achieved by expressing irrelevant assumptions, as long as this expressive behaviour is itself relevant".

Example 4: Data D1.2 (continues, after several exchanges)

Exchange		Moves
PE	D:	I feel sorry for the poor bloke if it does she'll scrape every penny out of him
	S:	Hmm
	J: (a)	Yeah, I bet. (8 secs) I can't believe people like that are actually married but to have children as well people you knew that well GOD. Incredible
IE	(b)	You didn't see my gorgeous bathroom did you ?
	S:	no I haven't seen that yet
	J: (c)	it's a lovely colour ((laughs))
SE	(d)	it's yellow
	S:	Ya ?
	D:	cork tiles

3.2.1 Maintenance for or challenge to aboutness

The construction and negotiation of aboutness in exchanges often makes use of specific techniques. One such technique is what was introduced in Section 2.2 as 'plane-change' (Sinclair 1981,1982,1991, 1992). Plane change signals a shift in discourse, from "saying something" to "talking about something" (1982:32). It is an act of encapsulation, often used strategically to introduce a new element in discourse or to mark exchange boundaries. In discourse tactics it is a common way of diverting the discourse from whatever response is prospected by the utterance preceding it. The data analysed for this and other studies suggests that, in the TF, the diversion generally results in a clear distinction between a maintenance of the aboutness constraints of a TF or a total disregard for the aboutness constraints. Example 5 below illustrates the former.

Plane change occurs in Exchange 2 (*Why do you ask? ...*) when, for some reason, ZJ does not answer the question of Exchange 1. What we see is the disruption of the structure of Exchange 1, as its anticipated response is not fulfilled, but is replaced by a new initiation. The new exchange, however, has incorporated the elements of Exchange 1 and supports the latter in the development of aboutness. What has happened here is that, although the challenge is not supportive of the preceding exchange, it supports the overall structure at the level of the higher unit of Topic-Framework. In Example 3, the exchange marked SE 2 (*sorry?*) also belongs to this category.

Example 5: Data: from field notes

Exchange		**Moves**
Exc 1	FH:	Are you working this evening Zan ?
Exc 2	ZJ: FH: ZJ: (a)	((smiles)) Why do you ask ? Phil and I thought maybe we come over for coffee ya I am
Exc 3	(b)	but come anyway, I need a break

The constraints of aboutness can also be totally disregarded, resulting in a challenge at the higher level of TF. In this case the attempt to generate an exchange is aborted. Plane change of course can be one of the techniques used in conversation to ignore the constraints of aboutness; other techniques still involve a general non-attendance to the prospection established by previous exchanges. In order to illustrate this use, let us go back to Example 3, which, for the sake of convenience, is reproduced below as Example 6.

The main issue at this stage of the conversation is 'what she sees in him', unresolved in two preceding exchanges. In SE 4, endorsed by SE 5, a counter challenge ensues in an attempt to resolve 'what he sees in her' instead. This is a conscious effort to steer the discourse in another direction by ignoring the constraints of aboutness. The interactive potential of SE 4 and SE 5 is yet again ignored by SE 6 which focuses back on 'why she is attracted to him', therefore attending to the prospection established by BE 3. As a consequence of a speaker's option to challenge being exercised, we have here two failed TFs because the aboutness criterion is not met. Coulthard & Brazil (1981:84) explain this phenomenon by suggesting that "a speaker can do anything he likes at anytime, but what he does will be classified as a contribution to the discourse in the light of whatever structural predictions the previous contribution (his own or another's) may have set up." And being irrelevant, they point out, is always one of the options available to speakers.

Example 6: Data D1.2

Exchange		**Moves**
BE 3	D:	But what does she see in him ?
SE 4	J:	What does he see in her ?
SE 5	S:	Ya exactly
SE 6	J: S: J: (e)	The bloke she married ? Ya He dribbles doesn't he ? Perhaps that's the attraction ((laughter))

Relevance is seen here in terms of a relation between an exchange, or more than one exchange, and a TF. For an exchange to be relevant it must be interpreted as being 'about' the TF. This, as noted in Section 3.2, does not exclude the possibility that an exchange is also relevant if the expressive behaviour is relevant.

3.3 Relevance as a property of structure in discourse

As pointed out in the introductory section of this paper, relevance in the Topic-Framework must be seen as a property of discourse whereby it is partly the position of the exchange in the structure that determines its relevance in terms of what it does in that structure. This claim brings us to the other key concept in relevance, namely prospection.

Sinclair (1985, 1991, 1992, 1993) suggests that there are two essential features of the interactive nature of language in general, namely, prospection and encapsulation. While prospection is seen as fundamental to the structure of the exchange in spoken discourse, whereby "each initiation prospects that the utterance following it will be interpreted under the same set of presuppositions as the initiation itself" (1992:83), encapsulation contains an act of reference to the I-R pair and is retrospective in nature. Going along the same lines established by Sinclair, I maintain that by examining the structural positions of the elements in a TF we can arrive at a description of how prospecting and prospected elements are distributed in the structure. There are, of course, constraints governing their behaviour, but it is beyond the scope of this paper to examine what these constraints are. Let us just note for the moment that an element can prospect another element in the structure and that the type of prospected element will vary according to the function of the prospecting element. The fulfilment of prospection can be considered as a fulfilment of relevance.

The next section seeks to demonstrate how the TF is realised through exchanges that carry out different functions in the structure by prospecting and encapsulating.

3.3.1 Prospection as a feature of structure

The notion of structure is a corollary of a dynamic model of discourse which has its origins in the seminal ideas presented in Sinclair et al (1972) and Sinclair & Coulthard (1975). The prospective features of spoken discourse were brought out when initiations were assigned predictive value according to how they preclassified what followed. This has developed over the years into a rigorous theory of discourse with different manifestations of prospection being highlighted. Many scholars have gone along with this view of discourse, leading to a proliferation of studies which describe the phenomenon: Tadros (1985) talks of prediction, Tsui's work (1986) is on the characterisation of utterances, Coulthard & Brazil (1981:83) describe the predictive power of the structural frame, Stubbs (1983:94) uses the term 'expectation', and proposes that hearers construct the coherence of what they hear by analysing structurally the discourse in which they are involved. Several terms have therefore been devised to describe a phenomenon similar to prospection. I am using the term according to the criteria that inform much of Tsui (1986) and Sinclair's (1981, 1991, 1992) theoretical framework on discourse.

Sinclair's (1991) approach to the relation between a sentence and the previous text captures the process of prospection in text. If we relate the notion of prospection to TF we note that, at the beginning of a TF, an Introductory Exchange prospects support from the exchange that follows it. The occurrence of the prospected exchange fulfils the interactive potential of the whole of the preceding exchange. The new exchange now assumes the status of 'current text' and can prospect ahead, thereby changing its status from a prospected to a prospecting exchange. The meaning of the previous exchange, following Sinclair (1992), becomes part of the shared knowledge that helps participants interpret the text at any point. With each prospection that takes place, the same process is repeated unless the exchange is challenged or stops prospecting by coming to a close. This completes the TF.

To sum up then, my position on the notion of prospection is that a TF must have as its beginning an obligatory exchange that introduces a topic potential. This exchange allows us to classify what follows, that is the prospected exchange, as being either:

(i) a supportive exchange that develops the budding topic, or

(ii) a non supportive exchange that ignores the interactional potential of its previous exchange, resulting in (a) the opening of a new topic, or (b) the supporting of the old one in terms of the overall structure of the TF.

We have so far defined the elements of TF in terms of a relation between a prospecting exchange and a prospected exchange. If we consider the position that such exchanges may have in the TF, we note that if an exchange prospects another exchange it cannot be final (except for the exchange that proposes closure) unless there is an unexpected dislocation of the discourse. If it is prospected it cannot be initial; it has to be preceded by another exchange. It is beyond the scope of this paper to discuss the specific positions that the different kinds of exchange take in the TF. In this context we can note that, although the possibilities of topics are unlimited, norms for how we conduct ourselves in conversation still constrain what we should talk about and at what stage this is appropriate. Thus, constructing a TF requires a well defined process beginning with an offer of topic potential, followed by a support in terms of topic development or elaboration and a consensus that there is sufficient convergence of knowledge that talk can now move towards a preclosing behaviour.

4. Concluding remarks

In this paper I have set out to examine the status of topic as a dynamic element in discourse. By proposing that the TF carries the patterns of topic I have attempted to account for the notion of relevance in the TF. I have also argued that the presumption of relevance is central to the TF structure because it accounts for how an exchange relates to the one before it, and the one it prospects. Using the notion of aboutness I have tried to illustrate the operating constraints within the structure. By adopting a prospective view of discourse, I hope to have shown that prospected elements are attempts to establish relevance, and a fulfilment of prospection must therefore surely be a fulfilment of relevance.

Author's Address
Language Centre, ITM, Shah Alam, Malaysia.

Notes

1. I would like to thank Dr Amy Tsui and Dr Hawa Rohany for allowing me to use their data.
2. I have adapted the term 'speech-in-action' from Cheepen (1988), who in turn took it from Malinowsky. The term was originally coined to describe utterances which are used "in connection with vital work". Cheepen draws parallels between the linguistic behaviour of Malinowsky's fishing activity and that of the interactional encounters she is concerned with. Some of her examples are *it might be hot* and *pass us the ashtray.*
3. In the model proposed (see Hazadiah 1991) the Maintenance exchange is prospected by the pair IE and SE when the IE is recognisably narrative initial. The IE realises a set of propositions and the SE establishes them. The role of the ME is to advance the narration to a given point in terms of its chronology, and to make a set of common presumptions which reflect the change from the preceding set. The Appraisal Exchange assesses the current relevance of the TF of a narrative extension and maintains its stability by ensuring that there is sufficient convergence between the narrator and the participants. Thus it does not propose a close. This category of exchange allows for the recursion of Maintenance exchanges.

References

Beekman, John & J. Callow. 1974. *Translating the word of God.* Grand Rapido, Michigan: Zondervan.

Beaugrande, Robert de & W. Dressler. 1981, 1988. *Introduction to Text Linguistics.* Singapore: Longman.

Brown, Gillian & G. Yule. 1983. *Discourse Analysis.* Cambridge: Cambridge University Press.

Cheepen, C. 1988. *The Predictability of Informal Conversation.* London: Pinters.

Coulthard, M. & D. Brazil. 1981. "Exchange Structure". *Studies in Discourse Analysis* ed. by M. Coulthard & M. Montgomery, 82-106. London: Routledge & Kegan Paul.

Grice, H. P. 1975. "Logic and Conversation". *Syntax and Semantics 3: Speech Acts* ed. by L. Cole & J. L. Morgan, 41-58. New York: Academic Press.

Halliday, M. A. K. 1961. "Categories of the Theory of Grammar". *Word* 17, 241-92.

Hazadiah M. D. 1991. "The Structure of Topic in Conversation with Special Reference to Malaysian Discourse". Unpublished Ph.D. thesis. University of Birmingham.

____________. 1992a. "The Exchange as an Element of Topic-Framework". Paper Presented at the 4th Systemic Conference, Nottingham, August 1992.

____________. 1992b. "What the Data Says about Real Conversation: Some Considerations for Instructional Materials Development". Paper Presented at the National Seminar, ITM, Malaysia, September 1992.

Philips, M. 1989. *Lexical Structure of Text (= Discourse Analysis Monograph 12).* Birmingham: University of Birmingham.

Schegloff, E. & H. Sacks. 1973. "Opening up Closings". *Semiotica* 7,4. 289-327.

Sinclair, J. M. 1981. "Planes of Discourse" *The Two Fold Voice: Essays in Honour of Ramesh Mohan* ed. by S. N. A. Rizvi, 70-89. Saltzburg: Saltzburg University.

__________. 1985. "On the Integration of Linguistic Description". *Handbook of Discourse Analysis* ed. by T. A. van Dijk, 13-28. London: Academic Press.

__________. 1987. "Mirror for a Text". Unpublished Manuscript, University of Birmingham.

__________. 1991. "Trust the Text". *Advances in Systemic Linguistics: Recent Theory and Practice* ed. by M. Davies & L. J. Ravelli, 5-19. London and New York: Pinters.

__________. 1992. "Priorities in Discourse Analysis". *Advances in Spoken Discourse Analysis* ed. by M. Coulthard, 79-88. London: Routledge.

__________. 1993. "Written Discourse Structure". *Techniques of Description: Spoken and Written Discourse* ed. by J. M. Sinclair, M. Hoey & G. Fox, 6-31. London and New York: Routledge.

__________, I. J. Forsyth, M. Coulthard & M. C. Ashby. 1972. *The English Used by Teachers and Pupils.* Report to SSRC. Mimeo: University of Birmingham.

__________ & M. Coulthard. 1975. *Towards An Analysis of Discourse: The English Used by Teachers and Pupils.* London: Oxford University Press.

__________ & D. Brazil. 1982. *Teacher Talk.* London: Oxford University Press.

Sperber, Dan & D. Wilson. 1986. *Relevance, Communication and Cognition.* Oxford:Basil Blackwell.

Stubbs, M. 1983. *Discourse Analysis. The Sociolinguistic Analysis of Natural Language.* Oxford: Basil Blackwell.

Tadros, A. 1985. *Prediction in Text* (= *Discourse Analysis Monograph 10*). Birmingham: University of Birmingham.

Tsui, A. 1986. "A Linguistic Description of Utterances in Conversation". Unpublished Ph.D. Thesis. University of Birmingham.

Wilson, J. 1989. "Conversational Matters: Towards a Theory of Conversation". Unpublished Ph.D. Thesis. Queens University, Belfast.

Interpreting Multi-act Moves in Spoken Discourse

Amy B. M. Tsui
The University of Hong Kong

Abstract
Since the proposal of a framework of linguistic description of spoken discourse by Sinclair & Coulthard (1975), subsequent studies based on this model have largely focused on the structure of 'exchanges'. Relatively little attention has been paid to 'move' and 'act', and even less to move structure and the relationship between acts in moves consisting of more than one act. On the basis of examining conversational data, this paper investigates the relationship between acts in multi-act moves and attempts to formulate rules which, it is proposed, govern the interpretation of these moves.

0. Introduction

Using classroom data as a starting point in formulating a descriptive model of spoken discourse which was largely based on Halliday (1961), Sinclair & Coulthard (1975) proposed that spoken discourse is describable in terms of a hierarchical rank scale. This rank scale consists of the following units in ascending order: act, move, exchange, transaction and lesson. Subsequent studies based on the Sinclair & Coulthard model have largely focused on the unit 'exchange' and its structure (see for example Coulthard & Montgomery 1981; Stubbs 1983; Berry 1987; Coulthard 1992). Relatively little attention has been paid to the units below an exchange, that is, move and act. So far, no follow-up work has been done on move structure, that is on the relationship between acts in moves which consist of more than one act. The aim of this paper is to examine the relationship between different kinds of acts in different types of multi-act moves and to propose rules which might govern the interpretation of these moves. The discussion is supported by examples from a corpus of conversations between native speakers of English.[1]

0.1 Multi-act moves

In the Sinclair & Coulthard model, a move is the smallest free unit of spoken discourse, and an act is the unit next below, which is the smallest unit. Drawing an analogy with morphemes and words in grammar, they argue that just as a word is the smallest free unit and can be made up of one or more than one morpheme, a move can be made up of one or more than one act.

Example 1 [Sinclair & Coulthard 1975:97]

T: *What cuts the piece of wood? Will the scissors cut the piece of wood?*
S: No.
T: Let's try. No, it won't.

In example 1, the responding move (*No*) consists of only one act, while the initiating and follow-up moves consist of two acts each. This type of multi-act move is also common in conversation, as in example 2, where all moves consist of more than one act each.

Example 2 [B:A:A:4:3]

M: Do you want me to pick you up? Are you are you in your office right now?
X: No, I'm I'm going to the h- I'm at the Great Hall. I have to go to the head's office.
M: Alright, maybe afterwards.

In accounting for the relationship among the acts in multi-act moves such as those given above, Sinclair & Coulthard (ibid) borrowed the concept of structure in the grammatical unit 'group'. The nominal group *the man next door* is made up of 'words', a word being a unit below group. The headword in this nominal group is *man*, which is obligatory; *the* is a modifier, *next door* is a qualifier, and both are optional. When a move consists of more than one act, one of them will similarly be the obligatory 'head act' carrying the illocutionary force of the entire move; the rest will be optional 'pre-head' and/or 'post-head' acts, depending on whether they precede or follow the head act. This description accounts for the structural relationship among acts but leaves the question of identifying the component acts unresolved. So, how do we determine which is the head act and which are the pre- and post-head acts?

0.2 The discourse value of acts

In answering the above question, Sinclair & Coulthard propose that the sequential location of the act is the most important factor in determining what act has been performed: "the discourse value of an item depends on what linguistic items have preceded it, what are expected to follow and what do follow" (ibid:34; and see also Schegloff & Sacks 1973).

Coulthard & Brazil (1981:83) explain the notion of discourse value by using the example of the adjective *cuddly* in the noun groups *the cuddly black cat* and *the black cuddly cat. Cuddly* belongs to two separate sub-classes of adjectives: qualitative and classifying. What determines which subclass it belongs to is the sequential location, that is, what precedes and follows it. If it precedes *black*, it is a qualitative adjective, and if it follows *black* and precedes *cat*, it is a classifying adjective. The discourse value of an act is similarly determined by what precedes and what follows it. The only difference is that spoken discourse takes place in real time. Speakers may start off by saying something and then change their minds. In contrast with written discourse, however, they cannot erase what they have said. What they may do, however, is change the direction of the discourse in the middle of the utterance. The addressee will then have to revise his/her interpretation of the uncompleted act in the light of what follows. Let us take F's utterance in example 3 for instance.

Example 3 [B:E:A:4:4]

F: What time are you leaving on on Monday next week?
D: Oh I'm not leaving till the evening, so I mean it's going to be a bit=
[
F: Alright, alright.
D: =complicated doing the packing in there if I'm free for the afternoon, we ah
F: Well, I'll take you to the airport as a as a ah ((laughs)) If you if you did it for me, I'll be very happy to take you to the airport.
D: Well that's very nice. Thank you very much. ((laughs))

D is a visiting professor and F has asked him to give a lecture to his students. F's highlighted utterance would be interpreted by D as an offer to take him to the airport, up to the point where F changes the direction of the discourse in the middle of the utterance by producing the conditional *if*. As soon as F produces the conditional clause, whatever precedes it is aborted. The uncompleted offer should be treated as though it has never been performed, and the act that F has performed is a contingent offer, not an offer. F's move is therefore a single-act rather than a multi-act move. F's contingent offer is responded to with an acceptance and thanking, hence implying D's acceptance of the condition that he gives a lecture to F's students.

Because of the real-time nature of spoken discourse then, participants have to interpret the discourse as it unfolds, and they constantly have to revise their interpretation in the light of what follows. The data examined in example 3 above represents an uncompleted act. In what follows, we shall look at acts which are fully performed and investigate how sequential location determines their discourse value.

0.2.1 Head acts and starters

In example 1 above, the teacher starts off with the interrogative *What cuts the piece of wood?*. This is an elicitation which expects a response from the pupils. However, what actually follows this elicitation is not a pause indicating the boundary of the initiating move, but rather another elicitation: *Will the scissors cut the piece of wood?*. Hence, the first elicitation is 'pushed down' or 'edited' (Coulthard 1981:13) and becomes a 'starter' whose function is "to provide information about or direct attention to or thought towards an area in order to make a correct response to the initiation more likely" (Sinclair & Coulthard 1975:40). The second elicitation, which is followed by a pause, and possibly by the teacher looking round the class for responses, realises the head act, which is obligatory and to which the pupils are expected to respond.

The following initiating move has different components presented in a different sequence from that in example 1. It consists of an elicitation followed by an informative and another elicitation, and yet the same 'pushing down' mechanism seems to be at work.

Example 4 [Sinclair and Coulthard 1975:35]

T: What about this one? This I think is a super one. Isobel, can you think what it means?
P: Does it mean there's been an accident further along the road?
T: No.

The first elicitation and the informative act as starters, and the second elicitation realises the head act.

On the basis of classroom data such as that in examples 1 and 4 above, Sinclair & Coulthard (1975:35) make the following generalisation of how the head act is identified in classroom discourse:

> ... while speaking the teacher produces a series of clauses classifiable as statements, questions and commands in situation. If the teacher then allows a pupil to speak, these clauses are seen as initiating, and have the discourse value of informative, elicitation, directive; if he follows one of these clauses

> with another it is pushed down to act as a starter.
>
> In any succession of statements, questions, and commands the pupil knows that he usually only has to respond to the final one, and only that has an initiating function.

While this generalisation does seem to apply to the examples that Sinclair and Coulthard give, there are many instances of multi-act moves to which it does not apply, even in classroom discourse. Let us take for instance the following piece of classroom data.

Example 5 [Sinclair and Coulthard 1975:67]

T: What's the next one mean? You don't see that one around here. Miri.
P: Danger falling rocks.
T: Danger falling rocks.

According to Sinclair & Coulthard, the elicitation would have been pushed down by the following statement to become a starter and the head act would have been an informative. Yet this is not the case. It is the elicitation which is the head act and the one to which the pupils respond. The informative, because it is preceded by an elicitation, is relegated to the position of a post-head, with the function of providing additional information to help the pupils respond to the elicitation (ibid:41). And indeed, the pupil did respond to the first rather than the second act. Moves with this kind of sequencing are also common in conversational data, as in example 6.

Example 6 [B:D:B:1:1]

M: Can you get a message to Larry? He's not home and he's not in the office.
H: Yes.

Here, we have a move consisting of a requestive and an informative. As in example 5 above, the first act is the head act and is not pushed down by the second act to become a starter. The informative becomes a post-head act providing a reason for the request. This can be seen from the fact that H responds to the request rather than to the informative.

The question that the above data raises is whether the identification of the head act is entirely up to the individual recipient or whether there are rules which govern this identification. The rest of this paper will attempt to address this question by focusing on acts in the initiating move. In the discussion that follows, four identification rules are proposed. It is important, however, to interpret 'rule' not as something which is inviolable in terms of what will actually occur in conversation, but rather, as Berry points out, as "invariant in that the existence of the rule will always be assumed for the purpose of interpreting what occurs ... The rules specify what is typical, thereby providing a basis for explanation of our recognition, interpretation and evaluation of the untypical" (1982:38-9). In other words, when the rules are violated, we will be able to recognise the violation and to provide an explanation for it.

1. Identifying the head act

In the ensuing discussion, we will deal with fully performed acts rather than with uncompleted acts. We will examine the relation among four major categories of acts occurring in the initiating move: requestives and directives which solicit non-verbal actions, and elicitations and informatives which solicit verbal responses.[2] Requestives give the addressee the option of not complying, whereas directives do not. Elicitations solicit a verbal response which provides the missing information, agreement, confirmation, and so on (see Tsui 1992a), whereas informatives solicit a verbal response which acknowledges the receipt of information (for a detailed description of these categories, see Tsui 1991, forthcoming).

1.1 The co-occurrence of elicitations

Let us start by examining moves which consist of more than one elicitation. In example 1, the first elicitation is a wh-interrogative and the second is a polar-interrogative. The first requires the pupils to produce a response from a closed set of possible responses. The second proposes one response from the set and asks the pupils to make a choice between *yes* and *no*. This type of sequencing occurs in classrooms when the teacher tries to help the pupils arrive at the right answer by narrowing down the choices available to them. In cases of this type, the first elicitation is pushed down and becomes a starter, and the second elicitation becomes the head act. This can be seen from the pupil's response, which addresses the head act, and the teacher's acceptance of it as an appropriate contribution to the discourse. Moves with this kind of sequencing of acts also occur in conversation, as the following example illustrates.

Example 7 [B:E:A:4:1]

A: *I'm just wondering are y- ah you're going to be here till what date? Till th- will you be will you be here on on next Monday or you'll be gone?*

B: Yes we leave on Monday.

A: That's too bad.

As in example 1, we have a second elicitation narrowing down the choices of the first elicitation by giving a specific date. The second elicitation becomes the head act to which B responds and the first becomes a starter. Here is a further example.

Example 8 [C:1:A:1:1]
H gave X back ten dollars that he borrowed from S.

H: I guess oh that's oh what's that? Oh what're you doing? Ah you're putting you want to give me a receipt?
S: No.
H: ((laughs)) Looks like you'll give me a receipt.

In example 8, the highlighted elicitations go from general to specific and the last elicitation realises the head act which is responded to.

On the basis of the regularities exhibited in the above pieces of data, we can formulate the First Identification Rule as follows.

In a multi-act move, when an elicitation co-occurs with another elicitation which is more specific, the more specific elicitation realises the head act.

Let us now consider the following piece of data in the light of this rule.

Example 9 [Sinclair & Coulthard 1975:71]

T: Can you say anything about these symbols?
S: They're all easy to draw.
T: Er Yes. Reasonably easy to draw. That's a good point, isn't it, because if they were very complicated it wouldn't be very easy to put them on whatever you wanted to put them on.
Can you think of anything that they would be put on? Would they just be written or painted or drawn on the sand?
S: They might be on the rocks.
T: Yes, they were mainly carved in rocks.

The teacher's second initiating move, which is highlighted, consists of two elicitations. The sequencing is similar to that in examples 1 and 7 in that the second elicitation is realised by a polar-interrogative, narrowing down the choices of correct answers. According to our first identification rule, it is the head act to which the pupil is expected to respond. However, the pupil responds to the pre-head act rather than the head act. And yet the teacher accepts the answer as an appropriate contribution to the discourse.

This is a case in which the identification rule is violated and the violation is accepted by the participants as in order. This can be explained by the fact that it is a common practice for teachers to ask a question and then follow it up with a series of more specific questions to help pupils arrive at the right answer. In example 9 above, the pupil responds to the initial elicitation instead of responding to the subsequent more specific one. The 'implicature' (Grice 1975) is that the pupil does not need the specific elicitation to help him arrive at the answer and that the teacher has probably misjudged his ability.

1.2 Elicitations and informatives

Let us now examine moves in which elicitations co-occur with informatives. In example 5 above, we saw that when an elicitation is followed by an informative, the elicitation realises the head act and the informative becomes a post-head act. In the following example, we have two elicitations followed by an informative.

Example 10 [B:G:A:1:5]

J: *Well, what's he referring to? Which of your researches?*
I mean you've done all kinds of researches.
[
L: I don't know. I don't know, I think they they really they don't like our attitude research which is published in refereed journals and so ah

In J's move, the second elicitation is more specific than the first. According to the first rule of identification, the first elicitation is relegated to a starter. The second elicitation is not pushed down by the following informative. Rather, it realises the head act to which L responds by specifying a particular piece of research, and the informative is relegated to the position of a post-head act.

Informatives may also precede elicitations in moves, as in the following examples.

Example 11 [C:2:A:1]

A: That Taipo Road is still blocked?
B: Yeah. *And aw we kept hearing the helicopter going over. Did you hear them?*
A: Mm-mm ((shakes head)).

Example 12 [B:C:A:1:2]

H: *See, he has a son at was in school last year ah does he have to re-apply?*
X: Ah yes, I think so.

In examples 11 and 12, we can see that, instead of acknowledging the receipt of information conveyed in the informatives, the addressees respond to the elicitations. The informatives function as starters which provide the context for interpreting the elicitations.

From the regularities exhibited in the above pieces of data, we can formulate our Second Identification Rule as follows.

In a multi-act move, when an elicitation co-occurs with an informative, the elicitation realises the head act.

1.3 Requestives and informatives

Requestives are acts which, if responded to positively, either solicit a non-verbal action from the addressee, or commit the speaker to a non-verbal action. They include acts such as requests for action, requests for permission, offers and invitations. Requests for action include those which solicit action from the addressee as well as those which commit the speaker to carry out the action (see Tsui 1991, 1992b, forthcoming).[3] Consider first the following examples (15, 16 & 17), in which requestives are followed by informatives.

Example 13 [BCET:A:B:51]

C: *Can I just use your lighter? I've run out of matches.*
B: Oh aye. Ahhhh ((gives C the lighter))
C: Ta.

In example 13, C asks B's permission to use his lighter. This requestive is followed by a statement of the reasons for using B's lighter. As we can see from the rest of the exchange, B responds to the requestive and gives C the lighter. The informative becomes a post-head.

Requestives are intrinsically face-threatening acts (Brown & Levinson 1987). By asking the addressee to do something (or not to do something), or by asking for the addressee's permission to do something, the speaker is imposing his or her own will on the addressee. As Goffman points out, "a request consists of asking license of a potentially offended to engage in what could be considered a violation of his rights" (1971:114). There are, however, various ways of minimising the face-threatening effects or the possible offence caused. One way is to provide a reason for making the request, hence demonstrating that the speaker is "fully alive to the possible offensiveness of his proposed act and begs sufferance" (ibid). Seen in this light, it is not difficult to see why the informative realises a post-head act with an auxiliary function in the move. Here are some further examples.

Example 14 [B:C:A:2]

J: *um I wonder if you might give my apologies. I'm - not going to make it tomorrow.*
[
K: Okay
K: Okay Jack, sure.

Example 15 [B:A:A:4:2]

X wants to talk about a research paper that she is doing jointly with M and asks if they could discuss the paper after lunch.

X: *What about could we talk about it after? I have to go to I have to go to ah the head's office.*
M: Alright, we'll we'll just we'll make a time after right after the ah lunch today alright? We'd just be we just say we have to go to class or something.
X: Okay.

In both examples (14 & 15), the highlighted requestives are followed by informatives. In both cases, as can be seen from the responses, it is the requestive which is the head act carrying the illocutionary force of the entire move, and the informative realises the post-head.

Let us now consider moves in which requestives are preceded by informatives. Consider the following piece of data, where H wants to change his lunch appointment with X to the following day

Example 16 [B:A:A:1:1]

1 H: Hey I forgot something. I have to go to lunch with Alice
2 ((laughs)) to see the ha you know the the the ah videotape=
[
3 X: ((laughs))
4 H: =of the of that show we did at the hotel?
5 X: Yup, yup.
6 H: So I I I already booked the ticket. So I have to go with her.
7 *I tried to talk her out of it out of it but she wants to go.*
8 ((laughs)) *So may be we have we have lunch tomorrow?*
9 X: Okay yeah it's yes it's fine, yes.

In the highlighted move, the requestive to get X to agree to change the lunch appointment is preceded by three informatives. However, rather than acknowledge the information given in the informatives, X responds to the requestive. Apart from an explicit solicitation of a response from H in line 4, where a rising intonation is used, X also withholds the acknowledgement of the news given up to the point when the requestive is made. This item of information is therefore clearly treated as a preface to something else, rather than as being of interest in its own right. Drew (1984:141) points out that "such withholdings of any appreciations/receipts of news display the recipient's treatment of the reportings as preludes to something arising from them". In other words, the head act is the requestive and the informatives become preheads, with the function of providing a detailed account of the events which lead to the postponement of the engagement so as to reduce the likelihood of causing offence. In the following piece of data, we have two informatives, one preceding and one following a requestive.

Example 17 [B:D:B:4:1]

H: Listen, I'm calling I'm calling about the we we have a um Public
Speaking Committee here at the university and they have public
lectures every so often and every year they have one in English.
I was wondering if you would be interested in giving y' know just
a short lecture. They want somebody who who's who's a good
speaker and who's prominent in the community to give a speech for
about a half hour to 45 minutes. May be they'd ask questions.=
[
X: Yeah.
=May be they wouldn't too, knowing them.
H: Well no, this y'know this group is a um what w I would prime and
ask questions.
X: No no no. Ah the the only thing about this ah is that ah I feel
hesitant ah to do it because basically I don't regard myself as
being in any way prominent or or ah qualified to make public
speeches um y'know I don't I don't regard as any part of my role
to to get up and make speeches for public consumption.

H's move in lines 1-7 consists of an invitation realised by *I was wondering if you would be interested in giving y'know just a short lecture*. It is preceded by an informative providing a context for the invitation (lines 1-3): that X is invited to give a public lecture organised by the Public Speaking Committee of H's institution. It is followed by another informative which gives reasons why X is invited (lines 5-7): that X is a *good speaker* and *prominent in the com-*

munity. The conversation then digresses a little on asking questions. But we can see from X's last move (lines 12-16) that instead of acknowledging the receipt of information conveyed, he addresses the invitation by saying that he is reluctant to accept it and by giving reasons for his reluctance. In other words, X identifies the head act in H's contribution (lines 1-7) as an invitation, and it is this that X addresses. This type of sequencing is by no means uncommon and is socially motivated. As Drew (1984:140) points out,

> Detailing the story behind an invitation can allay or mitigate its possible intrusiveness in the recipient's (unknown) plans, by forming an event as a sociability already planned independently of the recipient. Hence the speaker is not being presumptuous in expecting or depending upon the other's availability; though he may emphasise his desire for the recipient to attend ... and both may express disappointment if the recipient is unable to, the reporting displays that a declination will not jeopardize the event.

The following is another example of this kind of sequencing.

Example 18 [B:A:A:4:1]

X: Are are you ah going to be in your office for
M: Yeah I I think - now somebody's with me.
X: Yeah ah I'm thinking say around noon?
M: Yeah but y'know at noon I was going to ask you would this Professor Yang from Shanghai is here and we've invited him for lunch at Ch at ah Chung Ying at twelve to meet some of the staff. We're trying to=
[
X: mhm
M: =get hold of you ((laughs)). Would you be free?
[
X: mhm
X: What um
M: Just
[
X: Who's Yang?

In the above example, X rings M up to see if they can get together around noon. M's highlighted move starts off with *...at noon I was going to ask you would...* As we can see from the rest of the conversation, M was going to say something to the effect of "would you be free to join us for lunch?", which would have been an invitation. But M changes the direction of the discourse and produces two statements about a lunch meeting with Professor Yang. This is followed by an invitation to join them for lunch. The informatives would have realised the head act had M not continued to produce a further act. But once the invitation is produced, the two informatives become pre-heads which provide the context for the invitation. Their function is further confirmed by the fact that before M accepts or declines the invitation, she asks for more information about Professor Yang. Once the context is established, M accepts the invitation, as can be seen from the rest of the conversation reproduced below.

Example 19 [B:A:A:4:1]

X: Who's Yang?
M: Ah ah he's professor from um Shanghai Normal School or something. He's professor of English. He's been in here all morning. He's talking to Christine now. He's ah h- his contact is with Johnson. He's going=
[
X: uhuh
M: =to have lunch with us too. So we're=thinking of all going down to=
[
X: mhm
M: =Chung Ying at about twelve. I booked a table. Well simply because=
[
X: mhm
M: =he'd like to meet teachers and find out what we're doing first year English, the electives and stuff. *So if you're free.*
X: Okay.

One important feature of a requestive that distinguishes it from an elicitation and an informative is that a non-verbal action is involved. The feasibility of the non-verbal action is crucial in determining whether the requestive will be accepted or declined. Very often, therefore, before the requestive is performed, the speaker ascertains that the pre-conditions obtain. This is sometimes achieved by a pre-sequence, that is a sequence of turns which typically lead up to something else and which are treated by participants as being relevant in terms of what they foreshadow rather than in terms of what preceded them (Schegloff 1978:97). At other times, however, it is achieved in the same move in which the requestive is performed. The following is a piece of data in which an informative, co-occurring with a requestive, serves the function of establishing the pre-condition for the requestive.

Example 20 [B:E:A:3:3]

X: It's probably going to rain huh? Jesus, it hasn't rained since October=
[
H: I gu-
X: =thirtieth.
[
H: Yeah.
H: Yeah, but it's very dark here but it's it's not a thunderstorm, it's just kinda lousy drizzle rain which I don't have a bloody ah umbrella here.
X: Yeah, I got one here. You can borrow mine.
H: Oh yeah oh yeah, alright, do you think you can send it right through? ((laughs))

In the highlighted move, X is offering H his umbrella. One of the pre-conditions for the felicitous performance of an offer is that the speaker is able to carry out the proffered action. Therefore, before making the offer, X tells H that he has got an umbrella with him. In other words, the informative *I've got one here* serves the function of ascertaining that the pre-condition for making the offer obtains. The informative therefore realises the pre-head to the offer. Although, in this particular piece of data, the offer is facetious, since the participants are on the telephone and are miles apart, the sequencing is arguably no different from that found in a serious offer.

From the regularities exhibited in moves in which informatives and requestives co-occur, we can formulate our Third Rule of Identification.

In a multi-act move, when a requestive co-occurs with an informative, the requestive realises the head act.

1.4 Requestives and elicitations

We turn now to moves in which a requestive is followed by an elicitation.

Example 21 [B:F:A:1:3]

H: ...I just want to ask you can I pass it onto a colleague? Or do you want it back?
X: Sure, oh no, pass it on.

H starts off with a requestive asking for X's permission to pass the novel that X lent him onto a colleague. This is followed by an elicitation addressing one of the pre-conditions for getting permission from X, namely, that he does not want it back. X responds by giving H permission to pass the novel on. The head act of this move is therefore a request for permission, and the elicitation serves as the post-head whose function is to check up on the pre-condition. This is supported by the fact that even if X had simply said "Oh no, I don't need it back", hence ascertaining that the pre-condition obtains, X's response would still be interpreted as granting permission to H. Consider the following piece of data.

Example 22 [B:A:A:4:3]

M: Do you want me to pick you up? Are you are you in your office right now?
X: No, I'm I'm going to the h- I'm at the Great Hall. I have to go the the head's office.
M: Alright, maybe afterwards.

Here, M is offering to pick X up to go for lunch. But one of the pre-conditions for an offer is that there is a need for action. Therefore, immediately following the offer, M checks to see if X is in her office because if she is not, there is no need for the action. Again, the requestive is the head act and the elicitation is a post head whose function is to check up on the pre-condition. X responds by "reporting" (Drew 1984:135) where she is and what her present engagement is. That X's response is not to be interpreted as merely addressing the elicitation but as implicitly declining the offer can be seen from the fact that in the follow-up move, instead of acknowledging or commenting on the information given, M minimises the face damage caused by the offer being turned down. This is similar to reportings in invitation sequences in which, instead of rejecting the invitation, a reporting of events is given and the recipient is left to formulate the upshot of the reporting.[4]

The following is an example in which a requestive is preceded rather than followed by an elicitation.

Example 23 [B:C:A:3:2]

A: Will will you be will you be around ah say at 4:30? Shall I bring him round?
B: Ah yes, I'll be I'll be here.
A: Alright, I'll I'll just tell him to stop at 4:15 and we'll go down to meet you.

A offers to take a visitor to B's office. The elicitation tries to establish that the pre-condition for the offer obtains, that is, B is still in his office at four-thirty. Hence, the elicitation functions as a pre-head to the offer. As we can see from the data, B's response to the elicitation is interpreted by A as an acceptance of the offer. In other words, despite the fact that in both examples 22 and 23, the addressees explicitly respond to the elicitations rather than the requestives, the responses are to be interpreted in the light of the requestives rather than the elicitations.

From the data given above and taking into consideration the examples discussed in the previous section, we can re-formulate the Third Identification Rule as follows.

In a multi-act move, when a requestive co-occurs with an informative or an elicitation, the requestive realises the head act.

1.5 Directives and informatives

Finally, we shall examine the co-occurrence of directives with other major categories of initiating acts. Directives share an important feature with requestives in that they both solicit a non-verbal action. The only difference is that while requestives give the addressee the option of not complying, directives do not. Directives include those in which the action is for the benefit of the speaker, for example, 'order' and 'threat', as well as those in which the action is for the benefit of the addressee, for example, 'advise' and 'warn' (see Tsui 1991, 1992b, forthcoming).

Consider S's highlighted move below, where a directive which gives X some advice is preceded by an informative.

Example 24 [C:2:A:6:2]

S: *Anyway um I heard well I know because I went to Taiwan this summer. Don't try to exchange your money at a bank or at the airport but get your friend to take you to ah like a jewellery store and they'll exchange it a lot cheaper there.*
[
X: Ah
X: They have better rate.
S: Mhm
X: Good tip. Thanks.

We can see that X does not respond until the advice is given, indicating that the informative is understood as a preface to something else rather than given in its own right. That the illocutionary force of S's move is one of advice can be seen from the fact that X thanks S for the tip. Hence, the informative realises a starter and the directive realises the head act. The starter serves the function of providing the grounds on which the advice is given. In the following example, the sequence is reversed: a directive is followed by an informative.

Example 25 [B:E:A:3:4]

J: Okay I'll see you tonight then.
[
T: I'll take take take that and look at it
J: Oh and bring the Marlowe book. I'd like to see it.
T: Oh I ah with the what?
J: With the book by Marlowe.
T: Yeah, I got it set aside. ( ) so I'm going to bring that.
[
J: Oh.

In J's highlighted move, a directive reminding T to bring a book is followed by an informative which gives a reason for the directive. This kind of sequencing is similar to that in which a requestive is followed by an informative, and the social motivation is also similar. By providing a justification for soliciting the non-verbal action, the speaker demonstrates his sensitivity towards the infringement of the freedom of the addressee. The directive is therefore the head act and the informative the post-head act. As we can see from the rest of the conversation, it is the directive that is being attended to by T.

On the basis of the above observations, we can now formulate our Fourth Identification Rule as follows.

In a multi-act move, when a directive co-occurs with an informative, the directive realises the head act.

1.6 Directives and elicitations

In the following piece of data, a directive which makes a suggestion is followed by an elicitation asking for confirmation.

Example 26 [B:B:B:5:3]
J rang L to ask him to pass a message onto a colleague.

L: Why don't you try 400? You didn't try 400.
J: No, he said he'd be he was going to his office, so I thought=
[
L: Alright, I'll
J: =he'd be there.
L: Alright.

After getting the message that he has been asked to pass onto a colleague, L makes a suggestion that J try the latter on extension 400. This is realised by the directive *Why don't you try 400?*. This suggestion is followed by an elicitation asking for a confirmation. The elicitation serves the function of checking out the pre-condition for the directive, namely, that there is a need for the action. If J had already tried extension 400, then the directive would have been superfluous. As in responses to requestives, the reporting of the reasons for not trying extension 400 implicates a rejection of the suggestion. This rejection is in turn accepted by L. The head act of L's highlighted move is therefore the directive, and the informative is a post-head confirming the need for the suggestion.

Taking into consideration the data given in this section, we can re-formulate our Fourth Identification Rule as follows.

In a multi-act move, when a directive co-occurs with an informative or an elicitation, the directive realises the head act.

2. Further observations

In the process of formulating the identification rules, certain facts about the relation among acts in the same move seem to have emerged. First of all, in my corpus of spoken discourse data, elicitations co-occur with directives much less frequently than they do with informatives and requestives. This is probably due to the fact that directives expect compliance from the addressee. They therefore frequently co-occur with informatives, which expect acknowledgement, but not with elicitations, which provide the opportunity for the addressee, as it were, to answer back.

Second, there is not a single instance in my corpus of a co-occurrence of requestives and directives in the same move. This is hardly surprising since both solicit a non-verbal action and are mutually exclusive in the sense that, while the former give the option of non-compliance, the latter do not. What may occur, particularly in mother-child discourse, is that when a mother fails to get a child to comply with a requestive, she may then use a directive. But then these two acts occur in different moves and not in the same move.

Third, there seems to be a cline of elicitative force among the four major categories of initiating acts. Acts which solicit a non-verbal action seem to have a stronger elicitative force than those which solicit a verbal response. Hence, directives and requestives have a stronger elicitative force than elicitations and informatives. Elicitations in turn seem to have a stronger elicitative force than informatives, which require a mere acknowledgement of receipt of information. The strength of elicitative force appears to govern the identification of the head act. Hence, we can formulate an over-arching identification rule as follows.

> **In a multi-act move, if a requestive or a directive co-occurs with an elicitation or an informative, the requestive or directive realises the head act. If an elicitation co-occurs with an informative, the elicitation realises the head act.**

3. Concluding remarks

In this paper, I have attempted to follow up an aspect in the Sinclair-Coulthard model of linguistic description to which little attention has been paid, namely the structure of moves which consist of more than one act. I have investigated the relationship between the acts and have formulated rules for identifying the head act in these moves on the basis of examining conversational data. These rules are not assumed to be inviolable; the attempt to formulate them should be seen as no more than an exercise in capturing regularities observed in conversational data. It is hoped that these regularities can help us explain aspects of the structure of discourse which have not been addressed before.

Author's Address:
Department of Curriculum Studies, The University of Hong Kong, Pokfulam Road, HONG KONG.

Notes

1. The corpus of conversational data used consist of about thirty-six hours of face-to-face conversations (labels beginning with [C]) and telephone conversations (labels beginning with [B]) between native-speakers of English. The label [BCET] stands for Birmingham Collection of English Texts. The author wishes to thank the School of English of the University of Birmingham for allowing her to use their data.
The transcription conventions used are, briefly, as follows:
= contiguous utterances; [overlaps; ( ) indecipherable utterances.
2. The categories of acts used in this paper are different from those in Sinclair & Coulthard (1975). The former describe acts in conversational data, whereas Sinclair & Coulthard's categories describe acts in classroom discourse data. One major difference is that I make a distinction between directives and requestives, whereas Sinclair & Coulthard do not. I also identify sub-categories of the four major categories of acts.
3. In Tsui (1992b, forthcoming), five subclasses of requestives are identified.: request for action, request for permission, offer, invitation and proposal. If responded to positively,

the last subclass also commits the speaker to a future action. For example, the requestive *So why don't we arrange to get together maybe Sunday?* commits the addressee as well as the speaker to meeting up on Sunday, if complied with by the addressee. In this paper, proposal is subsumed under request for action.

4. Drew (1984:136) notes, however, that recipients may not formulate the upshot of the reporting. They may opt to talk more about the reporting. This would be a "withholding of sorts". Such a withholding leaves the outcome unresolved and provides the opportunity for the speaker to re-invoke the invitation. See examples 17 and 18 in which the invitation is extended again after the identity of the Professor is established.

References

Berry, M. 1982. "Review of M. A. K. Halliday (1978) *Language as Social Semiotic*". *Nottingham Linguistic Circular* 11, 1. 64-94.

_______. 1987. "Is Teacher an Unanalyzed Concept?" *New Developments in Systemic Linguistics* ed. by M. A. K. Halliday & R. Fawcett, Vol. 1, 41-63. London: Frances Pinter.

Brown, P. & S. Levinson. 1987. *Politeness*. Cambridge: Cambridge University Press.

Coulthard, M. 1981. "Developing the Description". *Studies in Discourse Analysis* ed. by M. Coulthard & M. Montgomery, 13-30. London: Routledge & Kegan Paul.

_______, ed. 1992. *Advances in Spoken Discourse Analysis*. London: Routledge.

_______ & D. Brazil. 1981. "Exchange Structure". *Studies in Discourse Analysis* ed. by M. Coulthard & M. Montgomery, 82-106. London: Routledge & Kegan Paul.

Drew, P. 1984. "Speaker's Reportings in Invitation Sequences". *Structures of Social Action* ed. by J. Atkinson & J. Heritage, 129-51. Cambridge: Cambridge University Press.

Goffman, E. 1971. *Relations in Public*. New York: Harper & Row.

Grice, P. 1975. "On Logic and Conversation". *Syntax and Semantics 3: Speech Acts* ed. by L. Cole & J. L. Morgan, 41-58. New York: Academic Press.

Halliday, M. A. K. 1961. "Categories of the Theory of Grammar". *Word* 17. 241-92.

Schegloff, E. 1978. "On Some Questions and Ambiguities in Conversation". *Current Trends in Textlinguistics* ed. by W. Dressler, 81-102. New York: Walter de Gruyter.

_______ & H. Sacks. 1973. "Opening up Closings". *Semiotica* 7,4. 289-327.

Sinclair, J. & M. Coulthard. 1975. *Towards an Analysis of Discourse*. London: Oxford University Press.

Tsui, A. B. M. 1991. "The Description of Utterances in Conversation". *Pragmatics at Issue* ed. by J. Verschueren, Vol. 1, 229-48. Amsterdam/Philadelphia: John Benjamins.

_______. 1992a. "A Functional Description of Questions". *Advances in Spoken Discourse Analysis* ed. by M. Coulthard, 89-110. London: Routledge.

_______. 1992b. "Socio-semantic Network Systems of Requestives and Directives". Paper presented at the Nineteenth International Systemic Functional Congress, held at Macquarie University, Sydney, 13-19 July 1992.

_______. Forthcoming. *English Conversation*. Oxford: Oxford University Press.

Theme and Prospection in Written Discourse

Anna Mauranen
University of Jyväskylä, Finland

Abstract
This paper explores the role of theme in the linear development of written discourse. Theme is seen in the dual role of providing the point of departure for the current sentence and establishing its relevance to the preceding sentence. I focus particularly on the prospections that each sentence makes in terms of the ensuing discourse, and the ways in which these prospections limit thematic choice. The data analysed consists of academic journal articles written in two unrelated languages, English and Finnish. Although the syntactic principles for ordering constituents differ radically in the two languages, it is shown that a functional model of theme and rheme can be applied to both, allowing us to see a basic similarity in the textual roles of the theme despite the differences in sentence structure.

0. Introduction

Theme is usually associated with the given element in a sentence, that is, the element which is shared between reader and writer, while rheme contains new and focal elements (see, for example, Halliday 1967, 1985, Daneš 1974, Fries 1981, Enkvist 1974). It is therefore natural in analysing text to assume that the theme part of a sentence is related to something that has already been established, whereas the rheme part is the most likely location for what takes the text in new directions. This assumption seems to underlie most thematic analyses, and is also the point of departure in the present paper.

However, there is little agreement among scholars as to the precise nature of theme. One source of variability is that this area of study is heavily biased towards Indo-European languages, particularly English, as pointed out by Hakulinen (1989) and others (for example Fries and Francis 1992). It is therefore important to develop analytical models of theme which cover a variety of language types, in order to explore the textual roles that themes can take. These roles should be approached from a functional perspective which transcends the syntax of a particular language.

It is also important to consider the nature of the relevance of theme to preceding discourse. The view is taken here that the part of a sentence theme which indicates what the sentence is about - the ideational theme in Hallidayan terminology - must bear semantic relevance to the preceding sentence. If this is not the case, either the ideational theme must be preceded by bridging material in the same sentence, which will be called here 'orienting theme', or the preceding sentence must provide the necessary interpretative framework for establishing the connection. Thus, a sentence prospects what happens next in the text, to a greater or lesser degree, and the theme part of the next sentence has to indicate the relevance of the sentence to the prospected development. The notion of prospection has been formulated by Sinclair (1983, 1992, 1993) to refer to the less-than-firm predictions that elements of discourse make in providing a framework of choice for what comes next. The role of prospection is tentatively probed here with respect to thematic choices in continuous text. The responsibilities of the theme of a sentence vary with respect to how much of the bridging between consecutive sentences has been done by the preceding sentence. That is, the connection provided by the theme must be appropriate to the preceding context.

This paper explores the thematic progression in paragraphs by English and Finnish writers, starting from the assumption that mutual relevance needs to be established between consecutive sentences and that the theme plays an important role in this process. It will be shown that texts in both Finnish and English support this assumption, despite major surface differences in the two languages, including different principles of constituent order in sentences.

A model of theme which fits both English and Finnish is first established, and hypotheses about thematic progression in text are formulated. The hypotheses are then tested on textual data, and finally questions concerning the role of prospection in thematic progression are discussed in the light of examples from these texts.

The data used here comprises 120 paragraphs drawn from 19 academic journal articles. The texts were drawn from the fields of medical science and economics. Seven of the source texts were written by native speakers of English, six by native speakers of Finnish in English, and six by native speakers of Finnish in Finnish. Forty paragraphs were sampled from each field, and grouped in sets of three (one native-English, one native-Finnish and one non-native English); each set was matched as closely as possible in terms of discipline and topic. All the sentences analysed are declarative: the discussion concentrates on declarative sentences throughout, since this is the mood that predominates in the scientific genre.

1. Theme in Finnish and English

Finnish is typologically different from English in many ways. The most relevant differences for thematic analysis are probably in the areas of word order and the role of the subject. Word-order in Finnish, or constituent order at sentence level, is relatively free: there are few syntactic constraints which would make particular grammatical permutations unacceptable. Word order is therefore determined primarily by discourse rather than syntactic considerations, or, as Vilkuna (1989:9) puts it: "...'free' with respect to Finnish word order means *discourse-conditioned*".

The generally accepted view of Finnish is that it was originally a non-SVX language, but has developed under the influence of surrounding SVX-languages into one which is predominantly SVX, although some disagreement seems to prevail as to how far this development has proceeded. The most authoritative surveys of written Finnish (for example Hakulinen et al 1980) render SVX as the preferred order, accounting for about 60% of written sentences.

A second important feature of Finnish from the point of view of thematic analysis is the role of the subject. The grammatical subject is a much less important and less obligatory constituent in Finnish sentences than it is in English: according to some surveys, about 30% of Finnish sentences do not have a subject constituent (Hakulinen & Karlsson 1979b). This is particularly common with passives, which never take a subject in Finnish.

A particular type of active generic sentence, known as the Missing Person sentence, is also relatively common. No person is indicated in these sentences. In principle, the Missing Person resembles in function the generic pronouns found in many Indo-European languages, such as *man* in German, *on* in French, and *you* or *one* in English, but a major difference is that there is no constituent with this meaning in Finnish. Finnish also lacks, at least in formal written varieties, equivalents of the dummy subjects *it* and *there*, which can conveniently be used as 'empty' themes in English.

When grammatical subjects do occur, that is when they are realised as constituents in Finnish sentences, the free word order allows them to be located virtually anywhere in the sentence without interfering with grammaticality. The notion of grammatical subject is a much investigated question among Finnish linguists, and its precise nature and extension are controversial. What is more important in the present context is that the tendency of the grammatical subject to coincide with theme (and also with the 'topic') in unmarked English declaratives is not similarly dominant in Finnish, because there are fewer subjects and these are not necessarily placed before the verb.

Such facts as these have caused problems in applying the concept of theme in Finnish linguistics and have led some Finnish linguists (for example Hakulinen 1989) to question the usefulness of the concept of theme in general and in particular the validity of thematic analysis for the description of Finnish. However, the analyses which reject the notion of theme altogether, or are at least highly critical of it, can in turn be questioned on the grounds that they assume, rather problematically, an inherent close association between theme and topic: theme has been equated with the topic of the sentence (for example Holman 1975), or with the topic of the discourse, or even both.

As far as the topic of the discourse is concerned, we would be making a very strong assumption if we said that it must be found in a particular syntactic slot in each sentence. In reality, elements which are connected with the discourse topic (as distinct from sentence topic) can occur anywhere in the sentence, as has been noted by many scholars (for instance van Dijk 1977, Fries 1981). Even more important is the fact that the discourse topic need not be expressed in every sentence at all. Whatever mechanisms readers and listeners resort to in deriving the topic of the discourse, it is not possible that they expect the same element to be expressed in every sentence in a text before they can arrive at an interpretation of what the discourse is about.

Assuming, then, that the requirement that theme should account for discourse topic can be dismissed, the more difficult problem of the relationship of theme and topic still remains at the level of the sentence. If theme is taken to be the topic of the sentence, and thereby to indicate what the sentence is about, it will be difficult to maintain a positional view of theme. The initial, or leftmost, constituent is frequently associated with functions other than topic in the sentence, and it has been argued that this cannot therefore be the theme (Hakulinen 1989, Vilkuna 1989). For instance, in example 1, both sentences have fronted material before the constituent which has the function of topic in the sentences. In S(1) the topic is the subject, while in S(2) it is not. (Italics indicate pre-topical thematic material; bold face indicates topic.)

Example 1

(1) *Kansainvälisissä pääomaliikkeissä* **verotuskäytäntö** on yhdenmukaistunut,
international capital movements taxation practice has become-uniform

mikä on merkinnyt sitä,	että ulkomaille	maksetut korkotulot on usein
which has meant it	that abroad	paid interest revenues have often

vapautettu lähdeverosta.
exempted withholding tax

(In international capital movements, taxation has become uniform, which has meant that interest revenues paid abroad have generally been exempted from withholding tax.)

(2) *Osingoista* **lähdevero**	sen sijaan	yleensä pidätetään.
dividends withholding tax	instead	generally collect

(To dividends, however, withholding tax is generally applied.)

The leftmost constituents are adverbials which correspond to prepositional phrases in English. In analyses such as Fries' (1981), these would be regarded as ideational themes because they have ideational meaning and are sentence-initial. In the analysis of English, they can be treated as fronted material, together with textual themes, interpersonal themes, and subordinate clauses preceding the main clause.

However, even though these initial constituents have ideational meaning, or propositional content (as most linguistic expressions do in any case), they are not what the sentences are about in the sense of containing the topics. S(1) is saying something about *verotuskäytäntö* ('taxation practice'), and the adverbial *kansainvälisissä pääomaliikkeissä* ('in international capital movements') provides a background to the message, not its topic. The sentence would have the same topic even if the adverbial were moved elsewhere. Similarly, S(2) is about *lähdevero* ('withholding tax'), and this would be the topic even if the adverbial *osingoista* ('to dividends') were moved elsewhere. If such fronted material as the adverbials above, subordinate clauses, or modal or reporting clauses, is taken to express topical themes in sentences, then the notion of topic becomes very different from what it appears to mean in simple declarative main clauses. In addition, texts which have perfectly good topical continuity would appear disconnected if such fronted elements were regarded as their topics.

Sentences like those in example 1 above are quite common in both English and Finnish. Moreover, it seems to be true of both languages that when topic, 'pure' topic as it were, is analysed as a semantic element in sentences (expressing what the sentence is about), it appears to have a preferred position, which is immediately before the finite verb (for Finnish, see for example Hakulinen 1989, Hakulinen & Karlsson 1979a). Therefore, the position of preverbal elements is not unimportant in Finnish. But obviously there is now more than one candidate for thematic position.

Because both the leftmost constituent and the constituent immediately preceding the verb appear to be important for theme in both Finnish and English, it seems useful to take the entire preverbal part of the sentence into consideration when comparing thematic choices in the two languages. If then we consider the preverbal section of a declarative sentence in each language, both similarities and differences emerge.

Finnish and English appear to share a preference for a topic-type nominal entity before a finite verb in declarative sentences. This is more commonly realised by the grammatical subject in English than it is in Finnish. In both languages, this element can be preceded by fronted material which serves a function other than expressing topic. This fronted element has been characterised for English as performing a framing function (Davies 1988), and for Finnish as expressing some kind of contrast (Vilkuna 1989), or indicating the function of the sentence (Hakulinen 1989). Such characterisations are sufficiently similar to warrant a general label of 'orienting' function, which is the term used here for the fronted material in both languages. The table below is a schematic representation of theme and topic in terms of functional roles and grammatical realisations in unmarked Finnish and English declarative sentences. (V(X) = verb optionally followed by other constituents.)

	theme		rheme
functional roles	'orienting'	'topical'	'core'/'new'
realisation tendencies	connectors adverbials complements modal clauses reporting clauses	Fin: nominal Eng: subject	V(X)

The following two examples illustrate how the above scheme renders the analysis of sentences in English, example 2, and Finnish, example 3. (In this and all subsequent examples, italics indicate orienting theme; bold face indicates topical theme.)

Example 2

Further, in some plausible situations, **such taxes** might even cause a gain in efficiency.

Example 3

Kilpailulakien osalta	**tämä**	synnyttää	kysymyksen
competition laws part	this	gives rise	question

niiden	yleistettävyydestä.
their	generalisability

(As to competition laws, the question of their generalisability arises.)

The two languages exhibit close similarities in the functional roles of the two thematic parts and the rheme, but less similarity in the grammatical realisations of these elements. Topic is more closely associated with a particular grammatical category, namely the subject, in English than in Finnish.

2. Some assumptions about thematic progression

In order to investigate the operation of themes and rhemes in a well-formed text, some background assumptions must be made clear.

The first assumption is that the role of theme in text as an essentially local organiser of discourse must be emphasised. Since theme-rheme configurations organise sentence constituents, they must do this with the needs of the immediate co-text in mind, so as to make the text progress smoothly. Relationships between adjacent sentences are therefore the most important, but continuity within paragraph boundaries can also be expected. Thus, the basic text unit felt to be relevant for theme analysis is the paragraph, at least in the genre of academic papers.

The second assumption is that the propositional relevance of a sentence to the previous one is primarily carried by the topical theme of the sentence. It is essentially a unit very much like the topical theme proposed here that Daneš (1974) seems to have had in mind when suggesting his patterns of thematic development. In Daneš' view, text accumulates a vast amount of information as it proceeds from sentence to sentence, and themes make principled selections from this otherwise rather overwhelming universe of information.

> it is evidently necessary to distinguish between the mass of information accumulated up to a certain point of text, and the portion of this mass contained (occurring) in the particular utterance following this point. This distinction involves a selection from the mass of known information for every utterance. We assume that this selection is determined, directly or indirectly, by the choice of the utterance theme (Daneš 1974:112).

Daneš, therefore, originally saw the role of theme in discourse as making choices among the information that accumulates in the text. It seems, however, that in this characterisation of the role of theme, these two ideas (the constant accumulation of information and the selections from this information which themes make in each utterance or sentence) are mutually contradictory. If all information or, to use the term adopted in this study, all propositional content, conveyed by the text accumulates all the time, then the selections that themes make at the outset of every utterance have no significance. In order for the choices to be significant, they must form the basis not only for the current utterance but also in some way for those that follow later. That is, a choice must have permanent consequences for the way the text is going, otherwise any later thematic choice would be possible so long as it had relevance to anything occurring earlier in the discourse. And this of course is not the case.

In order to formulate the meaning of theme, one possibility would be to adopt Sinclair's (1992) suggestion that at each stage of communication the text is contained in the current sentence. Sinclair's position is that each sentence contains the meaning of the previous one, by encapsulating it or by being prospected by it, and supersedes it as the current state of the text. The text can thus be likened to a mental representation which is developed by each sentence in turn, and which contains the meanings of all previous sentences as part of the knowledge that is used by the reader to interpret the text. Whether or not this representation is cumulative is not discussed by Sinclair, but in the light of what was said about Daneš' position, it is important to assume that it is not.

The third assumption, then, is that each sentence builds on the previous sentence, which represents the meaning of the text so far, rather than just on any sentence in the previous discourse. The selections that themes make can be seen as significant in a very similar way to that originally suggested by Daneš' - as selections from the available meaning elements that are made in order to provide a point of departure for the further development of the discourse. If the meaning of the preceding discourse is contained in the previous sentence, then the theme of a sentence should be relevantly connected to something in that sentence.

Having said this, it is important to note that the topical theme is not always directly relevant to the immediately preceding sentence. I have already pointed out that the topical theme is not always even realised as a constituent in Finnish. In such cases, intersentential relevance must be expressed by other means. The text can provide this by signalling that the relevant connection is there but not thematised, or that this is where a change occurs in the development of the text. Orienting themes serve an important function in such signalling.

Thus we may make a fourth basic assumption - that if the topical theme is not directly relevant to the preceding sentence, or if the topical theme is not realised in the text as in the case of Finnish, then there must be something in the text to warn the reader. The suggestion here is, then, that the default assumption in written discourse is that two sentences in a sequence are mutually relevant, so that the first contains something which will be followed up, and the second builds on the state that the first has provided. If the second involves a change of direction or a fresh start, this has to be signalled. This suggestion is basically in line with Sinclair's (1992) hypothesis that the state of the text at any point, that is, the sentence being currently processed, carries with it everything a reader needs to understand it. It encapsulates what has gone before, and often it also prospects forward and sets the scene for what follows.

The problem with this assumption is that studies dealing with theme in text, including Daneš', have been surprisingly vague about what counts as a connecting relationship between themes, or themes and rhemes. In considering meaning relations between any two sentences, it is normally possible to discern or imagine some relationship between them. There is thus an obvious danger of the analyst's reading meaning relationships into thematic relations arbitrarily. In view of this, it would be helpful to formulate some simple rules for systematically establishing interconnections between propositional sentence constituents. I have used studies of lexical cohesion (Hasan 1984, Hoey 1991) and reference chains (Källgren 1979, Näslund 1991) in order to draw up the following list of criteria for including relationships as links between thematic and rhematic constituents:

(1) identity, including coreferentiality and changes of word class (*aspirin - it, fetus - fetal*)
(2) synonymy, including instantial synonymy (*thromboxane production - thromboxane biosynthesis*)
(3) Partial identity (*thromboxane - thromboxane biosynthesis*)
(4) Possessive relationship (*dog - dog's / its*)
(5) Contrast, including antonymy and instantial contrast (*risk - safety; their results - our study*)
(6) Specification: hyponymy, co-hyponymy, meronymy (*drug - aspirin, income tax - property tax, placenta - cord*)

It is worth noting that relationships involving rhemes tend to be more complicated than those involving thematic elements only. Rhemes are usually complex, containing at least a verb, and often complements and adjuncts of various kinds. Nevertheless, in the analysis which follows, rhemes are treated as unanalysed wholes. Therefore notations like T1 = R2 must not be taken literally. They are shorthand expressions for situations where a theme constituent picks up a central propositional element of the preceding rheme.

3. Thematic progression: towards the formulation of hypotheses

Basically, Daneš' models of thematic progression assume that a paragraph beginning with a sentence in a theme-rheme sequence (T1 + R1) can be followed by two kinds of themes: the theme of the second sentence is either identical to or derived from the theme of the first sentence (T2 = T1), or it is identical with or derived from the rheme of the first sentence (T2 = R1). In addition, there are sentences whose theme picks up the entire proposition in the preceding sentence (T2 = T1 + R1).

Daneš's patterns are concerned with the role of theme in connecting current sentences with the preceding discourse, and therefore, in his description, rhemes do not play a retrospectively connecting role. He assumes identity relations between rhemes to be very rare (1974:122), but finally leaves open the possibility of finding patterning among rhemes as well. If we take his patterns as the basis of our hypotheses, then it is necessary for the sake of completeness to hypothesise the infrequency of two kinds of connections involving the rheme: rhemes are not normally derived from preceding themes or from preceding rhemes (R2 ≠ T1 and R2 ≠ R1). Daneš' thematic progression patterns can then be interpreted to make the following predictions concerning two adjacent sentences in a passage:

A. High probability options:
1: T1 -> T2
2: R1 -> T2

B. Low probability options:
3: T1 -> R2
4: R1 -> R2

We can also formulate hypotheses for text passages where the propositional content is divided into theme-rheme nexuses (S = T + R), assuming that the theme part is located before the finite verb, and that the theme consists of a topical element (topical theme, T) optionally preceded by an orienting element (orienting theme, OT). A more complete formula is therefore S = (OT) + T + R for English, and S = (OT) + T + R or OT + (T) + R for Finnish. The hypotheses are now as follows:

(1) The topical theme shares essential propositional elements with the preceding sentence, either with a) its theme, or b) its rheme, or c) with both.
(2) If no such propositional elements are realised in the topical theme, an orienting theme constituent must be present to mark the intended connection. This also applies to Finnish sentences where the topical theme may not be realised as a constituent.
(3) Rhemes do not contain shared propositional elements between adjacent sentences which serve to indicate the continuity between the sentences in the way themes do, as was discussed above.
(4) Native language texts follow principles (1)-(3) more closely than foreign language texts.

These hypotheses were tested by analysing the data separately for L1 English, L1 Finnish, and L2 English writers. All main clause themes within a sentence were included in the analysis, but if two themes within the same sentence did not follow the predicted principles, this was ignored, because the syntactic connection was held to provide a sufficiently powerful link in any case.

Examples of the relationships in hypotheses (1) and (2) are given below. Since the prediction of the relationship in hypothesis (3) is negative, i.e. that it will not be found, it cannot be illustrated, and hypothesis (4) makes a quantitative prediction which can only be tested on the data. *Hypothesis 1* is illustrated with three examples (4a, 4b, and 4c), following its subdivisions.

Example 4a The topical theme shares essential propositional elements with the theme of the preceding sentence:

(1) The study group consisted of pregnant subjects attending the obstetrics clinic at Vanderbilt University prior to 20 weeks' gestation.
(2) **All subjects** gave informed written consent and the study was approved by the committee for the Protection of Human Subjects at Vanderbilt University.
(3) **All subjects** were healthy at the time of admission into the study.

In this example, the relationship between the themes of S(3) and S(2) is one of identity. Moreover, these themes encapsulate all the information in S(1).

Example 4b The topical theme shares essential propositional elements with the rheme of the preceding sentence:

(1) Martin Faustmann is credited with developing the archetypal model used to optimize rotation length.
(2) **It** consists of a single even-aged forest stand on a uniform site owned by a perfectly competitive forest owner whose sole goal is rent-maximization.

This example also involves an identity relationship, realised as pronominal reference between the theme of S(2) and central noun group in the rheme of S(1).

Example 4c The topical theme shares essential propositional elements with both the theme and the rheme of the preceding sentence:

(1) *Studies in the United States and other countries consistently find that* women use a greater number of prescribed drugs than do men.
(2) **This** may in part reflect the higher rate of physician utilization among women, a tendency that increases the probability of obtaining a prescription.

The theme of S2 is a reference item which encapsulates the whole of the preceding sentence. Thematic reference items of this sort are typical of a thematic connection with a whole sentence, but lexical references with the same function also occur.

Hypothesis 2 states that if the topical theme of a sentence does not share essential propositional elements with the previous sentence, it is preceded by an orienting theme:

Example 5

(1) Although these hypotheses are plausible, much of the data on which they are based have serious limitations.
(2) *For example*, **most epidemiological studies assessing the use of prescribed drugs** rely on self-report even though patient reports on medication use tend to be unreliable.
(3) *It is also known that* **men** are more likely to under-report their use of psychotropic drugs, which may lead to some exaggeration of the sex differential.

In this example, the relationship between S(1) and *epidemiological studies* in S(2) is somewhat indirect, and is much strengthened by *for example*. In S(3), *men* is not predicted or anticipated by anything in S(2), but the reporting clause which serves as an orienting theme establishes a connection between the previous sentence and the current sentence.

Some additional remarks need to be made here. First of all, certain theme types appear to be inherently acceptable. Self-reference, i.e. reference to the discourse itself, seems to be an acceptable theme without much restriction. It marks a change in orientation or a 'plane change' (Sinclair 1983) when it appears in the middle of a paragraph, but does not need an orienting theme before it. This is illustrated in example 6, from an L2 English text:

Example 6

(1) For example, a fifth of the women were using drugs for high blood pressure, and the use of allergy and thyroid drugs was also high.
(2) **This study** did not include women who had a hysterectomy before their menopause, and because these women are sicker than other women, our data may underestimate the real frequency of drug use among mid-age women in the study area.

Similarly, *it* and *there* are acceptable as themes almost anywhere, with or without orienting themes. In the next example (7), the orienting theme is long and heavy, and the empty theme is merely holding a place in the syntactic construction:

Example 7

Since corporate capital income is subject not only to corporate taxes but also to personal taxes and property taxes, **there** is a strong presumption in the literature, e.g. Feldstein, that there is too little corporate capital, with large efficiency costs.

Because the subject element *there* in such cases has little meaning in terms of the propositional content, sentences where it appears after an orienting theme are not too far removed from the Finnish constructions with OT + R. As noted before, these occur with impersonal constructions, i.e. where no surface subject is present.

In order to illustrate how the analytical method works in a text passage, it may be applied to a passage from an L2 English text:

Example 8

(1) **Several women** maintained their patterns of using prescribed drugs in the two surveys done three years apart.
(2) *However*, **short-term therapies** were also common.
(3) *In the case of pain relievers and psychotropic drugs*, **that** was to be expected,
(3b) *but* **even some women using cardiovascular drugs, mainly drugs for high blood pressure**, dropped their therapy during the follow-up.
(4) **This instability in drug use patterns** have both health care and research implications.
(5) **Short-term use of prescribed drugs for chronic diseases** suggests either overprescribing or poor follow-up and patient compliance.
(6) *Also*, **more women** are exposed to drugs than any cross-sectional inspection would suggest,
(6b) *and* **this** amplifies the number of such adverse effects which depend

more on number of exposures than on their duration.
(7) *If a drug is commonly used*, **the characteristics of users and non-users** are difficult to study in cross-sectional inspections, because the classes overlap each other.
(8) *In further studies*, **it** would be interesting to compare women using drugs consistently over time to those using them only for short durations.

In the light of the above analysis, the passage appears relatively coherent. S(2) shifts the attention to a topic which contrasts with that of S(1), marking the shift by means of the orienting theme *however*. In S(3), the topical theme, *that*, is a reference to the previous sentence. The theme of S(5) can be seen to be connected to the rheme of S(4) by specification. Sentences (6) and (7) have orienting themes which relate them to sentences (5) and (6) respectively. Moreover, the inherently acceptable but topically uninformative theme *it* in S(8) is preceded by an orienting theme, which is connected to the topic of studies in this field in the preceding sentence.

The analysis of the 120 paragraphs gives strong support to the hypothesised patterns: L1 English texts had 13 deviations in all, L1 Finnish texts 12, and L2 English texts 54. This suggests that the principles in the hypotheses cover the possibilities of normal paragraph development fairly well. The text group which stands out from the others is the non-native English texts: they have over four times more deviations from the hypothesised patterns than either of the native language texts. That is, exceptions to the patterns were mostly produced by the L2 writers, as predicted by hypothesis 4 (for a closer analysis of the results, see Mauranen 1992).

4. Theme and prospection

In addition to the Finnish L2 writers' expected and frequent departures from the patterns, the analyses brought up some interesting features which contribute to the currently emerging view of thematic progression. One of these concerns the influence of prospection on themes.

After certain sentences it is difficult to determine whether the following topical theme is relevant to the preceding sentence according to the principles of meaning connections laid out above, because the first sentence allows almost anything to follow. Example 9 is a case in point:

Example 9

(1) *When we take uncertainty into account*, **the results** change dramatically.

(2) **Investors** would now require that the expected after-tax real rate of return on an investment at least equal the after-tax real risk-free interest rate, which we denote by r^a_z, plus enough to compensate for the risk in the return of the investment.

In S(1), the non-individuating topical theme (the results) and the anticipation of a drastic change in the rheme create an expectation of discontinuity rather than continuity. Apart from the need to stay within the general topic area of 'the results', the writer is quite free to choose the next theme. The topical theme which follows cannot therefore very easily be regarded as unmotivated, even though it does not adhere to the principles of meaning connection. However, unlike example 9, which places only minor restrictions on the next sentence, most sentences restrict their successors much more clearly. Example 10 illustrates this:

Example 10

(1) This argument that the taxation of corporate income is non-distorting is entirely different from that of Stiglitz.
(2) **Stiglitz's argument**, developed in a certainty setting, relied on the possibility of 100 percent debt finance for marginal investments.

In this example, the rheme of S(1) requires that an explanation follows as to how Stiglitz's argument is different from the one mentioned. This is probably more obvious in the context of the text so far than when the two sentences are considered alone, because at this point the reader knows that Stiglitz and his argument are being mentioned for the first time.

The difference between examples like 9 and 10 can be interpreted as one of prospective specificity, since 9 provides fewer specifications of the next possible theme than 10 does. Despite the difference in specificity, though, both sentences make a strong prospection in the sense that they distinctly limit the choices available for the next sentence.

Prospection as understood here is a broader category than prediction in Tadros (1985), where prediction is contrasted to anticipation. According to Tadros, prediction involves a commitment on the part of the writer to a later linguistic event, whereas anticipation involves guesses on the part of the reader as to what is to follow (1985:6). The distinction is a useful one, and basically analogous to the distinction between intended and unintended information. Intended information in text is of the kind that the writer conveys or intends to convey, whereas unintended information is any information a reader or listener can figure out from the form or content of a message (see Dahl 1976: 37-8). Unintended information, and with it, anticipation, will be excluded from consideration here, and the focus is on prospection as recoverable from the text and as in principle intentionally provided by the writer. However, prospection is taken to subsume both prediction in Tadros's sense, that is, firm predictions, as well as prospection in the sense in which it is used by Sinclair (1983), that is, less-than-firm predictions.

Within the domain of prospection, it is possible to discern a cline of prospective strength. Prospective indicators in a sentence may determine what follows to a greater or lesser extent, the limiting cases being firm prediction and no prospection at all. Our assessment of the success of a text is partly dependent on whether or not it fulfils its prospections. If a text violates its own prospections, especially strong ones, it is felt to be problematic. For instance, the following example from the L2 English data does not fulfil its prospection and therefore appears to break a promise:

Example 11

(1) Part of this difference may reflect population differences,

(1b) but part of it may result from different methodologies.

(2) *In this study* **current drug taking** was asked

(2b) *but* **an alternative "less than once a week"** was given.

(3) This wording may have prompted sporadic drug users to report their use.

(4) Quite likely taking of a nonprescribed drug is not an important event and it is easily forgotten, but the fact that they sometimes take them may be better remembered.

Here *different methodologies* in S(1b) prospects a comparison, but as can be seen, the text continues by developing only one alternative. A reader would have expected an explanation of how the alternatives differ from the one chosen, as is done in a very similar example from another text (L1 English):

Example 12

(1) Previous studies have reported thromboxane formation to be increased, unchanged, or decreased in pregnancy.

(2) **The difference between studies** may in part reflect the analytic approaches that were used.

(3) **A major difference between this and previous studies** was that thromboxane biosynthesis was determined non-invasively with the use of methods that avoid sampling and analytic errors.

(4) *In many studies* **thromboxane formation** is determined by measurement of thromboxane B2 in plasma.

The prospections in examples 10 - 12 can all be characterised as both strong and specific, whereas the prospection in example 9 is strong but unspecific. An unspecific prospection relaxes the requirement that a theme should keep within the limits of the semantic relationships outlined above. Example 13 illustrates both possibilities in a sequence:

Example 13

(1) What mediates the increase in platelet activity in pregnancy is unknown.
(2) **Formation of thromboxane A2, the major cyclooxygenase product of arachidonic acid in platelets and a potent platelet aggregant**, is reported to be increased in pregnancy.
(3) **A possible source of increased thromboxane formation in pregnancy** is the placenta, which has been shown to generate thromboxane in vitro.

In this extract, S(1) makes a strong but unspecific prediction, in other words it predicts its successor rather liberally: almost anything in the general topic area would be read as a potential cause of the increase in platelet activity. There is no need for the theme of S(2) to ensure its relevance to S(1) by adherence to rules of meaning connections. S(2) in turn prospects nothing at all about its successor. The relevance of its topical theme must therefore be shown in that sentence, S(3). S(3) does this by encapsulating S(2) as a nominalisation in its theme. In this way, sentences (2) and (3) also seem to fit in with the principles of text development outlined by Sinclair (1992), which hold that sentences can fulfil prospections made on them by previous sentences, or encapsulate previous sentences. S(1) in example 13 also illustrates how the prospection made in a sentence may concern the discourse function of its successor: the next sentence must be interpretable as an explanation.

Example 13 also shows an instance of a very weak prospection, or perhaps a zero prospection, in S(2). It fulfils one, without making a prospection itself. A similar case is S(3) in Example 12, which fulfils part of the prospection made by S(2), but it does not prospect the next sentence, which fulfils the rest of the prospection in S(2). This is basically very similar to the instances Daneš (1974) calls 'split rheme': a particular rheme makes two specific prospections, which are fulfilled one after the other.

When an unspecific prospection is made, the need for an orienting theme is greatly reduced (or in fact orienting themes are made redundant). Nevertheless, an orienting theme can still be present even when quite a strong prospection has been made. One role for orienting themes when the topical theme ascertains the relevance to the previous sentence is an attitudinal or modal one, as in example 14:

Example 14

(1) *Although prostacyclin can be generated by the placenta,* **the source of the increment in prostacyclin biosynthesis during pregnancy** is uncertain.
(2) *It is possible that* it represents a vascular response to platelet activation.

Here, the introductory modal clause in S(2) expresses the writer's attitude towards the proposition in the following clause. At the same time, it enhances the relevance of S(2) to S(1) by picking up the uncertainty predicated in S(1). The role of orienting themes in such situations would thus appear to be both attitudinal and textual: in example 14, the orienting theme of S(2) corroborates the prospected discourse function of the main proposition of S(2) as a possible cause of the phenomenon discussed. However, since this function has been prospected, it is not necessary to thematise it for the relationship to be understood. Thus, the choices of orienting themes appear to result from a complex interaction of factors in textual development. How prospections of different kinds and strengths influence the choices of orienting and topical themes in different genres would be worth further investigation.

5. Conclusion

This paper has argued that theme can be compared in two unrelated languages in terms of a simple functional model. When the analyses are brought together in a common sentence-level description, it becomes possible to carry out a comparison of the textual behaviour of the theme independently of differences in syntactic rules.

From the textual analysis of thematic development it seems that the mutual relevance of two consecutive sentences in a text is largely established by the theme. To see this, it is useful to keep two kinds of sentence themes separate: orienting themes and topical themes. However, the role of the theme in ensuring the connection between sentences varies with the prospections made by the previous sentence. If the previous sentence either allows almost anything to follow it, or establishes the textual role or function of the next sentence so clearly that practically anything will be interpreted in the light of this function, the theme of a sentence need not manifest very explicitly a specific kind of relationship between the sentences, but has a wider range of options which satisfy the need for a connection between adjacent sentences.

Author's Address:
Language Centre for Finnish Universities, University of Jyväskylä, P.O. Box 35, 40351 Jyväskylä, FINLAND.

References

Dahl, Ö. 1976. "What is New Information?" *Approaches to Word Order* ed. by N. E. Enkvist & V. Kohonen, 37-50. Åbo: Publications of the Research Institute of the Åbo Akademi Foundation.

Daneš, F. 1974. "Functional Sentence Perspective and the Organization of the Text". *Papers on Functional Sentence Perspective* ed. by F. Daneš, 106-28. Janua Linguarum Series Minor 147.

Davies, F. 1988. "Reading between the Lines: Thematic Choice as a Device for Presenting Writer Viewpoint in Academic Discourse". *The ESPecialist* 9, 1/2. 173-200.

Enkvist, N. E. 1974. "'Theme Dynamics' and Style: an Experiment". *Studia Anglica Posnaniensia* ed. by J. Fisiak. Vol. 5. 127-35.

Fries, P. 1981. "On the Status of Theme in English: Arguments from Discourse" *Forum Linguisticum* 6. 1-38.

______ & G. Francis. 1992. "Exploring Theme: Problems for Research". *Occasional Papers in Systemic Linguistics* (Volume 6) ed. by Dirk Noël, 45-59. University of Nottingham.

Hakulinen, A. 1989. "Some Notes on Thematics, Topic, and Typology". *Text and Discourse Connectedness, Proceedings of the Conference on Connexity and Coherence, Urbino, July 16-21, 1984*. 53-64. Amsterdam: John Benjamins.

___________ & F. Karlsson. 1979a. *Nykysuomen lauseoppi* (= Modern Finnish syntax). Helsinki: Suomalaisen kirjallisuuden seura.

______________________. 1979b. *Kvantitatiivinen tutkimus suomen morfosyntaksista ja sen tekstuaalisista tekijöistä* (= A quantitative study of morphosyntax and its textual determinants in Finnish). Unpublished Manuscript. Department of Finnish and General Linguistics, University of Turku.

______________________, & M. Vilkuna. 1980. *Suomen tekstilauseiden piirteitä; kvantitatiivinen tutkimus* (= Features of Finnish text sentences, a quantitative study). Helsinki: Publications of the Department of General Linguistics, 6. University of Helsinki.

Halliday, M. A. K. 1967. "Notes on Transitivity and Theme in English" (Part 2). *Journal of Linguistics* 3. 177-274.

_______________. 1985. *Introduction to Functional Grammar*. London: Edward Arnold.

Hasan, R. 1984. "Coherence and Cohesive Harmony". *Understanding Reading Comprehension* ed. by J. Flood, 181-219. Newark, Delaware: International Reading Association.

Hoey, M. 1991. *Patterns of Lexis in Text*. Oxford: Oxford University Press.

Holman, E. 1975. "Teema/reema -rakenne ja suomen kielen sanajärjestys" (= Theme/rheme structure and Finnish word order). *Virittäjä* 79. 279-88.

Källgren, G. 1979. *Innehåll i Text.* Ord och Stil. Språksamfundets skrifter 11. Stockholm: Studentlitteratur.

Mauranen, A. 1992. *Cultural Differences in Academic Rhetoric. A Textlinguistic Study.* Unpublished PhD thesis, University of Birmingham. Näslund, H. 1991. *Referens och koherens i svenska facktexter.* (= "Reference and Coherence in Swedish LSP Texts"). Uppsala: Institutionen för Nordiska Språk vid Uppsala Universitet.

Sinclair, J. 1983. "Planes of Discourse". *The Twofold Voice: Essays in Honour of Ramesh Mohan* ed. by S. N. A. Rivzi, 70-89. Salzburg: Universität Salzburg.

________. 1992. "Trust the Text: the Implications are Daunting". *Advances in Systemic Linguistics: Recent Theory and Practice* ed. by M. Davies & L. Ravelli, 5-19. London: Frances Pinter.

Tadros, A. 1985. *Prediction in Text.* Discourse Analysis Monograph No. 10. Birmingham: English Language Research, University of Birmingham.

van Dijk, T. 1977. *Text and Context. Explorations in the Semantics and Pragmatics of Discourse.* London: Longman.

Vilkuna, M. 1989. *Free Word Order in Finnish. Its Syntax and Discourse Functions.* Helsinki: Suomalaisen kirjallisuuden seura.

Professional Conflict
Disagreement in Academic Discourse[1]

Susan Hunston
Cobuild, Birmingham

Abstract
This paper brings together two approaches to Discourse Analysis in what will hopefully be a mutually beneficial interaction. One is an examination of the way in which propositions in a discourse are evaluated, and how this evaluation may be open to analysis. The other is an examination of the role of discourse, in this case academic research articles, in the construction of knowledge by particular academic discourse communities. I thus hope to provide what Fairclough (1992) calls a description and an explanation of a text. The focus of the paper is conflict in academic discourse. The texts discussed are academic research articles in which the writer's knowledge claims are presented as being in conflict with another researcher's knowledge claims.[2]

0. Introduction

The social groups whose discourse is the subject of this paper are those concerned with particular academic disciplines - biochemistry, linguistics and history - and the texts which will be discussed are the research articles associated with those disciplines, in which knowledge claims are presented and argued for. The social role of the research article is to participate in the gradual process by which the ideas of one group of researchers become accepted as part of the general knowledge-bank of the social group as a whole (Latour & Woolgar 1979).

The view of knowledge assumed in this paper, as something which is essentially constructed by texts, owes much to the work of sociologists of science (see Myers 1990 for an overview). Like them, I have chosen to focus on areas of conflicting knowledge claims. Investigations of the competition between different paradigms (Gilbert & Mulkay 1984; McKinlay & Potter

1987) and the negotiation of the presentation of controversial programmes (Myers 1990) are valuable in that they can make apparent processes which are often hidden. In this paper, the texts discussed are academic research articles in which the writer's knowledge claims are presented as being in conflict with another researcher's knowledge claims.

The paper deals in turn with three main topics: conflict relevance, the presentation of knowledge claims, and conflict resolution. My central concern is to describe the linguistic strategies by which the writer attempts to achieve the required outcome: persuasion of the readership to accept the writer's claim above all opposition. In addition, I shall compare the conflict presentation in the research articles of three disciplines in order to throw light on the choices available and to link the strategies chosen with the value-system of each discourse community. For the sake of simplicity, I shall distinguish between the knowledge claim made by the writer under discussion (Proposed Claim) and that made by opposing writers (Opposed Claim).

1. The texts

The texts discussed in this paper are taken from three discipline areas: biochemistry, sociolinguistics and history, representing Sciences, Social Sciences and Humanities respectively. Two texts are taken from each area. Details of the texts are given in Appendix 1. A brief account of the topic of each text, and the nature of the conflicting claims, follows.

S1 is a report of an experiment to measure the dichroism of chromatin, a substance found in cells, in order to deduce its structure. The results and the consequent description of chromatin are in direct contradiction of similar experiments carried out by the writers of S2 and other papers. Basically, the writers of S1 find the dichroism of chromatin to be positive, while the writers of S2 find it to be negative. S2, like S1, is a report of an experiment, the aim of which is to draw conclusions about the structure of chromatin by conducting measurements of its dichroism. The dichroism is found to be negative.

L1 is an account of an experiment carried out on pre-school children in the United States to ascertain how many of them, of what gender and class, would use *thank you* spontaneously. Conclusions are drawn concerning the age at which children acquire sociolinguistic knowledge. Both results and conclusions conflict with experiments carried out under different conditions by other researchers, and with Piaget's theories. L2 presents itself as a direct challenge to a monograph about phonological variation in Chipewyan written by two other writers (Scollon and Scollon), which has been generally accepted as valid by other workers in the field.

H1 reports a study of the voting patterns of the electorate in Shrewsbury before and after the Reform Bill of 1832 and argues that the Bill had a greater reforming effect than is currently believed by many historians. H2 directly challenges the accepted view among French historians that the organisation Croix de Feu was not a fascist one. The paper argues for parallels between the Croix de Feu and the parties led by Hitler and by Mussolini.

Figure 1. Type of data

	Experimental		**Non-experimental**	
	Physical world	*People*	*People*	*Records*
New	S1, S2	L1		H1
Re-worked			L2	H2

Some kind of pre-theoretical classification of the papers is tempting but it is difficult to produce categories which satisfactorily divide the texts according to consistent parameters. Intuitively, however, the divisions presented in Figure 1, based on the type of data presented and whether it is new or re-worked data, seem valid. S1, S2 and L1 present new, empirical data. H1 also presents new data, but this is the result of the observation of historical records, not experimentation. L2 and H2 both re-work and challenge existing data. We might then predict two possible sets of groupings: along discipline lines (S1 and S2; L1 and L2; H1 and H2) and along article type lines (S1, S2 and L1; L2 and H2; H1). One question to be answered now is what groupings the analyses of the conflict presentations produce.

2. The relevance of the conflict

Bearing in mind that writers always exert a choice over how they present their material, we might ask why writers choose to report views which conflict with their own. A clue to the answer may be found in Swales' (1990) description of the writers of experimental research articles finding a 'niche' for their knowledge claims by pointing out gaps in existing research, or in Barton's (1992) observation that the problematisation of a topic at the beginning of an essay carries a high cultural value. The conflict provides a context within which the writer's Proposed Claim may be seen as relevant, thereby justifying the writer in putting it forward. In addition, the conflict is used to present the Proposed Claim in the best possible light, at the expense of the Opposed Claim.

The Introduction sections of the papers offer evidence for a binary division between those who express the context - the conflict - in terms of a lack of knowledge and those who express it in terms of the need to correct an erroneous Opposed Claim. In both cases, the strategy chosen allows the Proposed Claim to be seen positively, in that it fills a gap in knowledge or corrects a misconception. Behind this lies the value assumption that lack of knowledge, or lack of 'correct' knowledge, is bad. Examples 1, 2 and 3 illustrate the 'lack of knowledge' context:

> **Example 1: S1** The results of those studies left some unresolved discrepancies, which we seek to clear up in this paper. Specifically, McGhee et al ... found a limiting dichroism of about -0.2 ... Our experiments ... showed a limiting dichroism of +0.06 ...

> **Example 2: L1** The results of research on the use of *thank you* by preschoolers are inconsistent. ... Additional research is necessary to determine whether preschoolers are aware of the contexts in which *thank you* is appropriate ...

> **Example 3: H1** The question is, therefore [consequent upon different opinion], not how reform miraculously transformed English electoral politics from mire into meadow, but rather how reform affected an already broadly participatory political system. In a system already familiar to some extent with parties, issues, and partisanship, what was the impact of the Great Reform Act on electoral behavior?

The evidence may be summarised as:

(i) lexis identifying lack of consistency in current knowledge (*unresolved discrepancies*, *results...inconsistent*, *disagreements*) and/or lexical contrast indicating such a lack of consistency (for instance the contrasting symbols '+' and '-' in example 1);

(ii) either a commitment to fill the gap in the current paper (*clear up in this paper*) or an expression of need for future research (*additional research is necessary*), including the use of interrogative forms (*what was the impact of the Great Reform Act on electoral behaviour?*)

It is notable that the knowledge-gap itself may be characterised either as a problem, as in example 1 (*discrepancy)*, and example 2 (*inconsistent)*, or as a question, as in example 3. In these texts, only the empirical papers present it as a problem.

In the two papers which re-interpret and challenge existing data, the problem is presented not as the lack of certain knowledge but as the presence of an 'incorrect knowledge claim', as examples 4 and 5 illustrate.

> **Example 4: L2** However, neither of these claims [made in a monograph by Scollon and Scollon] can be substantiated from what I have learned during my recent work in the community. Because the Scollons's book has been hailed as "excellent" and an "original contribution" (ref), I feel it is incumbent upon me (as it is unlikely for anyone else to have direct access to the data) to review their claims ...

> **Example 5: H2** Given its dominant role on the French Right, it is a matter of some consequence that most French historians insist that the *Croix de Feu* was not fascist. ... Indeed, by formally excluding the *Croix de Feu* most historians of French fascism would seem to have talked themselves out of a subject ... The history of fascism in France in the interwar years ought to be about as lively as the study of fascism in the United States ... In fact, however, it has become a veritable growth industry yielding an increasing number of scholarly studies on all aspects of French fascism - except the *Croix de Feu.*

The indication of a false claim may be explicit, as in example 4, or implicit, as in example 5, where the falseness accompanying the exclusion of the *Croix de Feu* from the category 'fascist' is implied by the conjoining of such items as *a matter of some consequence* and *insist that*, and the ironic contrast between *should have talked themselves out of a subject* and *a veritable growth industry.*

It is interesting that no examples of 'incorrect knowledge claim' papers are found in the biochemistry texts. In fact, in the physical sciences, it appears that papers which specifically address themselves to the correction of error are treated as a separate genre and printed in a separate section of the journal, such as the 'Technical Comments' section of *Science* (Beaton, forthcoming). It is as if such discussions are deemed to occur prior to and subsidiary to the presentation of findings in the main section of the journal. This in turn mirrors the role of the debate among competing theories as perceived by members of the discipline: disagreement is superseded by discovery, or, as Gilbert and Mulkay (1984) put it, "Truth Will Out".

In History, however, there is no bar to the presentation of new claims as constituting a direct attack upon existing ones. The fact that the writer of H2 does not acknowledge openly that an attack is taking place suggests that the attack is within the normal course of events for the discipline. The writer of L2, by contrast, goes to some pains to state the necessity for her attack upon the Scollons (*I feel it is incumbent upon me*), and her qualifications to do this (*as it is unlikely for anyone else to have direct access to the data*), as if such an attack is unusual within the discipline.

Thus, there are two options apparently open to writers when presenting a conflict: 'lack of knowledge' and 'incorrect knowledge claim'. In addition, when an incorrect knowledge claim is challenged, there are variations in the degree of deference and justification with which this is done. All these choices represent alternative views of how knowledge develops.

3. The management of conflict: representing knowledge claims

Having discussed how the conflict as a whole is presented within the context of the research article, I now turn to the question of how the two conflicting knowledge claims - the Proposed Claim and the Opposed Claim - are managed in such a way as to present a choice between them but to influence that choice in favour of the Proposed Claim. In doing so I shall first draw attention to some strategies that are used in a similar way in all the articles, and then point out some differences between the empirical articles (S1, S2, L1) and the non-empirical articles (L2, H1, H2).

3.1 Common strategies

3.1.1 Differentials in status

The first strategy is the creation of a differential in the status of the two claims. The status assigned to a proposition indicates where it is located in terms of the process of knowledge construction, for example, whether it is an observation, an experimental result, an interpretation or a conclusion. The status of a proposition may be indicated prospectively, by a reporting clause for instance,[3] or this may be done retrospectively, by an anaphoric noun (Francis 1986). Sinclair's concept of 'encapsulation' (Sinclair 1992a & 1993) may be used to explain what happens when a stretch of discourse is summarised by a status noun. The noun effectively re-evaluates the preceding discourse in terms which determine how the discourse may then proceed. It is very easy, for example, to proceed to countermand a conclusion; one simply has to adduce contradictory results, as in example 6.

Example 6: S1 It has been proposed that in the 10-nm fiber observed at low salt concentration, the nucleosomal disks are arranged with their diameters nearly parallel to the fiber axis. However, *this conclusion* is not supported by the flow dichroism studies of Tjerneld and Norden (1982).

To contradict previously-published results, without accusing the writers of fraud, is altogether a more difficult matter, as will be shown below, and in most cases results are not contradicted, but are treated as factual. Indeed, to avoid the problem, a writer wishing to contradict a previous researcher will often re-evaluate the Opposed Claim in terms of an interpretation rather than as a set of results, as the writer of L1 does with respect to Piaget (see example 8 below).

A writer presenting conflicting claims, then, has the option of giving them different statuses. Mulkay (1985) comments on the use of a series of opposites, such as *fact* versus *opinion* and *observed value* versus *true value* in the arguments conducted in scientists' private letters. Examples 7, 8 and 9 provide illustration of this strategy from research articles (my emphasis).

Example 7: L2 (fact versus belief) I have dwelt on the *ts* --> *tš* shift to emphasize two *facts*: (1) this shift is not irregular and unpredictable, contrary to what the Scollons have led us to *believe*; and (2) two phonological systems coexist, a point the Scollons completely missed.

Example 8: L1 (theory versus finding) *The present results* can also be used to address Piaget's *claims*. *Piaget argued* that children under the age of seven years, especially between the ages of three and five years, find it difficult to accommodate the perspectives of their listeners. *The results of the present study, however, indicate* that children between the ages and 3½ and 4½ years do adapt to differences in listener status and say 'thank you' more frequently to adults than to peers.

Example 9: L1 (experiment-specific result [*found*] versus generalisable result [*indicate*]) *Greif and Gleason (1980) also found* that preschool-aged boys and girls were equally likely to say 'thank you' spontaneously. *The present results indicate* that preschool-aged girls say 'thank you' spontaneously more frequently than boys.

3.1.2 Modification of status

In examples 8 and 9 a pattern emerges which is used even when there is no actual differential in status: a named researcher is used as the source of the Opposed Claim, as the subject of a projecting clause for example; while a non-human entity, usually a piece of research or its results, is used as the source of the Proposed Claim. In Halliday's terms, for the Opposed Claim, the subjectivity of the judgement is made explicit, while for the Proposed Claim, its objectivity is made explicit (Halliday 1985:336). In this way, the status is *modified* by its named source. Modification may also be done by the status-assigning verbs, which may weight the Proposed Claim more positively than the Opposed Claim. Examples 10 and 11 illustrate these contrasts.

> **Example 10: S1** .. *McGhee et al ... found* a limiting dichroism of about -0.2 and did not observe saturation of the orientation at increasing electric field. *Our experiments* ..... *showed* a limiting dichroism of +0.06, with saturation of orientation at voltages above 15 kV/cm.

> **Example 11: H1** *Gash's characterization* of the act itself is undoubtedly correct; Whig reforms fell well short of radical, much less revolutionary. Even so, *the behavior of Shrewsbury's electorate in both political debate and at the hustings demonstrates* that if the Reform Bill was too weak to tear England's political fiber, the experiences associated with its passage profoundly altered the electoral behavior of many voters.

This strategy unambiguously creates a difference between the Proposed Claim and the Opposed Claim in favour of the former, but does so in a way which implicitly denies the intervention of the personal, in keeping with an ideal 'objectivity of discovery'. Firstly, both the Proposed Claim and the Opposed Claim are apparently presented in parallel terms. Secondly, the Proposed Claim's apparent virtue is that it does not arise from an individual but from research findings. The writer's presentation of self is of a non-judgemental reporter, in spite of the judgement implicit in the contrast between, for example, *McGhee et al found* and *our experiments showed*. The act of persuasion is, as Latour and Woolgar (1979) comment, rendered invisible.

In the above examples, there is a certain even-handedness in the treatment of claims, in that both the Opposed Claim and the Proposed Claim are presented as contingent upon some form of evidence or 'hearsay' (a term used in Chafe 1986). In example 10, for instance, the Opposed Claim is contingent upon the words of McGhee et al while the Proposed Claim is similarly contingent upon the inscriptions (graphs, tables, and so on) which are the outcome of *our experiments*. This contingency in itself modifies the status of the claims: both are the results of single sets of experiments rather than undisputed or self-evident facts.

The weighting of one claim over the other can, however, go one stage further. The source of the Proposed Claim can be left unidentified, as in examples 12 and 13 below, so that the truth of the claim is not contingent upon anything. (However, *surely* in example 12 does represent contingency upon a consensus, as will be discussed below.) Alternatively, the source can be expressed in a phrase introduced by *as*, as in examples 13 and 14, making the Proposed Claim contingent upon something which has already been agreed. In the following examples, OC = Opposed Claim, PC = Proposed Claim.

Example 12: H2

OC Pierre Milza insists that "the sociology of the *Croix de Feu* movement" excludes it from the ranks of French fascism (ref). Relying on the figures of Janine Bourdin, he establishes that a quarter of the *Croix de Feu* was drawn from the upper middle class ... The *Croix de Feu*, Milza concludes, was therefore "less plebeian and more bourgeois than fascist organizations."

But which fascist organizations?

PC Overrepresentation of the middle and upper middle classes and under-representation of the factory laborer were surely characteristic of the social composition of both Italian fascism and nazism (ref).

Example 13: L2

OC They [Scollons] correctly observed that the alternation of the alveolar and the alveopalatal series is most conspicuous among the Chipewyan speakers in Fort Chipewyan.

PC But they failed to understand both the origin of this variation and the nature of this phonological variation in the process of "normal" language change and language death.

PC As indicated in the previous section, the unsystematic alternation of the coronal obstruents in the speech of semi-speakers is primarily due to the coexistence of the two affricate systems.

Example 14: S2

PC As can be seen from Figure 1, the dichroism measured for the filament form of all chromatins studied is strongly negative. ....

OC On the other hand, Tjerneld et al (ref) have reported that the flow dichroism of unfractionated rat liver chromatin at low ionic strength is positive.

In these examples, then, the validity of the Opposed Claim is modified by its expressed contingency, while the validity of the Proposed Claim is not modified in this way. As with the other strategies discussed in this section, the Proposed Claim is, with an appearance of objectivity, given greater weight.

3.2 Differences between empirical and other papers

So far, we have seen how writers in different disciplines share strategies for presenting knowledge claims differentially. We now turn to points of difference which exist between the empirical papers, that is, those which present new experimental data (S1, S2, L1) and the non-empirical ones (L2, H1, H2). My argument here will draw on Sinclair's observation of the essentially interactive nature of discourse (Sinclair 1981). In particular, I shall argue that the degree to which such interaction is made explicit differs between the two sets of articles. Broadly speaking, in the non-empirical articles, language is used which appeals more overtly to shared attitudes and values than is common in the empirical articles. As this is a very complex issue, I shall take up only three aspects of it here: the choice of an attitude-indicating verb, adverbial modification, and the appeal to consensus.

3.2.1 The use of attitudinal verbs

The first point is the occurrence of attitude-indicating verbs, such as *show*, *point out*, *argue*, *insist*, and *understand*. As a preliminary to the discussion, it is first necessary to propose a tentative classification of such verbs. Numerous parameters for such a classification are possible (for example Thompson & Ye 1991). The categories below owe something to Thompson and Ye's classification in making salient the issue of whose attitudes towards the epistemological status of the proposition are implied, but they also take account of the fact that a reporting verb may make implications about conflicting attitudes of current writer and original researcher. I tentatively propose the following seven classes.

Class 1. Verbs of reporting (human subject): *find*, *observe*, *report*. No information about attitude is given.
Class 2. Verbs of subjective interpreting (human subject): *conclude*, *believe*, *suspect*, *interpret*, *suggest*, *characterise*. Information is given about the attitude of the subject of the verb.
Class 3. Verbs of objective interpreting (non-human subject): *show*, *indicate*, *demonstrate*, *establish*, *characterise*, *suggest*. Information is given about the attitude of the writer.
Class 4. Verbs of arguing (human subject): *note*, *point out*. Information is given about the attitude of the writer.
Class 5. Verbs of arguing (human subject): *claim*, *argue*, *admit*, *deny*, *insist*, *acknowledge*, *concur*, *declare*. A potential difference of attitude between writer and subject of verb is implied.
Class 6. Verbs of cognition (human subject): *realise*, *be aware*, *understand*. When grammatically positive, information is given about the attitude of the writer. When grammatically negative, a difference of attitude between writer and subject of verb is implied.
Class 7. Verbs of reaction: *find significant*, *be struck/impressed by*, *con-*

vince, persuade. A potential difference of attitude between writer and original researcher is implied.

Figure 2: Number of tokens in each verb class, by text

Text Verb class	S1	S2	L1	L2	H1	H2
1	4	5	4	2	-	-
2	2	1	1	5	-	1
3	1	-	3	1	3	1
4	-	4	-	1	-	-
5	-	-	1	3	3	11
6	-	-	-	1	-	2
7	-	-	-	-	-	5
total 1-4	7	10	8	9	3	2
total 5-7	0	0	1	4	3	18

The essence of my argument here is that whereas verbs from classes 1-4, such as *show* in example 10 above, simply imply a single judgement regarding the certainty of the following proposition, verbs from classes 5-7, such as the negative *realise* in example 15 or *convince* in example 16, involve a great deal more complex interaction of judgements.

Example 15: L2 They [Scollons] apparently *did not realize* that a sound shift or merger in a given position and a given lexical item does not necessarily eliminate the segments involved from the (underlying) system.

Example 16: H2 One of the things that *convinces* Pierre Milza that the Croix de Feu could not have been fascist is "the links that seem to attach it to certain leaders of the liberal Right".

In particular, the verbs in classes 5, 6 and 7 (for example *claim*, *not realise*, *is convinced by*) all imply a state of disagreement between the writer and the original researcher. Thus, the presence of these verbs indicates a greater subjectivity and a more heavily argumentative style than does the presence of the verbs in classes 1-4. As figure 2 shows, the empirical articles (S1, S2, L1) form a natural grouping, with verbs predominantly from classes 1-4. The figure also shows H2 to be the most 'argumentative' paper, with verbs predominantly from classes 5-7. Papers L2 and H1 fall between the two extremes.

3.2.2 Adverbial modification

The degree to which the status modifiers are used to indicate interaction between writer and reader can be increased by their further modification by adverbials. The effect of negation upon class 6 verbs has already been noted. What is particularly interesting, however, is that in the non-empirical articles, adverbs are sometimes used to modify verbs which apparently cannot be modified in this way in the empirical articles. Two examples of note are *correctly observe* in example 13 (page 123) and *persuasively demonstrate* in example 17 below.

> **Example 17: H1** Yet O'Gorman *persuasively demonstrated* that, despite its flaws, the electoral system played far more than a marginal role politically.

In each of these examples, the modification appears to contradict the 'meaning' of the verb, at least as it is used in the empirical articles. In those articles, an *observation* is non-interpretative, and cannot therefore be either correct or incorrect, and a *demonstration* occurs as a result of the convincing nature of experimental data, not as a result of human rhetorical skills. In modifying these verbs in this way, the writers of L2 and H1 are revealing an attitude towards knowledge that is very different from that exhibited by the writers of S1, S2 and L1. Alternatively, one could say that the meaning of the lexical items *observe* and *demonstrate* are different in the different types of article.

3.2.3 Appeal to consensus

We turn thirdly to the appeal to consensus. Certain grammatical items make a specific appeal to an implicit, shared consensus of opinion between writer and reader and, again, the presence or absence of these distinguishes S1, S2 and L1 from L2, H1 and H2. Some of these grammatical items are illustrated in example 12, quoted in section 3.2.1 above, and example 18. They include rhetorical questions, as in example 12, adverbs such as *certainly*, *no doubt*, or *surely*, also in example 12, and the representation of the source of knowledge claims as generally assumed knowledge, as in example 18.

> **Example 18: H2** ... *it can hardly be disputed that* the mobilization of a radical Left preceded that of a radical Right ... Moreover, *no one would contend that* the action of the Croix de Feu on the night of February 6, 1934 had anything to do with "responding to an attempted revolution."

My argument, then, is that what might be felt intuitively as a difference between an impersonal, report-like style and a more interactive, argumentative style is at least in part explicable in terms of the lexical and grammatical features associated with the presentation of knowledge claims within the various discourse communities.

4. The resolution of the conflict

The conflict, as we have seen, is presented in terms of a differential between knowledge claims. It is resolved by amassing positive value to the Proposed Claim and/or negative value to the Opposed Claim. Value is the evaluation of the claim as being 'good' or 'bad', that is, of positive value to the discipline (the claim is probably right) or of negative value to the discipline (the claim is probably wrong). In academic writing, value is often not given explicitly, using attitudinal lexis, but implicitly, by stating the grounds on which the judgement is made.[4] One example is:

> **Example 19: L1**
> **OC** Greif and Gleason (1980) also found that preschool-aged boys and girls were equally likely to say 'thank you' spontaneously.
> **PC** The present results indicate that preschool-aged girls say 'thank you' spontaneously more frequently than boys.
> **PC + fit to other data** This finding is in accord with other research which supports sex differences in language development (refs).
> **PC + fit to explanation** Moss (1974) has suggested that girls receive more social training from adults than do boys and this may result in faster acquisition of the more social aspects of language.
> **PC + fit to explanation** Sex differences in the spontaneous use of 'thank you' by preschoolers may also have a cultural basis, with Western culture placing more emphasis on the use of politeness routines such as 'thank you' by females.

In this example, a single result (OC) is contrasted with a generalised one (PC). The Proposed Claim *preschool-aged girls say 'thank you' spontaneously more frequently than boys* is given positive value on two grounds: that it fits, or is consistent with, results of experiments carried out by other researchers; and that it fits, or is consistent with, an explanation. The Opposed Claim is given no corresponding positive value, but neither is it given negative value - as a finding it remains unchallenged. The Opposed Claim thus remains unchallenged as an accurate result, though one which is isolated and not in accord with the generalisation which the paper accepts.

This examination of value can reveal the value-system and the worldview of the discipline (Hunston 1993). The value-system is shown by the nature of the grounds upon which claims are given value. In this case, as in the physical sciences, consistency with data and with explanation is of high value, suggesting that knowledge is perceived as fitting into a network of compatible data, hypotheses, assumptions. It is also notable that, unlike the physical sciences, results which lead to incompatible generalisations are nonetheless acceptable: there is no assumption that the 'physical world' - in this case, child speakers - will be homogeneous. This discussion of example 19 isolates three important areas:

(i) where the conflict lies, i.e. where in the process of knowledge construction from observation to model, human intervention can be seen to occur;
(ii) the grounds on which positive or negative value is given;
(iii) the outcome, in particular, how strong the dismissal of the Opposed Claim is.

To show how the disciplines may differ in each of these areas, three further examples (20, 21 and 22) will be taken.

Example 20: S2

PC As can be seen from Figure 1, the dichroism measured for the filament form of all chromatins studied is strongly negative.

OC1 On the other hand, Tjerneld et al (ref) have reported that the flow dichroism of unfractionated rat liver chromatin at low ionic strength is positive.

OC2 (PC + artefact)[5] Both Tjerneld et al (ref) and Yabuki et al (ref) have attempted to resolve this discrepancy by suggesting that the chromatin filament could be distorted by the applied electric field.

OC2 no fit to data We have reported previously (ref) that we could find no evidence that the electric field induced any chromatosome reorientation relative to the adjacent spacer DNA.

PC + fit to other data Furthermore we observe negative dichroism at low electric field strengths where the orienting force is of the same order of magnitude as the reported flow experiments.

PC + fit to other data Finally, our own flow experiments on chicken erythtocyte, sea urchin sperm, and rat liver chromatin filaments give negative dichroism (ref).

OC1 no fit to data We have attempted to mimic the results of Tjerneld et al (ref) by working at high chromatin concentrations or by addition of polyamines or even RNA, but the sign of the flow dichroism has remained consistently negative.

OC1 + artefact It has been suggested that a positive chromatin dichroism is often associated with aggregation (ref).

In terms of where the conflict lies, example 20 shows the most difficult of problems in Science: results of apparently replicated experiments which nonetheless conflict. In terms of grounds for value, as in example 19, the Proposed Claim is supported by other data. The Opposed Claim is given negative value on the grounds that what has been 'seen' in experimental terms is a distortion, an artefact (*aggregation*) of the experimental process (penultimate sentence). The outcome of the conflict is that, by contrast to example 19, the Opposed Claim and Proposed Claim cannot co-exist. Neither, however, can the Opposed Claim be said to be simply a mistaken reading. The writers must find an explanation which neither accuses the proponents of the Opposed Claim of falsifying results nor allows those results to remain as valid. The claim that the opposed researchers have inadvertently created an artefact fulfils this requirement.

Example 21: L2

OC + fit to PC They [Scollons] correctly observed that the alternation of the alveolar and the alveopalatal series is most conspicuous among the Chipewyan speakers in Fort Chipewyan.

OC no fit to PC But they failed to understand both the origin of this variation and the nature of this phonological variation in the process of "normal" language change and language death.

PC1 As indicated in the previous section, the unsystematic alternation of the coronal obstruents in the speech of semi-speakers is primarily due to the coexistence of the two affricate systems.

PC1 + fit to other situations One might wonder how such diversity exists in a small community of some 1,500 residents, not all of whom speak Chipewyan.

But such a phenomenon is not a rarity in Athapaskan.

In addition to a much more complex case of coexisting subsystems in Slave (ref), I know of three other similar cases among the Northern Athapaskan languages. ......

OC no fit to PC2,3 I have dwelt on the ts-->tš shift to emphasize two facts: (1) **PC2** this shift is not irregular and unpredictable, contrary to what the Scollons have led us to believe; and (2) **PC3** two phonological systems coexist, a point the Scollons completely missed.

OC no fit to PC4 Furthermore, this change has nothing to do with Cree
PC4 + fit to data influence, as evidenced by similar changes elsewhere.

In example 21 the conflict lies, not in the data, about which there is no disagreement, but in the interpretation of the data in terms of conflicting model systems. The grounds for value are in part those found in examples 19 and 20, particularly consistency to data and to explanation. In addition, there is a category of grounds which has a special significance for sociolinguistics: 'fit to other situations'. In other words, the writer's Proposed Claim is given credence by not being unique, but having analogies in other speech communities. But the writer of L2 uses another evaluative device too: claiming that the Opposed Claim is to be given positive or negative value based solely on whether it is consistent with the Proposed Claim (S3, S4, S9). What this does is to resolve the conflict through the presentation of a more convincing argument, rather than through an amassing of empirical data. The outcome of the conflict is that, while the data leading to the Opposed Claim is allowed to be valid, the Opposed Claim itself is totally demolished.

Example 22: H2

Why, exactly, have historians of French fascism excluded the Croix de Feu from consideration?

OC1 Some are struck by the fact that Colonel de la Rocque does not fit the typical fascist profile.

OC1 + problem The assumptions here are more than a little problematic.

OC1a + internal consistency La Rocque is very different from Mussolini or Hitler, although social origins may be the least important of the differences.

OC1a no internal consistency But Hitler and Mussolini were also very different from one another.

OC1b no clear criteria Moreover, it is not clear to what degree the fascist qualities of a movement can be derived from its leader.

OC1b no internal consistency Certainly it would be quite wrong to suggest that ex-army officers with social contacts among the bonne bourgeoisie are refractory to fascism.

Such elements were notoriously present in National Socialism and, above all, Italian fascism.

OC2 Pierre Milza insists that "the sociology of the Croix de Feu movement" excludes it from the ranks of [fascist organisations] ...

OC2 no clear criteria But which fascist organizations?

OC2 no internal consistency Overrepresentation of the middle and upper middle classes and underrepresentation of the factory laborer were surely characteristic of the social composition of both Italian fascism and nazism (ref).

OC2 no fit to records Moreover, as Milza admits, by 1937 the PSF made significant breakthroughs in "the working-class milieu": it claimed 19 percent of its membership was drawn from industrial workers, 16 percent from peasants, 24 percent from white-collar workers, 15 percent from small businessmen, and 26 percent from the liberal professions (ref).

In example 22, the conflict lies in how particular groups of people are to be classified. The classification into 'fascist' and 'non-fascist' is presented not as an automatic or natural one, but as one that is made by the historians themselves. The grounds upon which the classification is given value are the clarity of its criteria and the consistency with which the classification is applied. These are features which belong to an argument, not to a model based on empirical data. Fit to data in the form of records is, however, another grounds for evaluation. It is also notable that the writer in this example uses a simple attitudinal judgement in the third sentence, where the Opposed Claim is labelled as 'problematic' before the grounds for this evaluation are given. As with example 20, the outcome of the conflict is that the Opposed Claim is totally invalidated.

In sum, I have tried to show in this section that an investigation into which knowledge claims are presented as conflicting, and the grounds upon which knowledge claims are evaluated, is a useful method of revealing similarities and differences in the resolution of conflicts in the different disciplines. These differences are very revealing of the process of discovery, or knowledge construction, as perceived by each discipline. The place in the process of discovery where the conflict is seen to lie indicates what part of that process is seen as legitimately open to human intervention. The grounds upon which the claims are evaluated indicate the system of values of the discipline.

In general, there appears to be a progression from the 'report-like' biochemistry papers, through the linguistics papers to the 'argument-like' history papers. The report-like papers present conflict as occurring at the experiment-design stage of discovery, with accurate, consistent data leading naturally to correct interpretation. The argument-like papers deal with conflict in classification, evaluated mainly in terms of the internal consistency of the classification.

To a certain extent, the papers do divide satisfactorily along discipline lines. In the science texts, there is an assumption that the physical world is homogeneous and that observation of that world leads, inevitably, to correct interpretation. Differences in observation cannot be tolerated and are reinterpreted in terms of distortion by the experimental process itself. Conflict is, to a large extent, presented as an unfortunate by-product of human fallibility, not part of the discovery process itself.

In the sociolinguistics texts, on the other hand, it is not assumed that speakers will be so homogeneous that investigation of one group will lead to data that can be reliably generalised to all groups. Neither is it assumed that observation will lead inevitably to a model. The relating of data to model - the interpretation - is where human intervention and therefore conflict occurs. Resolution can be brought about by amassing data as evidence, but also by the construction of the most convincing argument.

In the history texts, although new data can be brought as evidence to support Proposed Claims against Opposed ones (as in H1), there is no necessity for this to be so, and the conflict can be resolved by forcefulness or reasonableness of argument alone (as in H2). As regards the discovery process, it is assumed that human intervention is crucial in selecting, assessing and interpreting evidence, and that it is this intervention, and its concomitant conflict, which constitutes the discipline itself.

The disciplines themselves are far from homogeneous, however. The two linguistics texts, in particular, show significant differences, with the empirically-based L1 having more in common with the science texts and the argumentative L2 showing a greater similarity with the history texts. This is true in terms of where the conflict lies and in the grounds for value. In particular, there is a difference between L2 with its greater use of attitudinal language and L1 with its science-like reliance upon implicit value.

5. Conclusion

This discussion of instances of conflict in academic discourse has illustrated writers' exploitation of linguistic resources, particularly those associated with evaluation, as part of the process of influencing their discourse community in favour of one knowledge claim over another. Further study is clearly necessary, however, as many of the knowledge claims made in this paper require validation from data of other kinds. As an example, the proposed classes of verbs used to modify the status of claims would need evidence from collocational studies to bear them out.

The comparisons between the six texts have shown several different possible groupings. For example, the investigation into how the conflict was made relevant distinguished those papers where earlier findings were re-worked (L2, H2) from the others, which presented new findings. The way that knowledge claims are represented divides the empirical papers (S1, S2, L1) from the non-empirical (L2, H1, H2). The site of the conflict and the grounds for value that are invoked divide the papers along discipline lines, but more attitudinal language is used in the papers which challenge earlier findings.

The variety of possible groupings suggests that the notion of genre in academic discourse, and assumptions about what is 'the same' genre and what is 'different' must be treated with the utmost caution. This is not to say that ideas of genre can be ignored, however. On the contrary, this study also suggests that the differences between texts is something that corpus-based research needs to be sensitive to. It has been suggested, for example, that even the meaning of certain words such as *observe* or *demonstrate* depends on the specific type of text they are used in.

The benefit of taking as a topic those research articles which express conflict is that they call attention to the linguistic strategies employed by writers to support their own claims at the expense of competing claims. These strategies in turn reveal the assumptions regarding the nature of knowledge held by each discipline.

Author's Address:
Cobuild, Westmere, 50 Edgbaston Park Road, Birmingham. B15 2RX.

Notes

1. An earlier version of this paper was read at the conference 'Discourse and the Professions', Uppsala, August 1992. This version was written while the author was a lecturer at the University of Suurey.
2. A knowledge claim is an item in a research article which the writer puts forward to be added to the sum of knowledge agreed upon by the community of that discipline.
3. For full details of how status may be assigned, see Hunston (1989).
4. For more details of my analysis of evaluation, see Hunston (1989; 1993; forthcoming). For other treatments of evaluation, see Winter (1982) and Hoey (1983).
5. 'Artefact' here is used in the scientific sense of a phenomenon occurring as a result of the experimental process itself.

References

Barton, Ellen L. 1992. "Evidentials and Rhetorical Structure". Paper read at the conference *Discourse and the Professions*, Uppsala, August 1992.

Beaton, Bryan. Forthcoming. *The Discourse of Technical Comments*. Unpublished MA Thesis: University of Surrey.

Chafe, Wallace. 1986. "Evidentiality in English Conversation and Academic Writing". *Evidentiality: The Linguistic Coding of Epistemology* ed. by Wallace Chafe & Johanna Nichols, 261-72. New Jersey: Ablex.

Fairclough, Norman. 1992. "Introduction". *Critical Language Awareness* ed. by Norman Fairclough, 1-29. London: Longman.

Francis, Gill. 1986. *Anaphoric Nouns (Discourse Analysis Monographs No 11)*. Birmingham: English Language Research, University of Birmingham.

Gilbert, G. Nigel & Michael Mulkay. 1984. *Opening Pandora's Box*. Cambridge: Cambridge University Press.

Halliday, M. A. K. 1985. *An Introduction to Functional Grammar*. London: Arnold.

Hoey, Michael. 1983. *On the Surface of Discourse*. London: Allen & Unwin.

Hunston, Susan. Forthcoming. "Evaluation and Organisation in a Sample of Written Academic Discourse". *Advances in Written Discourse Analysis* ed. by Malcolm Coulthard. London: Routledge.

____________. 1993 "Evaluation and Ideology in Scientific Writing". *Register Analysis: Theory and Practice* ed. by Mohsen Ghadessy, 57-73. London: Pinter.

____________. 1989. *Evaluation in Experimental Research Articles*. Unpublished PhD Thesis: University of Birmingham.

Latour, Bruno & Steve Woolgar. 1979. *Laboratory Life: The Social Construction of Scientific Facts*. Beverley Hills: Sage.

McKinlay, A. & J. Potter. 1987. "Model Discourse: Interpretative Repertoires in Scientists' Conference Talk". *Social Studies of Science* 15. 593-630.

Mulkay, Michael. 1985. *Word and the World: Explorations in the Form of Sociological Analysis*. London: Allen & Unwin.

Myers, Greg. 1990. *Writing Biology: Texts in the Social Construction of Science*. Madison: University of Wisconsin Press.

Sinclair, John McH. 1993. "Written Discourse Structure". *Techniques of Description* ed. by J. McH. Sinclair, M. Hoey & G. Fox, 6-31. London: Routledge.

____________. 1992a. "Priorities in Discourse Analysis". *Advances in Spoken Discourse Analysis* ed. by Malcolm Coulthard, 79-88. London: Routledge.

____________. 1992b. "Trust the Text". *Advances in Systemic Linguistics: Recent Theory and Practice* ed. by M. Davies & L. Ravelli, 5-19. London: Pinter.

____________. 1981. "Planes of Discourse". *The Two-Fold Voice: Essays in Honour of Ramesh Mohan* ed. by S. N. A. Rizvi, 70-89. Saltzburg: University of Saltzburg.

Swales, John. 1990. *Genre Analysis: English in Academic and Research Settings*. Cambridge:Cambridge University Press.

Thompson, Geoff & Ye Yiyun. 1991. "Evaluation in the Reporting Verbs Used in Academic Papers". *Applied Linguistics* 12. 365-82.

Winter, Eugene O. 1982. *Towards a Contextual Grammar of English: The Clause and Its Place in the Definition of Sentence*. London: Allen and Unwin.

Appendix 1: Texts Referred to.

Biochemistry: (S1). Yabuki et al. 1982. "Orientation of nucleosomes in the thirty-nanometer chromatin fiber". *Biochemistry* 21. **(S2)**. McGhee et al. 1983. "Higher order structure of chromatin: orientation of nucleosomes within the 30 nm chromatin solenoid is independent of species and spacer length". *Cell* 33.

Sociolinguistics: (L1). Becker & Smenner. 1986 "The spontaneous use of thank you by preschoolers as a function of sex, socioeconomic status, and listener status". *Language in Society* 15. **(L2)**. Cook. 1991. "Linguistic divergence in Fort Chipewyan". *Language in Society* 20.

History: (H1). Phillips & Wetherall. 1991. "The great reform bill of 1832 and the rise of partisanship". *Journal of Modern History* 63. **(H2)**. Irvine. 1991. "Fascism in France and the strange case of the Croix de Feu". *Journal of Modern History* 63.

2. Corpus Studies: Theory and Practice

A Corpus-Driven Approach to Grammar
Principles, Methods and Examples

Gill Francis
Cobuild, Birmingham

Abstract
This paper outlines the approach to grammar currently being implemented at Cobuild, Birmingham. It begins by pointing out some of the major differences between a data-driven approach and that embodied in traditional descriptive grammars. It goes on to define the characteristics of a new grammar which gives due consideration to lexis and phraseology, as well as to the meanings which are typically encoded by particular syntactic structures. A working method for compiling this type of grammar is suggested, and in the final section this is illustrated by an investigation of a particular structure, the appositive *that*-clause. Access to large corpora reveals aspects of this structure which have not been captured before. It is argued that corpus-driven studies will lead to similar discoveries in other areas of language description.

0. Introduction

In this paper I will attempt to explain and illustrate the approach to grammar which is currently being developed at Cobuild, Birmingham. We have recently started the compilation of a large descriptive grammar of English, using the corpora of the Bank of English.[1] John Sinclair is the guiding force behind this enterprise, and the grammar will embody many of the views on language that he has been formulating over the years.

Our approach differs fundamentally from that of the descriptive grammars currently available. Perhaps the best-known and most authoritative of these is Quirk et al's *A Comprehensive Grammar of the English Language* (1985). I will begin by indicating some of the ways in which a new descriptive grammar should, and will, differ from this very well-established work.

First, we are providing only real examples, a practice that has always been followed by Cobuilders in the compilation of dictionaries and other reference works (e.g. Sinclair 1987, 1990). The fundamental principle behind this is that the evidence of the ways in which language is really used is available in plenty, and there is no longer any need to invent example sentences in the time-honoured way. I am currently consulting corpora amounting to 120 million words of running text, and while the provision of authentic untampered-with examples is the simplest and most obvious way of using this evidence, it is nevertheless crucial.

Quirk et al, incidentally, had access to small corpora, including the Survey of English Usage, but these were apparently not used extensively, and the examples given are in some cases idealised from the corpus and in others clearly invented. Opening the book at random, we find sentences like:

> (1) Walter played the piano more often in Chicago than his brother conducted concerts in the rest of the States. (ibid:1132).
>
> (2) I've never seen a dog more obviously friendly than your cat. (ibid: 1135)

While real corpora often provide some very odd examples indeed, their oddness rarely has this unmistakable ring of artificiality. One of the reasons for the artificiality of examples like the above is that they are given as possible variants of other sentences, in order to contrast one item, structure or system with another in terms of what is possible and what is not. Examples 1 and 2 are in the section on comparative clauses: example 1 illustrates the fact that while single-variable comparisons are the most common, multi-variable comparisons are possible. But the number of variables involved is precisely what makes the sentence unnatural: Walter and his brother are being compared in terms of their frequency of performance, but since their fields of expertise and their geographical coverage are also different, it is difficult to think of any motivation for the comparison which would be meaningful in genuine interaction. Example 2 is contrasted with **I've never seen a more obviously friendly dog than your cat* (ibid: 1135)[2], which if said would imply that the *cat* in question is actually a *dog*: neither variant, however, is likely to occur. These two examples are not ungrammatical sentences, but they do not meet basic standards of conversational relevance: they are unnatural precisely because they point to possibilities rather than realities. The descriptive basis of the grammars which accommodate them is essentially contrastive.

This brings us to a second feature of our grammar which will distinguish it from traditional descriptions like that of Quirk et al: it is our aim to compile, as far as is possible, a *non-contrastive* grammar, which shows how each item or structure is used in its own right, rather than as compared with members of the same or a contrasting paradigm. The development of this standpoint is high on the agenda.

A third defining characteristic of the grammar we are compiling is that it will be a *data-driven* grammar of English. It might be supposed that this is a principle held by corpus linguists in general, but that this is not the case is evidenced by arguments like that of Aarts (1991:45), who suggests that the corpus linguist make up his [sic] grammar on the basis of his "intuitive knowledge of the language and whatever is helpful in the literature", and that when he "has the conviction that his grammar is tolerably reliable and complete" he tests it on the sentences of a corpus. Incidentally, the fact that Aarts refers to 'the sentences of a corpus' while we prefer to talk about 'x million words of running text' points to a fundamental difference in our attitudes towards data and invokes very divergent linguistic traditions.

In our approach, the data comes first. The corpus is the major informant, providing the raw information we need in order to describe the language, and intuition is considered to be of secondary importance. Intuition may be useful to linguists in a number of ways, but for the purposes of saying exactly how language is used, it is notoriously unreliable. Even when people are describing their own language practices, there is often a difference between what they think they say and what they actually do say. Moreover, there are multiple aspects of language use which are simply not accessible to introspection: Louw (this volume), for example, discusses the prevalent but often unconscious patterning which he calls 'semantic prosodies'. The corpus has a lot of surprises in store for us at every turn, and contains some threats to our accepted views of language, but it is the only reliable authority and must be treated with respect. It may be used as the testing-ground for theories only if we are ready to abandon our theories at any moment and posit something new on the basis of the evidence.

A fourth major characteristic of the Cobuild approach to grammar is that it is essentially a *lexical* approach: we are attempting to compile a lexicogrammar in the true sense of the term. Basically, most grammars are interested in lexis only insofar as it is useful, indeed indispensable, for the illustration of syntactic structures. The assumption is, or appears to be, that grammar and lexis are quite distinct, and that grammar is an activity which is concerned solely with the description of syntactic structure. Thus they see the clause as consisting essentially of Subject, Verb, Object, Adjunct and so on, each of whose internal structure is similarly described in terms of its smaller constituents. All these constituents are seen as slots into which lexical items can be fitted, subject to certain obvious restrictions: for example, the modifier slots before the head of a nominal group are typically filled by members of the class of adjective or noun, and the qualifier slot by a prepositional phrase. The implication is that so long as you choose a word from the right class to fill the slot, then the result will be a grammatical utterance. But while it may indeed be grammatical in the narrowest sense of the term, it is unlikely that it will be natural or acceptable unless lexical factors are taken into account.

Before going on to consider the implications of a lexical approach to grammar, I will give two examples, from different areas, of the traditional treatment of the relationship between syntax and lexis as represented by Quirk et al's *Comprehensive Grammar*.

1. The treatment of lexis in a traditional descriptive grammar

1.1 Introductory 'it' as Object

The first example concerns a grammatical structure which is lexically restricted. This is where the introductory, or non-referring pronoun *it* is the Object of a verb, and is followed by an adjective or noun group. These citations were selected at random from a full concordance of the structure when *it* precedes an adjective:

```
policies because they often find it difficult to explain why a r
ass unemployment, many will find it harder than usual to relax i
ist, who, unlike Genscher, finds it easy to make friends among B
orers in this country, has found it almost impossible to get a g
counted by Lord Graham who found it necessary to the point of hi
 put Gwynned first. But he found it hard, too, being wed to John
want to pressure him, and I made it clear that I was not angry.
cent. Dealers said the move made it clear that the Spanish autho
ut his party's policies, he made it plain that Labour planned to
his strategy would at least make it possible to conceptualise mo
ess ahead of the match have made it inevitable that French rugby
inflationary pressures and makes it less and less likely that th
gainst a rape conviction, making it likely that he will remain i
area of refurbishment, by making it almost impossible for them t
```

Quirk et al mention this structure, taking it to involve the extraposition of a clausal Object. They state (1985:1393) that "When the object is an *ing*-clause in SVOC and SVOA clause types, it can undergo extraposition; when it is a *to* infinitive clause or a *that* clause, it must do so". Thus the sentence *You must find it exciting working here* is considered to be an extraposed variant of *You must find working here exciting*, while the sentence *I made it my prime objective to settle the matter* has no such 'non-extraposed' alternative. In other words, they treat the structure entirely in terms of syntactic operations, interesting only insofar as it is, or is not, a transformation of some other 'original' clause (though when there is no 'original' variant available, it is difficult to see how this is relevant). They give no indication as to its key lexis, while using it, perforce, in their examples.

In fact, as the concordance given above indicates, the structure occurs with an extremely restricted range of verbs, of which *find* and *make* are by far the most frequent, accounting for over 98% of all the citations of the structure in the corpus. There is a mere scattering of occurrences with *think* and *consider*, and a few more fixed phrases like *see to it*. Moreover, when *it* is followed by an adjective, the range of adjectives involved is fairly restricted: *find* occurs predominantly with *difficult*, *hard*, and *easy*, and *make* occurs most frequently with *clear*. *Clear* and *likely* do not occur with *find*, and so on. These lexical patterns are closely connected with the communicative function of the structure, which is to present a situation in terms of how it is evaluated, putting the evaluation straight after the verb. The ways in which we typically evaluate situations, using this particular structure, are stereotyped, showing that our concerns tend very often to be with how difficult or easy life is made for us, how predictable things are, and how well we understand what is going on. Such facts about lexico-syntactic interdependence and its related communicative functions become obvious only when one interrogates a corpus larger than those to which Quirk et al had access.

It should be pointed out that the sort of co-selection involved here goes beyond Firth's notion of 'colligation' (in Palmer 1966:181). This involves the relations between grammatical abstractions, which in this case could be stated as a verb, introductory *it*, and an adjective. What we have here though is a highly-restricted choice of verb, introductory *it*, and an adjective with one of a narrow range of meanings. While there is colligation at work here in the broadest sense, the most interesting aspect of this and many other structures is the blend of colligation and collocation, where the collocational possibilities involve not individual words, but semantic sets of words and phrases.

1.2 Appositive 'that' -clause qualifiers

A second area worth mentioning here is nominal-group qualifiers (post-modifiers) and in particular, the restrictive appositive *that*-clause in sentences such as *Which French leader made famous the remark that the British are a nation of shopkeepers?* Of the head-nouns found with this type of qualifier, Quirk et al (1985:1260) say this: "the head of the noun phrase must be a general abstract noun such as *fact*, *idea*, *proposition*, *reply*, *remark*, *answer* and the like", and follow it with a couple of examples.

This begs a number of questions, none of which is addressed by the authors. What is the range of 'general abstract nouns'? What are the criteria for recognising them? And, assuming we can find out exactly what they are, is this type of qualifier used with all these nouns? If not, what are the restrictions: what semantic considerations apply? *Type* and *thing* are presumably both general and abstract, but neither of them occurs in this environment. And finally and most important, what is the function of this type of qualifier in relation to its head-noun: what meanings does it typically encode? I will return to these questions later in this paper, and try to answer some of them.

2. A lexical and phraseological grammar

It is clear that the role of lexis, and hence of meaning, is not given sufficient consideration in traditional grammars. In many cases there is no mention at all of the fact that a structure may be lexically restricted other than by considerations of word-class and related factors. Thus in the case of introductory *it*, say, there is no attempt to link this to the lexis of evaluation and hence to communicative function. In other cases these grammars do give some lexical exponents of structural elements, in particular in the area of verb complementation, where lists of verbs are given for many of the patterns described. The attention paid to verbs and the lexis associated with them is symptomatic of the fact that verbs have always been treated as the privileged word-class in grammatical study. Even here, though, we have no way of knowing whether the lists are intended to be exhaustive, or, more seriously, whether the verbs listed are actually used in the given pattern with any frequency.

The approach to lexis embodied in these grammars cannot help us very far in our attempts to understand the range of meanings that are typically encoded in language and the syntactic resources that are typically exploited to grammaticalise these meanings. This is because their authors seem to assume that the study of grammar has strict limits, and lexis, on the whole, lies outside them. Thus, implicitly, they ignore the fundamental fact that syntax is driven by lexis: lexis is communicatively prior. As communicators we do not proceed by selecting syntactic structures and independently choosing lexis to slot into them. Instead, we have concepts to convey and communicative choices to make which require central lexical items, and these choices find themselves syntactic structures in which they can be said comfortably and grammatically.

The grammar which we are compiling will attempt a rather different approach to lexis. It is an approach which recognises the importance of syntagmatic patterning and dependent lexical choices in the structuring of language. In other words, we take the view that syntactic structures and lexical items (or strings of lexical items) are co-selected, and that it is impossible to look at one independently of the other. Particular syntactic structures tend to co-occur with particular lexical items, and - the other side of the coin - lexical items seem to occur in a limited range of structures. The interdependence of syntax and lexis is such that they are ultimately inseparable, and it becomes merely a methodological convenience to regard them as different perspectives from which to view language use.

2.1 The role of phraseology

It should be clear from the above discussion that the co-selection of syntax and lexis often involves not just one word but a series of them; each choice systematically constrains choices in the immediate context and in the development of the discourse as a whole. The co-occurrence of *make* and *find* in introductory *it* structures is an illustration of this: we have seen that the selection of one of these verbs followed by introductory *it* tends to involve a simultaneous choice within the *easy/difficult/impossible* parameter, the *unlikely/likely/inevitable* parameter, or the *clear/plain/obvious* range of meanings.

These co-selections are so frequent and stereotyped that they have to be seen as located somewhere on the phraseological cline between the fixed idiom and the echo of a very fuzzy pattern (see Sinclair 1991a:110ff). At one extreme are idioms like the much-cited *it's raining cats and dogs* and *it never rains but it pours*. Here, once you have chosen this way of evaluating a situation, all internal choices have been pre-empted: this is in effect a single-choice chunk and can be seen as a multi-word lexical item. To alter any of the elements would involve a very marked choice indeed. Further along the scale, there are less fixed idioms, in which some but not all of the lexico-syntactic choices have been preempted. Take the concept of *putting on a brave face*:

```
ing still to put on a brave face, even though they this week ann
palpable. He put on a brave face, of course, but there was too l
er. Lex was putting a brave face on the loss of the Volvo conces
ren Pritchard, puts a brave face on it. We are always hopeful th
nisters are putting a brave face on the prospect of weak recover
deavoured to put on a brave face about the rebellion, saying the
nd to those who put a brave face on exceedingly difficult circum
 Others are putting a brave face on the future, but we conclude
e ANC, trying to put a good face on the defeat of its surrogate,
e fashionable to put a good face on the violence, and they will
```

Here you require the verb *put*, but it can be in any tense, and the clause can be finite or non-finite, active or passive. You also require the particle *on*, but its position is not fixed: it can either come directly after the verb (*put on a brave face*), in which case there is no Object, or it can introduce a prepositional phrase (*put a brave face on the future*). Finally, though the adjective is usually *brave*, there are some citations with *good*, but whether this is really a variant or a different idiom is not absolutely clear. So while the main lexical choices are pre-empted, there is some variation in the form of this phrase.

Further along the cline still are semi-prepackaged phrases which allow for a good deal of variation. If you want to say emphatically that you simply do not know something, one way of doing it is to say *I haven't the faintest idea*, or something very similar, as the following citations[3] illustrate:

```
them. Labour still had not the faintest idea how a free enterpri
cottish electorate has not the faintest idea what it wants. The
tten very well knows, haven't the least idea what the word means
possession and no-one had the slightest idea what to do with it.
ir Lordships did not have the slightest idea what in particular
? After all, I didn't have the foggiest idea how a telephone wor
e realized that the hadn't the remotest idea of how to find out
rved in it, nor have they the slightest conception of the great
the Americans lacked even the slightest conception of the great
s few appeared to display the slightest notion of what cricket i
ation doesn't seem to have the foggiest notion of how to turn it
using, we still don't have the foggiest notion. However, we do k
revealed that they had not the remotest notion what sort of a wo
hat I mean?" Nope. Haven't the foggiest. They shared a brief lau
```

In this pattern there is no single lexical item which is essential; the only essential elements are the' superlative' marking of the adjective and the definite article *the* which accompanies it; these are the criteria for recognising this as a pattern. The pattern always has some negative orientation, usually achieved by *have* with negative polarity, but also by *no-one* or *nothing*, or by a verb like *lack*. There is a small range of head-nouns, comprising *idea*, *conception* and *notion*, though zero is also possible when the adjective is *foggiest* (last citation), showing that the predictability of this collocation is so strong that the head-noun has disappeared altogether (see Sinclair 1991:497). There is also a narrow range of adjectives: *faintest*, *slightest*, *least*, *foggiest* and *remotest*. Even within these limits, however, variation is restricted: there is no citation of *the faintest conception*, for example, in 110 million words. The open choices here are the Subject, the time orientation, the thematic development, and what it is, precisely, that the Subject is totally ignorant of. As Winter (forthcoming) points out, these open choices are what makes every clause unique, though each unique choice may of course constrain other choices, and so on, ad infinitum.

Finally, at the other end of the scale there are lexico-syntactic interdependencies which are so nebulous that they leave behind only an echo of the co-selection constraints governing their selection, and are discernible only when one examines a large number of citations of a lexical string. An example of this is the range of choices following the string *be a case of*. In itself, this is a fixed phrase, though the tense of the verb *be* varies, there may be modal expressions involved, and the polarity may be negative. What is interesting is the patterning in the nominal groups which follow. The following citations are a random selection of the multi-word concordance *be a case of*:

```
ious condition, this was a case of deja vu. He was shot and crip
here. However, it is not a case of either or but both and. Emil
ps for Graeme Hick it is a case of Hope deferred maketh the hear
is stage, so it's simply a case of keeping fingers crossed that
ia today. So it has been a case of grab a quote and run, especia
line on the plate, it is a case of carry on bowling. Such pain o
es over it.  It was just a case of me staying up tall. Although
. For it may not just be a case of having egg on your face, it c
on Francis Bacon are not a case of the king's new clothes. If Ba
 and that it is not just a case of pulling oneself together. Oft
s an easy option. It was a case of use it or lose it, says Bob T
no good whatsoever. It's a case of, if you can't beat them, join
```

The phrase *be a case of* is a preface, and is used for comparing the situation which has been established in the discourse with another that is familiar to the reader, often a proverb, aphorism, or other fixed expression. In other words, it is a way of packaging the immediate situation by summarising it with reference to something established as a typical or recognisable aspect of collective human experience. The point of reference is often so familiar that the speaker or writer does not consider it necessary to provide the whole story: *the king's new clothes*, for example, is a reference to the well-known fairytale, and the reader is expected to understand the relevance of the comparison. Sometimes, too, the preface is used to alter the text of the proverb or fixed expression, thereby tailoring it to fit a new situation: an example of this is *grab a quote and run*, which is presumably a variant of *take the money and run*. A strong additional tendency is that the expression to the right of *be a case of* tends to involve a negative comparison of some kind: there is often a parallel being drawn between two unfortunate circumstances or events.

Not all of these observations apply to every citation of the structure, and sometimes none of them apply, but the pattern of co-occurrence is highly probable. This then is the freer end of the phraseological cline, where we can say that a fixed phrase has a strong tendency to constrain the choice of another fixed phrase, but there is an enormous range of possibilities open to the speaker or writer as to which fixed phrase is chosen, and the extent to which it is adapted to the speaker's or writer's needs.

Our approach to grammar is one that takes account of all these co-selection tendencies in the language, and attempts to restore all such evidence to its rightful place in a description of how meaning is made in text.

3. Item and environment: a methodological convenience

At this point I would like to say something about the methods that are being explored in order to investigate the nature of lexico-syntactic interdependence and the role of phraseology. Basically, the method that seems to work best involves moving from environment to item and back to environment again. As has already been implied, the two are really one and the same, but viewing them as separate makes it possible to view the patterning of language from different perspectives.

3.1 From introductory 'it' to the grammar of 'possible'

Starting with the environment, then, I begin with a structure like introductory *it* as Subject or as Object. In an earlier study (Francis 1993) I looked specifically at the structure involving introductory *it* followed by an adjective and an adjectival complement clause (or, as Quirk et al see it, an extraposed Subject) in such sentences as *It is unlikely that he will race again this season*. The adjectives that occur as clausal Complement group themselves into broad semantic sets according to the nature of the evaluation involved: it is possible to classify them and arrive at a lexico-syntactic profile of this particular structure. One of the most frequent of the adjectives is *possible*, an interesting one since it evaluates in terms of either likelihood or ability, according to the type of clause following it. A *that*-clause encodes likelihood, as in the sentence *It is possible that the zoo will have to close*. A *to*-infinitive clause encodes ability, as in the sentence *It is possible to hire towing cars and trailers if you require them*.

This observation about the meanings of *possible* prompted a move from the lexical environment of the introductory *it* structure to the item *possible*, with a view to specifying other structural environments with which it is typically associated. It emerged that the assessment of possibility is a recurrent aspect of human communication: *possible* has a wide range of environments which make it unique among adjectives. For example, it is used in the following structure:

```
gue, have made the best possible start in the Refuge Assurance L
cy requires the highest possible level of predominantly shared c
the dark until the last possible moment unless the government is
```

```
decision until the last possible moment, up to half an hour befo
nd duties to the lowest possible level. John Paul II is perhaps
cs.I have the strongest possible belief in him and so does my su
 festival to the widest possible audience. From the beginning we
unication to the widest possible audience was the essence. It wa
 loss came at the worst possible time for Kleinwort, where 1990
sport routes. The worst possible outcome of that meeting would b
e for the smallest area possible - the constituency. They are el
amateur in the best way possible by beating them. It is clear th
```

This pattern involves the definite article *the*, a 'superlative' adjective, and a head-noun. The last two citations show that *possible* may occur after the head-noun as well as before it. The point is that this structure is not available to other adjectives in general: the only other adjective I found here is *imaginable*, occurring after the head in such citations as *...in the simplest and clearest language imaginable.*

Another structure which is associated with *possible* is the *as...as possible* structure, which is found with no other adjective, as far as I can ascertain:[4]

```
sion they may make. As early as possible in his life, train your
ow and it was checked as far as possible in the case of those no
eping their child as healthy as possible so that he or she can w
She tried to visit as little as possible, but there were times w
a man who liked sex as often as possible, and although he was sh
athe. Try to stay as relaxed as possible, it will help breathing
fensive on the Somme as soon as possible in the New Year. Though
him if he could come as soon as possible to help with the prepar
scenes described, as totally as possible. This does require a go
e over as much of the course as possible. I also appreciated tha
```

There is a wide range of adjectives, adverbs and quantifiers occurring to the left of *as possible*, of which the adverb *soon* is by far the most frequent: *as soon as possible* is a single-choice unit to the point that it has its own acronym: ASAP.

The citations given above do not exhaust the unique grammar of *possible*: there are other predictable environments like *where/wherever possible*, *when/whenever possible*, and *if possible*, a group of mini-clauses which occur frequently only when the adjective is *possible*. This notion of a unique grammar has far-reaching implications. If we take any one of a huge range of the more frequent words in English, and examine its citations *en masse*, it will emerge that it, too, has a unique grammatical profile, which certainly cannot be encapsulated by calling the word in question an adjective or a noun or a preposition. The generalisability of this observation thus has the potential to threaten the received position on word-classes, since in many cases an aspect of the grammar of a word is that it is versatile and may well cross more than one accepted word-class boundary. One of the most exciting aspects of corpus-based studies is that classes and categories of all kinds can be reformulated anew, from the bottom up, as suggested by the evidence.

This case of *possible* is just one example of the item-environment method, approached from the 'item' perspective, but it shows a way forward for exploring language in terms of the interdependence of lexis and syntax. Finally in this paper, I will take the grammatical structure referred to in section 1.2 above, the appositive *that*-clause qualifier, and attempt to show how it would be dealt with in a truly lexical grammar. It should be pointed out that this study as it stands is preliminary and incomplete: my intention is simply to give a brief sketch of the approach.

4. Appositive *that*-clause qualifiers: a lexical approach

It was pointed out in section 1.2 above that Quirk et al's treatment of appositive *that*-clauses is too vague and begs a number of questions. It seems to me that a thorough-going lexical approach to this, as to any, syntactic structure, must attempt to provide the following information:

i) We need to state the lexical range of head-nouns which have this kind of qualifier, and specify the semantic groupings which these nouns suggest. A complete study would also specify the nature of the restrictions, since clearly not all 'general abstract nouns' have this type of qualifier, but this aspect of the topic will not be discussed in this paper.

ii) We need to specify the range of meanings of this structure, stating the nature of the relationships between the head-noun and the *that*-clause, and how they relate in turn to the semantic classification of the head-nouns.

iii) We need to state the phraseological constraints which operate here: some head-nouns have appositive *that*-clauses only when there are specific lexico-syntactic co-selections involved.

In the following sections I will attempt to deal very briefly with each of these questions in turn.

4.1 Lexical range and semantic sets

Preliminary research shows that the number of head-nouns which have this type of qualification is in the hundreds: there are perhaps four hundred nouns, or broad senses of nouns, which have appositive *that*-clause qualifiers with any regularity at all, and perhaps a further two hundred which have them only rarely.[5] There may also be a number of nouns for which we can cite a single occurrence of the structure, in which case further evidence would be needed before they could be added to the list. It will be the task of a comprehensive grammar to indicate at least the most frequent of these: the larger the grammar, the more information it will be able to give.

Grouping such nouns semantically is a two-way process. It is up to the grammarian to begin somewhere by recognising typical, generally-accepted types of meaning and assigning lexical items to provisional places. After that it is the lexis itself which refines the groupings and defines them more precisely: we have in effect a thesaurus in the process of compiling itself from the bottom up.

We can start by grouping these nouns into six broad sets, as shown below. It should be noted that in some cases only one sense of the noun is involved: a complete description would include information about senses.

i) nominalisations of illocutionary processes, like *allegation*, *announcement*, *assertion*, *assurance*, *contention*, *implication*, *observation*, *recommendation*, *stipulation* and *suggestion*:

> It was almost impossible to refute Haig's *contention* that victory could only be gained by defeating the Germans on the Western front.

ii) nouns encoding language activity of some kind, or its results, like *axiom*, *caveat*, *dictum*, *formulation*, *generalisation*, *legend*, *maxim*, *message, point* and *proviso*:

> She is aware of granny's ancient *maxim* that you catch more flies with honey than with vinegar.

iii) nouns encoding mental states vis-à-vis particular issues, like *assumption*, *belief*, *conviction*, *doctrine*, *hypothesis*, *idea*, *knowledge*, *notion*, *opinion* and *view*:

> I have an unhappy *conviction* that none of us served him as he was entitled to be served.

iv) nouns encoding thought processes or their results, like *analysis*, *assessment*, *bet*, *calculation*, *conclusion*, *decision*, *inference*, *judgement*, *realisation* and *speculation*:

> It was open to an industrial tribunal to draw an *inference* that the discrimination was on racial grounds.

v) nouns encoding feelings and attitudes, like *anger*, *annoyance*, *astonishment*, *bitterness*, *determination*, *disappointment*, *expectation*, *fear*, *sorrow* and *surprise*:

On the negative side we have experienced overwhelming *disappointment* that man so often uses his will for selfish and destructive purposes.

vi) general nouns, like *advantage*, *basis*, *chance*, *complication*, *danger*, *discovery*, *fact*, *possibility*, *requirement* and *sign*:

This business had the added *advantage* that it could be run from home.

Of these sets, the last is the largest and is as yet largely unanalysed. Further research into the corpora may well suggest ways to classify these nouns, and input from a detailed study of what is often known as 'sub-technical vocabulary' would also be very useful.

It should be remembered that for the purposes of semantic grouping the appositive *that*-clause is just one syntactic structure. As we build up a record of all the recurrent lexis across a range of structures, the semantic sets will become ever more clearly delineated. We can ask, for example, what is the range of noun-heads in the Subjects of clauses with *that*-clause Complements (*the point is that...*). It is not exactly the same as that given above, since many of the nouns, particularly most of those in set (v), reject this environment. Facts such as these motivate further investigation into the ways in which we categorise lexical meaning.

4.2 Meaning

In the majority of cases, the appositive *that*-clause is the lexical realisation (Winter 1982) of the head-noun, or as Downing and Locke (1992:469) put it, "*that*-clause qualifiers represent a proposition or situation as equivalent to the head noun". This is the case in examples 1-3:

(1) In May 1991 his *forecast* that there would be no recovery in house prices in 1991 turned out to be the most accurate among the analysts.
(2) The *attitude* that it doesn't matter what you look like so long as you are clever and capable of doing the job does not work.
(3) The Chinese drafters of the Basic Law thus found themselves in the *dilemma* that they either had to question the Chinese constitution, or else to emasculate the Sino-British agreement and erode Hong Kong confidence.

Here, the *that*-clauses qualifying the head-nouns tell us respectively what the *forecast*, *attitude*, and *dilemma* are: it expands the noun and lexicalises it. In example 3, two dependent *to*-clauses are required within the *that*-clause in order to lexicalise the either-or semantics of *dilemma*.

This identity or equivalence relationship, however, is by no means the only relationship holding between a head-noun and its qualifying *that*-clause. Examples 4-9 illustrate a second type of relationship:

(4) He was in conference at the time, and picked up the receiver with a mild *annoyance* that his secretary had let the call through.

(5) Kemp's first reaction was amused *astonishment* that an intelligent woman should be so ignorant...

(6) ...old people can claim housing benefits, but there is always *anxiety* that some of the most needy old people will fail to take advantage of their rights.

(7) He expressed *concern* that the shop may be in distress.

(8) Howe expressed his *confidence* that the fall of Philadelphia would decide the issue once and for all.

(9) He felt a surge of *happiness* that she could not hurt him any more.

Unlike examples 1-3, the function of the *that*-clauses qualifying the nouns *annoyance*, *astonishment*, *anxiety*, *concern*, *confidence* and *happiness*, all from set (v) above, is not to supply their lexical realisation. These are all words encoding feelings and are in some sense lexically complete: they do not necessarily require expansion or identification. Rather, what the *that*-clause supplies is the *cause* of the feeling. I can find no reference to this kind of semantic relationship in any of the descriptive grammars available to me, although Downing and Locke (1992:464) list *anxiety*, *eagerness* and *happiness* as de-adjectival nouns which have an appositive *that*-clause.

The question is whether it makes sense to call these clauses appositive *that*-clauses at all, since there is no apposition involved in the usual sense in which the term is used. But if we no longer see them as appositive, then we need to refer to their meaning in another way. We certainly cannot call them restrictive relative clauses of any kind. Perhaps the corpus has thrown up, here, evidence of the frequency and range of a qualifying clause which has not previously been considered worthy of a name: the best term for it could be 'causal *that*-clause', for want of something more imaginative.

Finally, there are some nouns whose semantics again preclude the function of simple lexical realisation in the sense in which it was applied to examples 1-3 above. Examples 10-12 illustrate other head-qualifier relationships, less easily definable than those discussed above:

(10) Finally a waiter asked if he would like fresh coffee, a *hint* that the table was needed.

(11) This is a *reminder* that to live a full life you need not only daily bread, but a worthwhile cause to live for.

(12) The *denial* that unemployment is related to crime, for example, pervades the crime prevention initiatives discussed here.

In example 10 the apposition is between the waiter's request and the nominal group whose head-noun is *hint*: these are the two elements which are being presented as equivalent. The function of the *that*-clause in relation to *hint* is to specify what was being hinted *at*. In example 11, too, there is an equivalence between the Subject, *this*, and the head-noun *reminder*, and the function of the *that*-clause is to identify not the *reminder* itself, but what it is a *reminder* of. Example 12 is different: the *denial* is not identical with the proposition *that unemployment is related to crime* but rather its exact opposite: there is a reversal of polarity involved here.

These examples by no means cover the full range of meanings that appositive *that*-clauses may have in relation to their head-nouns. They should be sufficient, however, to show that identity, equivalence or even inclusion are not the only meanings that this type of clause encodes. These facts cannot be ignored in any grammar which is concerned with meanings as well as with the mechanics of syntactic structure.

4.3 Phraseological constraints

Several head-nouns have appositive *that*-clauses only in very restricted lexico-syntactic environments, or phraseologies of various degrees of fixedness. This can be exemplified by looking at the unique grammatical profile of the noun *reason*.

There are two sides to the causality equation in which *reason*, as a crucial member of the metalanguage (Winter 1977) frequently features, and both of them require lexicalisation. There is the *reason* itself, and there is the event or situation for which the *reason* is being given. The second of these is typically lexicalised in the qualifier of *reason*, often in a clause introduced by *for* as in the sentence *What was your real reason for visiting Russia?*, and in a variety of other structures. One of these is a *that*-clause - usually not an appositive *that*-clause in which the *reason* is lexically realised, but a relative clause where again it is the event or situation for which the *reason* is being given which is lexicalised, as shown in the following concordance. These are not standard relative clauses: in them the relative pronoun *that* can be roughly glossed as *for which*, but they are relative clauses nevertheless. (Proof of this is the deletability of *that*, as in the last four citations.)

```
mes in Braille. The reason that this has been so successful is t
 But the underlying reason that the diplomats will leave Kuwait
hat the fundamental reason that he made the decision was to ensu
And that's the main reason that people want to go there. There's
he unconscious. One reason that so few people ever have big idea
he surface. And the reason that it does this is that it's got t
```

```
Clint Eastwood. The reason Burton made these films was no profou
d to an end and the reason everyone works. Making more money is
ed properly, is the reason I am so certain that even people who
ng in her head. The reason I killed my wife the reason I killed
```

The point I wish to make here is that there is a unique phraseological environment in which *reason*, like the 'general nouns' in set (vi) above, is qualified by an appositive *that*-clause in which it is lexically realised. The following concordance shows this pattern:

```
inism for the simple reason that it hasn't worked. It hasn't w
sions for the simple reason that he had usually passed out, be
lated for the simple reason that they are untranslatable. Vari
date, for the simple reason that he is ahead in the polls; tha
eland for the simple reason that the gates have long been open
ntial for the simple reason that 95 percent of parents educate
areas for the simple reason that they are not profitable at th
orld, for the simple reason that they are outnumbered. My expe
lants for the simple reason that they taste better than lean,
 for the very simple reason that he knew nothing about it. He
```

Basically, in order to have an appositive *that*-clause qualifier, *reason* must be the completive of a prepositional phrase introduced by *for*, and it must be modified by a definite article and an adjective, of which *simple* is by far the most frequent. The only other modifiers found more than once are *good* and *very* (*for the very reason that*), though there were a number of single occurrences which included *decent*, *excellent*, *interesting*, *obvious*, *legitimate* and *practical*. The other occurrences of *reason* with a *that*-clause are accounted for by the fixed phrase based on another of it senses: *it stands to reason that*.

This is what is meant by a phraseological constraint: in this case the fixed phrase *for the simple reason* licenses a particular kind of qualifier and hence a particular kind of information structuring in the clause. The collocation is so predictable that the phrase can be seen as a single-choice chunk with a variable at *simple*; in fact it no longer makes much sense to think of *reason* as the head-noun within a nominal group.

There are many other nouns which have an appositive lexicalising *that*-clause only when there is some sort of phraseological constraint. *Proviso* occurs only in the environment *with the proviso that/subject to the proviso that*. *Advantage* occurs only in the phrases *have the advantage that/with the advantage that*. *Danger* requires the existential process clause *there is a danger that*. *Difference* requires the environment *with the crucial/key/important difference that*. The short concordance below illustrates the restricted range of environments in which *complication* has an appositive *that*-clause:

```
. There is also the added complication that the site is currentl
 Then there was the added complication that Dena was the wife of
e and there was the added complication that Wellby recognised Do
```

```
performers with the added complication that not only physical bu
ement, and with the added complication that the lives, physical
There is also the further complication that he did not write the
There is also the further complication that vitamins and mineral
```

As for the most frequent of all head-nouns with appositive *that*-clause, *fact*, the fixed phrase *the fact that* has taken on a general role as all-purpose nominalisation device in cases where a noun or nominal group is required by the grammar of the preceding elements, especially complex prepositions like *due to*, *in spite of* and *in view of* , which cannot be followed directly by *that*-clauses. In this role it has become so delexicalised that it introduces propositions that are not *facts* at all:

(13) I think the *fact* that Eileen might have been carrying his child when she died aroused emotions in Dermot which were slow to surface.

While no grammar can hope to specify all of these restrictions, there must be some discussion and exemplification of their nature and scope.

4.4 Indefinite articles and plural forms

Finally, there are several other pieces of syntactic information which need to be supplied and related to the function of the appositive *that*-clause. Quirk et al, for example, state (1985:1261) that it is "normal, but by no means invariable" for its head-noun to be modified by the definite article, and that plural heads are "rare with appositive postmodification".

As to the first of these claims, the corpus does not confirm that the definite article is the norm. In the case of some nouns, definite and indefinite articles are equally distributed, and, more important, some nouns are always modified by an indefinite article when they have a *that*-clause qualifier. Two examples are *pact* and *prayer*, where the definite article never occurs:

(14) But today they'll have made a *pact* that each must be courageous for the other's sake.

(15) I quickly dialled the number and said a silent *prayer* that Rusty would forgive me.

As to the second of Quirk et al's claims, that plural heads are rare, this is also challenged by corpus evidence. Number, like definiteness, depends on the individual headnoun. Some which are frequently found in the plural are *allegations*, *charges*, *claims*, *demands*, *denials* and *doubts*:

(16) The trouble appears to have arisen out of *demands* that several senior police officers should stand down.

These facts show that we must be wary of over-generalisations such as that the definite article is 'normal' and that the plural is 'rare': the truth is that the unique grammar and phraseological regularities of the head-noun includes specifications as to the type of determiner involved.

5 Conclusion

In the previous sections I have attempted to indicate the kind of information which would be given about the appositive *that*-clause, its lexis, its phraseology, and its meanings. The same approach would be applied to any grammatical structure. The 'item-environment' method described in section 3 provides a tool for building up a grammatical description gradually, allowing the compiler considerable freedom of movement. There it was shown how the lexical range of adjectives occurring with introductory *it* led to an exploration of the grammar of *possible*. This could then branch off in any of several directions: towards a study of related adjectives, or towards cognates like *possibly* and *possibility*, or towards the comparative structures suggested by the grammar of *possible*, to name but three. Similarly, the examination of appositive *that*-clauses could profitably lead to an exploration of the grammar of some of the more frequent noun-heads, of which *fact* and *reason*, for example, promise rich findings. There is no constraint upon the order in which a comprehensive grammar of this kind is built up, so long as all the information collected is stored in such a way as to be retrievable and fully cross-referenced.

The end result will be that we will be able to specify all major lexical items in terms of their syntactic preferences, and all grammatical structures in terms of their key lexis and phraseology. And as we build up and refine the semantic sets associated with a structure, we move closer to a position where we can compile a grammar of the typical meanings that human communication encodes, and recognise the untypical and hence foregrounded meanings whenever we come across them. The size of the task ahead is daunting, but the goal is ultimately achievable.

Author's address:
Cobuild, Westmere, 50 Edgbaston Park Road, Birmingham, B15 2RX, U.K.

Notes

1. The Bank of English is the name given to the new corpus-building initiative at Cobuild. It is expanding rapidly and currently comprises almost 200 million words of running

text, offering a balanced representation of a wide variety of spoken and written language.

2. As we approach the goal of a non-contrastive grammar, the asterisk of the 'impossible sentence' will become a thing of the past.
3. This is not a standard concordance, since there is no node word common to all the citations. It is a 'stepped' concordance: I concordanced *idea* and found this meaning with the adjectives *faintest, least, slightest, foggiest* and *remotest.* I then concordanced all these adjectives in order to find head-nouns combining with them to form the same meaning, and found *conception* and *notion.* I then made a random selection of the results.
4. At the moment I cannot be sure whether there are in fact any other adjectives which occur in the same structures as *possible*. This is because of the constraints of concordancing which make it impossible to look for structures which contain no single word to operate as the node. We will be obtaining tagged corpora in the near future, and this will greatly facilitate the process of looking for lexico-syntactic co-occurrences.
5. This is a very rough estimate and may need to be adjusted when the study of this structure is further advanced.

References

Aarts, Jan. 1991. *English Corpus Linguistics: Studies in Honour of Jan Svartvik* ed. by Karin Aijmer & Bengt Altenberg, 44-62. London & New York: Longman.

Downing, Angela & P. Locke. 1992. *A University Course in English Grammar*. London: Prentice Hall.

Firth, J. R. 1968. "A Synopsis of Linguistic Theory, 1930-55". *Selected Papers of J. R. Firth 1952-1959* ed. by F. R. Palmer, 168-205. London: Longmans.

Francis. 1993. "Corpus-Driven Grammar and its Relevance to the Learning of English in a Cross-Cultural Situation". *English in Education: Multicultural Perspectives. Essays in Honour of J. M. Sinclair*. Singapore: Unipress.

Louw, Bill. "Irony in the Text or Insincerity in the Writer? The Diagnostic Potential of Semantic Prosodies". This volume.

Quirk, R., S. Greenbaum, G. Leech and J. Svartvik. 1985. *A Comprehensive Grammar of the English Language*. London and New York: Longman.

Sinclair, J, ed. 1987. *Collins Cobuild English Language Dictionary.* London: Collins.

________, ed. 1990. *Collins Cobuild English Grammar*. London: Collins.

________. 1991a. *Corpus, Concordance, Collocation*. Oxford: Oxford University Press.

________. 1991b. "Shared Knowledge". *Georgetown Round Table on Languages and Linguistics 1991. Linguistics and Language Pedagogy: State of the Art* ed. by James E. Alatis, 489-500. Washington D.C.: Georgetown University Press.

Winter, Eugene. 1977. "A Clause-Relational Approach to English Texts". *Instructional Science* 6,1 (special issue). Amsterdam.

________. 1982 *Towards a Contextual Grammar of English.* London: George Allen and Unwin.

________. (forthcoming) *Meaning and the Metalanguage of English.* Birmingham, English Language Research, University of Birmingham.

Irony in the Text or Insincerity in the Writer?
The Diagnostic Potential of Semantic Prosodies

Bill Louw
University of Zimbabwe

Abstract
One consequence of the advent of large corpora has been their increased potential for revealing consistencies in the influence of collocation on the behaviour of particular linguistic forms. Irony relies for its effect on a collocative clash which is perceived, albeit subliminally, by the reader. In order for a potential collocative clash to attract the ironist's interest, there must be a sufficiently consistent background of expected collocation against which the instantiation of irony becomes possible. A consistent aura of meaning with which a form is imbued by its collocates is referred to in this paper as a semantic prosody. Semantic prosodies have been largely inaccessible to human intuition about language and they cannot be retrieved reliably through introspection. Evidence is emerging that departures in speech or writing from the expected profiles of semantic prosodies, if they are not intended as ironic, may mark the speaker's real attitude even where s/he is at pains to conceal it. This paper explores the question of the diagnostic capability of semantic prosodies and suggests what role they may play in the composition and grading of suasive language and in 'radical stylistics' in the future.

> *Happy (trying to quiet Willy)*: Hey, Pop, come on now ...
> *Willy (continuing over Happy's line)*: They laugh at me, heh? Go to Filene's, go to the Hub, go to Slattery's, Boston. Call out the name Willy Loman and see what happens! Big shot!
> (Arthur Miller, *Death of a Salesman*, Act I)

0. Introduction

As irony is to form the subject of this paper, one can perhaps be forgiven for allowing certain agreeable ironies in the history of stylistics to contextualise this study.

In 1970, Donald C. Freeman edited a collection of articles under the general title *Linguistics and Literary Style*. In his brief introduction, Freeman (ibid: 6) alludes to Bernard Bloch's definition of style and criticises it:

> Another difficulty in the work of the 'style as deviation' school of linguistic stylistics is its definition of the norm from which an author's style is supposed to differ in certain ways. For example, Bernard Bloch defines style as 'the message carried by the frequency distributions and transitional probabilities of [a discourse's] linguistic features, especially as they differ from those of the same features in the language as a whole.' *This definition is a chimera. The 'frequency distributions and transitional probabilities' of natural language are not known and never will be, and even if they could be ascertained they would constitute no particularly revealing insight into either natural language or style.* (emphasis added)

Ironically, all the supposed impossibilities on this fairly comprehensive agenda have been made possible in the past twenty-two years by the author of a paper written in 1966 and reprinted in Freeman's collection, Professor John Sinclair.

Sinclair's research in lexis and collocation over the 26 years can now furnish evidence that the description of vocabulary patterns is both necessary and possible. He writes, for example (1987; 1991) of the patterning associated with the verb *happen* and the phrasal verb *set in*. *Happen* is habitually associated with unpleasant events. The irony in the quotation from Arthur Miller with which this paper begins hinges on the fact that the reader knows that an evaluation like *Big shot*! is unlikely to follow *See what happens*! Of *set in*, Sinclair says the following (1987:155-156):

> The most striking feature of this phrasal verb is the nature of the subjects. In general they refer to unpleasant states of affairs. Only three refer to the weather; a few are neutral, such as *reaction* and *trend*. The main vocabulary is *rot* (3), *decay, malaise, despair, ill-will, decadence, impoverishment, infection, prejudice, vicious (circle), rigor mortis, numbness, bitterness, mannerism, anticlimax, anarchy, disillusion, disillusionment, slump*. Not one of these is desirable or attractive. The subjects of *set in* are also ... largely abstractions: several are nominalisations of another part of speech.

This is effectively the first computationally derived 'profile' to appear in print of the phenomenon which Sinclair has begun to refer to as a 'semantic prosody' (personal communication 1988), applying the term 'prosody' in the same sense that Firth (in Palmer 1966:40) used the word to refer to phonological colouring which was capable of transcending segmental boundaries. The nasal prosody in the word *Amen* would be an example: we find that the vowels are imbued with a nasal quality because of their proximity to the nasals *m* and *n*.

In the same way, the habitual collocates of the form *set in* are capable of colouring it, so it can no longer be seen in isolation from its semantic prosody, which is established through the semantic consistency of its subjects.

It is difficult to establish whether the existence of this phenomenon was perceived by traditional semanticists before the advent of linguistic computing. Bréal (1897) seems to have come closest to understanding it. He referred to transference of meaning which is the product of habitual collocation as 'contagion'. Ullmann (1963:185) refers to Bréal's explanation in these words:

> Among its widespread semantic ramifications, Latin 'persona' has become a negative particle in French. This change is due to purely *linguistic* reasons: contiguity with the negative particle *ne* has 'infected' this word, in the same way as *rem, passum, punctum,* by a process which Bréal has termed 'contagion'.

Bréal, as the term 'contagion' suggests, seemed to sense that there is in the language a preponderance of pejorative examples of this phenomenon. A further detail in which Bréal and Ullmann were correct is that 'contagion' is a general linguistic phenomenon which pervades every type of language. In other words, no amount of genre-based or register-based study in this century could ever have revealed its presence. However, Bréal offered too few examples for us to infer that semantic prosodies, and in particular their potency and proportions, might be accessible to our intuition. They are essentially a phenomenon that has been only revealed computationally, and whose extent and development can only be properly traced by computational methods.

1. Semantic prosodies in Larkin's poem's

1.1 Semantic prosodies in Larkin's 'First Sight'

Let us turn now to consider a semantic prosody Philip Larkin's poem *First Sight,* which Sinclair analysed in 1966, in the paper referred to above:

First Sight

Lambs that learn to walk in snow
When their bleating clouds the air
Meet a vast unwelcome, know
Nothing but a sunless glare.
Newly stumbling to and fro
All they find outside the fold,

Is a wretched width of cold.

As they wait beside the ewe,
Her fleeces wetly caked, there lies
Hidden round them, waiting too,
Earth's immeasurable surprise.
They could not grasp it if they knew,
What so soon will wake and grow
Utterly[1] unlike the snow.

The last line is particularly interesting, as is illustrated by the concordance for *utterly* drawn from the original 18 million word corpus at Cobuild. There are 99 citations altogether, and every third one is reproduced here:

```
 1 nothing. The farmers were utterly against the union and utterl
 2 rly against the Union and utterly against the Wages Board. Now
 3 f it everything seemed so utterly altered that I felt illogica
 4 But Io has no wind and is utterly arid. Most likely this erosi
 5 rst thing we'd ever seen, utterly blackened now, the skin on t
 6  in his diary: "Whitehall utterly burned to the ground, nothin
 7  In my experience it gets utterly confused. And there are seve
 8 s ambitious wife, are not utterly convincing. Miguel Fernandes
 9 nfident, well-trained and utterly dedicated to the idea of win
10 feet. Its hopes appeared utterly demolished in 1956, when Mr
11 outwards from the centre utterly destroying everything in its
12  the island the view was utterly different. The filmy enchant
13 . I think it would be an utterly different kind of programme
14  probability's sake, not utterly disconfirming the tale of a
15 ng which could be called utterly effete, self-centred and poi
16 them. The idea was to be utterly exhaustive, to examine the n
17  how kind he was, oh how utterly good and what trouble he too
18 ivor which worked for me utterly. I mean I'm not really inter
19  to reveal a man who was utterly insensible to the gravity of
20 monies latent in her was utterly jumbled by the clumsiness an
21 s listeners but which is utterly meaningless to himself and t
22 iver. When you think how utterly obsessed by self-denial Engl
23 our not uniquely evil or utterly out of line with custom and
24 oops would have found it utterly ridiculous. Dean had worked
25 e, but only if they feel utterly secure, and this is a contra
26 er words, of course, are utterly stupid, "a hard day's night"
27 he nearest town. "I'd be utterly terrified to go up, wouldn't
28 lways making? We fail so utterly to reward responsible marrie
29 ore 1914. Little of this utterly transformed scene can be tra
30 and illusion, then it is utterly unreasonable to suppose that
31 rrassment. Posy had been utterly unsympathetic when she heard
32 d I feel I fail to do so utterly. Why? The stipulation of the
33 d of ours and destroy it utterly. "You can work miracles", sa
```

If we study the collocates to the right of *utterly* we find a phenomenon similar to that identified for *set in*. The concordance shows that *utterly* has an overwhelmingly 'bad' prosody: there are few 'good' right-collocates. Indeed a 'bank' of the bad ones would be easy to establish and would be bound to have its own unique profile.

If one applies this knowledge to Larkin's poem, one discerns at once that the poet's intuition was keenly aware of the presence of a bad semantic prosody on *utterly*. The catch-all word *unlike* after *utterly* in the closing line of the poem allows the cumulative force of the bad prosody, in all its disagreeable variety, to be mustered from our prior knowledge as readers and then ranged against the deceptive homogeneity of the new born lambs' first problem ever: snow's *vast unwelcome*. The poet is content to hint at the myriad cruelties of the world the lambs have just entered and the burden of that hint is borne by the word *utterly*. Sinclair (1966:141) comments on the power of the line *Utterly unlike the snow* in critical language which the evidence of semantic prosodies can now corroborate: "It is the traditional 'brush off' structure: 'Something I'd prefer not to talk about' 'Nothing you won't know all about in time'".

Sinclair makes the point that any attempt to spell out these sinister implications would of necessity be couched in language which paints in too much context. In other words, in order to bring a prosody to the fore, it would have to be laboured upon, and this would be unacceptable for the same reason as the "rarity, in love poetry, of lines like: Her smile was not in the least like the grin of a decomposing vampire" (ibid:141)

Let us suggest another prosody in the poem: Larkin's depiction of waking. As corpora at Cobuild have grown since the first 18 million word corpus, we find that we are able to extract examples of the sinister sense of *wake* followed by *and*, as used by Larkin. The following citation is drawn from a more recent 37 million word written corpus, now incorporated into the Bank of English:[2]

> ... ancestor lying asleep with his head on his arm who was one day to *wake and* gobble up everything that lives.

It is becoming clearer that the great 'sleeping giant' in the application of Sinclair's work to stylistics lies in matching texts against corpora: this supplies the 'bottom up' *textual evidence* for what has until now been regarded as 'top down' *prior knowledge* in the act of reading (Louw 1989).

1.2 A semantic prosody in Larkin's 'Days'

Sinclair's stated position has long been that the pursuit of independent word meaning has been as illusory as it has been sustained. Testimony to this can be readily demonstrated by concordancing for two or more consecutive words chosen from a literary text, rather than just one word, as is often the practice.

The correlation between repeatable aspects of text process and repeatable events outside the text is far more striking than our intuition could ever have predicted. Here is another example, this time from Larkin's poem *Days:*

Days

What are days for?
Days are where we live.
They come, they wake us
Time and time over.
They are to be happy in:
Where can we live but days?

Ah, solving that question
Brings the priest and the doctor
In their long coats
Running over the fields.

The line *Days are where we live* purports to offer happy associations, but it is a line which leaves the reader with inexplicable feelings of melancholia. It appears to foreshadow the theme of death with which the second half of the poem becomes preoccupied and from which it never escapes. Whatever it is in the reader's prior knowledge that accounts for this, it must be verifiable as a piece of text process in our experience of the whole language. In more than two thirds of the following concordance from the 18 million word corpus, *days are* is followed by words like *gone*, *over*, and *past.* Days are not so much *where we live* as where we *have* lived and where we are likely, possibly sooner rather than later, to die:

```
1 t it yourself the prices these days are absolutely astronomica
2 ite 'The world is wide, no two days are alike, nor even two ho
3 ays are gone whenel. But those days are almost twenty years go
4 gless extinction when the grey days are done but who are reaso
5 o men for unequal pay. But the days are gone whenel. But those
6 or do I. The big beer drinking days are gone. They drank becau
7 nd cry for peace. My political days are good and over. I'm not
8 ople making these things these days are making money out of th
9 erage trawler when its fishing days are over - as the Morning
10 it was before. Those good old days are over because trout fis
11  Lourenco Marques. Alas those days are over. What did he die
12 o walk means that his babying days are over. The stroking cea
13 ade me regret that my dancing days are over. Rudolph couldn't
14 fate of Czechoslovakia. These days are over, and that is what
15 ng after me Grandad's working days are past walk along with m
16  a black black sky. But those days are rare and usually to be
17 f I had a striking clock. The days are stretching out again a
18 ness and constancy of country days are the very qualities tha
19 e the only movies I see these days are these nights, on the l
20  finances of old people these days are very much better than
21 rate that situation. The hard days are with us and they are c
```

The proportions of this semantic prosody can be further verified from the larger 37 million word corpus. In it the term *days are* appears 104 times and corroborates the profile which emerges from the concordances given above.

2. Irony and semantic prosodies

The remainder of this paper has four objectives:
(i) It will seek to demonstrate the role of semantic prosodies in making certain forms of irony possible.
(ii) It will argue that if irony can be shown to be a phenomenon foregrounded by virtue of the fact that it runs counter to a semantic prosody, then this must provide one criterion for the existence of the semantic prosody in question.
(iii) It will argue that where forms are retrieved which run counter to a semantic prosody but which cannot be shown to have a deliberate ironic *intention*, an alternative explanation might be inferred from the utterance's context of situation. It may be possible to uncover details about the attitude of speakers or writers which indicate that their utterances are *insincere*. The paper will propose that semantic prosodies, used in this way, help to resolve some of the problems which Grice (1978:124) poses about irony, and thus they could eventually make a contribution to the debate on felicity conditions or pragmatics in general. Where a speaker or writer's insincerity cannot be demonstrated from the context, other hypotheses may be entertained, for example that the speaker is young or is not using his or her first language .
(iv) It will suggest that semantic prosodies are a powerful component in suasive writing and that they might be pressed into service for the purpose of producing data-driven suasive writing. For example, this section will propose that semantic prosodies can be used to grade advertising copy in order to assess its 'sincerity'.

It is now worth returning to the concordance for *utterly* (see section 1.1 above) with some of these objectives in mind. If one moves through the concordance lines in search of exceptions, i.e. any apparently 'good' collocates instead of 'bad' ones to the right of *utterly*, one finds only four in the entire KWIC concordance drawn from the original 18 million word corpus:[3]

```
1 nfident, well-trained and utterly dedicated to the idea of win
2 h how kind he was, oh how utterly good and what trouble he too
3  it up - "I think it's oh utterly grand of you to give us all
4 t. Edward III, placid and utterly venerable, his face flowing
```

These citations of *utterly* are an excitingly clear example of Mukařovský's (1932) notion of foregrounding, demonstrated computationally. All of the forms listed here carry a fairly obvious ironic intention. Citation (1) (*confident, well-trained and utterly dedicated*) may appear at first to be a dubious candidate, but the wider context, when accessed, makes it plain that the subjects are Nazi storm-troopers and that their 'profession' is extermination! Citation (2) is taken from a telephone conversation in the spoken corpus in which the participants are criticising someone else. The irony in (3) is self-evident and the irony in (4) becomes more evident once one realises that the reference is to the king who is, at the time of utterance, dead.

Examples of this kind begin to assist us in determining criteria for recognising semantic prosodies. It is only because the prosody on *utterly* is as consistent as it is that it admits the possibility for irony. Prosodies are undoubtedly the product of a long period of refinement through historical change and even though new prosodies may be in the process of being formed, they cannot be *used* for the purpose of instantiating irony until their prosody predominates sufficiently strongly.

It will be obvious so far that semantic prosodies must furnish one of the most compelling arguments for building and using larger and larger corpora. Their precise characteristics will only be revealed and be seen to stabilise once the content of a prosody can be collected extensively. The nature of the exponency of irony through exceptions to a prosodic trend will, equally, only begin to gel once one has a reasonably large collection of what is a foregrounded phenomenon and therefore an extremely rare one. The original 18 million word corpus has disclosed only the tip of the iceberg in indicating the prevalence and extent of prosodies in the language. The investigator cannot but yearn for more and more evidence as the Bank of English increases in size and generic range.

Consider for example the following short passage from the novel *Small World* by David Lodge:

> The modern conference resembles the pilgrimage of medieval Christendom in that it allows the participants to indulge themselves in all the pleasures and diversions of travel while appearing to be austerely *bent on* self-improvement. (emphasis added)

One finds, upon comparing this with the contexts of *bent on* in the original 18 million word corpus, that its irony is made possible by a small but consistent semantic prosody. Although this corpus rises only to a modest total of ten citations for *bent on*, the seven cited below are sufficiently negative to suggest the consistency of background that Lodge needed in order to create the desired ironic effect:

```
1 them were so bent on defending themselves and on distinguishin
2 ment is hell bent on destroying British Leyland, aided and abe
3 side, seemed bent on getting down my collar and up the trouser
4 of the crowd bent on harrying the speakers, often for a laugh
5 nt. They are bent on 'improving' and perfecting existing weapo
6 n or persons bent on mayhem had not so far chosen to resort to
7 ated figure, bent on the same routine. Thereafter every Mamous
```

The remaining three entries for *bent on* in the original 18 million word corpus follow:

```
8 nergies were bent on convincing her and enlisting her support
9 ryone seemed bent on hurrying to some particular business and
10  resolutely bent on the rescue to which they had been called,
```

These last three citations deserve comment. We note, for example, that they are drawn from the same literary source and that they are written neither by an American nor a British author. Their frequency in this single source, which raises the frequency of *bent on* in the corpus as a whole, is suggestive of singularity of style rather than any common core; moreover, their use seems blander than the prosody which persists in the other entries.

The larger the corpus, the greater the insight it affords us into the nature of the prosody itself and the potential within it for further instantiations of irony. Some measure of the ability of larger corpora to establish a profile of a semantic prosody can be seen in the concordance for *bent on* taken from the 37 million word corpus. There are 103 citations, and again, every third one is reproduced here:

```
1 in a society hell bent on achievement. Mutable thinkers don't
2 werful enchanter, bent on bringing Arthur to his ruin this dev
3 iding donkeys all bent on business, they were forcibly impress
4 cter development. Bent on change, even to the point of shatter
5 r world she seems bent on conquering. Well, I suppose that you
6 overnment is hell bent on demanding greater and greater protec
7 of Yoller's wood, bent on destroying all survivors before purs
8 he people who are bent on doing good they can be the danger, s
9 stic savagery and bent on engulfing and drowning trapped men a
10 sonal safety and bent on escaping not only the enemy, but the
11  seen seems hell bent on expiating the great Fascist guilt. H
12 marriage, he was bent on finding a beauty if, while out walki
13 one else is hell bent on getting fit, I might as well join in
14  pass but he was bent on going that way, the shortest way, to
15 , who really was bent on helping his fellow man, and who was
16 e mine, patently bent on its early despatch and eager to witn
17 hy was I so hell bent on locking myself up in a hotel in orde
18 er as if she was bent on making me see the Host clearly, then
19 le making youths bent on mischief coming along my previous pa
20 y. He seems hell bent on owning all of Colorado.' Fancy findi
21 d and fast as if bent on placing an irrevocable distance betw
22 came, but people bent on recreation had a much narrower idea
23 e Derby, who was bent on resigning, that in the interests of
24 ish battleships, bent on revenge for the sunken Hood in the s
25 lear you're hell bent on ruining yourself". And he walked out
```

```
26 s they were hell bent on seeking the comfortable security of
27 n of individuals bent on some purpose on which they all agree
28  the 7th is hell bent on success it can be totally destroyed,
30 he fools were so bent on taking over the national government
31 f dead but still bent on the most terrible revenge. This is t
32 atchful eyes are bent on these 700 acres in the sea. There is
33 ut why he seemed bent on tripping up the deal with all that m
34  Tartar cavalry, bent on villainy. After which much happened
```

This concordance shows that the pursuits that people are *bent on* are almost always negative or unpleasant in some way.

Once one has this much detail of a semantic prosody, one also has the potential for beginning to profile the forms of irony which it makes possible. Here are some examples of irony which, like Lodge's example, are also built upon the form *bent on*. They are included in the concordance of *bent on* in the 37 million word corpus[4]:

```
1 am presently only bent on courting you. "Very well then listen
2 he people who are bent on doing good they can be the danger, s
3  and one twitcher bent on glory crossed the razor backed rock
4 lish countrywoman bent on good works and the continuance of tr
5 r, who really was bent on helping his fellow man, and who was
6 as they were hell bent on seeking the comfortable security of
7  to an apprentice bent on sincerity this multiplicity of inter
```

The length of the concordance span often needs expansion in order for one to appreciate the subtlety and complexity of the prosodies, but in most cases it is the collocate immediately to the right of *bent on* which signals that irony may be intended.

In common with so many areas of language study for which Sinclair's work has revolutionary implications, the theoretical debate on irony has sought to content itself with the study of made-up examples and with made-up contexts. Grice (1978:124) names everything from pragmatics to tone of voice in search of the perceived irony involved in *He's a fine friend*. He assumes far too readily that the expression, as a repeatable event, has no anchor in the conventions of the language apart from tone of voice:

> Even if, however, there is no specifically ironic tone, it still might be suggested that a contemptuous or amused tone, when conjoined with a remark which is blatantly false, conventionally indicates that the remark is to be taken in reverse. But the suggestion does not seem to me to have much plausibility ... (i) To be ironical is, among other things, to pretend (as the etymology suggests), and while one wants the pretense to be recognised as such, to announce it as a pretense would spoil the effect. (ii) What is possibly more important, it might well be essential to an element's having conventional significance that it could have been the case that some quite different element should have fulfilled the same semantic purpose; that if a

contemptuous tone does in fact conventionally signify in context that a remark is to be taken in reverse then it might have been the case that, e.g. a querulous tone should have been used (instead) for the same purpose. But the connection of irony with the expression of feeling seems to preclude this; if speaking ironically has to be, or at least appear to be, the expression of a certain sort of feeling or attitude, then a tone suitable to such a feeling or attitude seems to be mandatory, at any rate for the least sophisticated examples.

Less sophisticated examples than Grice provides would be difficult to find. It is worth making two observations about his position in the light of semantic prosodies. First, an ironic tone of voice could be used to create a sense irony on almost any piece of language. The criterion for identifying it should be authenticity of usage rather than speculation. The instantiation of irony through a prosodic clash would relieve one of the responsibility of providing a particular tone of voice. Novels like *Small World* are crammed with irony of the most sophisticated kind (an area Grice chooses not to examine) and no reader ever expects phonetic diacritics to be present in the text of a novel to assist them in getting the tone of voice right.

Second, it should never be assumed that corpora have nothing to contribute by way of verifying phenomena, even if the instances for verification are based on made-up examples. In this citation for the phrase *fine friends*, from the 18 million word corpus, tone of voice is patently not the device which brings its irony to our notice:

> Six ninety-five at Ohrbach's. The only piece of clothing she had bought since she came to New York. "Will I shame you in front of your *fine friends*?" she said, "A dozen of my *fine friends* will come up to you tonight and ask for your telephone number, he said."

Even from this small extract we can infer the irony of *fine friends*. Its presence seems to be confirmed by the echoing repetition of the authentic example and in some ways by the word *shame*, acting as it does, as a foil to *fine*. We find no reference to repetition in Grice. Here are some further citations of *fine* in the environment of *friend(s)*, this time from the 37 million word corpus:

> Genteel sufficiency, eh, my *fine friend*. What more can one ask? Superfluity, perhaps.

> And now all you *fine friends* of John, all you fine and decent people. Her sarcasm curdled the words. You're all here to pay your respects to her.

> I damn nearly fired off a piece to the Trib putting the record straight. Bond laughed. He said, "you're a *fine friend*, Felix. When I think of all the trouble I've been to to set you a good example all these years."
>
> Timidly he held out his hand. "Mister" he whined, "You von't turn me down like all my *fine friends*! Gott! Dey none of dem don't know me anymore!

The prosodic profile in the case of *fine* seems to be that it has a good prosody when it appears as an attributive adjective before any noun representing a professional class, such as *actress*, or objects such as *china*, works of art and the like. It seems to begin to entertain the possibility of irony when it appears before words which refer to family membership or friendship. Even the term *greeting* seems to be close enough to intimate human concerns to provoke usage in an ironic sense, as this citation, also from the 37 million word corpus, demonstrates:

> He blew out his lips. That' s a *fine* greeting after all this time.

The term *woman* seems to be sufficiently generic to suffer calumny too:

> The judge even reprimanded Sue for not having made friends with this *fine* woman who was trying to take her child away.

If semantic prosodies and the concomitant supportive forms which seem to surround them were to be systematically profiled by the computer, there can be little doubt that candidates for ironic interpretation can be collected and used to assist in the deeper analysis of prosodic profiles. Collocative clash, the use of inverted commas, and repetition would constitute only three of a growing bank of descriptors invoked by such filters. They might begin to identify entries such as the following, and other descriptors could be added to the filter as corpora are enlarged:

> You know, Joey, you're going to make some faithful woman a hell of a *fine* fugitive!
>
> And as for that bloody housemaster horror, who told you that love between men was a *fine* ideal so long as it remained ideal!
>
> The most serious change of all is one attributed to John Dean. He said that President Nixon was not only involved in covering up the Watergate escapade but actually congratulated him on 'the *fine* job' he'd done. This is extremely serious, since it immediately implies that the President too had lied and thereby obstructed justice.

3. Prosodic clash and possible insincerity

It has been established so far that ironists, that is, people who mean the opposite of what they say, are amply served by semantic prosodies in their endeavour to be ironic in text. Instances of their irony can, as we have seen, be recovered from corpora. However, something else that can be recovered from corpora is exceptions to prosodies which are apparently not intended as ironic by their encoders. What are we to make of them?

If we take the not indefensible view that an ironic statement is a deliberate 'device', to borrow a term from literary criticism, then we can safely say that the ironist wittingly 'writes the device' in order to achieve the ironic effect. In the cases we have been examining, this involves deliberate foregrounding of an exception against the prosody, which in Mukařovský's (1932) terms has become 'automatised' through semantic change over a period of time. Prosodies, therefore, are reflections of either pejorative or ameliorative changes that have run their course sufficiently to allow irony to be instantiable against them through the deliberate creation of an exception to the trend.

However, if we find ourselves confronted with forms that are not intended as ironic, but which are still 'devices' in the sense that they run counter to a well established prosody, how are we to determine their status? To say that they are 'slips of the tongue' is no longer entirely satisfying when we consider the extent to which the evidence in corpora suggests that they are indicative of something other than this. Consider the following example, recorded in a live interview on Zimbabwe Television. This is a transcript of an interview with Richard Francis, Director General of the British Council, conducted by Masimba Musariri and broadcast on 24 January 1989:

> *Musariri*: The British Council is concerned with education and culture et cetera. Is the British Council concerned with anything else?
> *Francis*: Well, they cover a multitude of sins. I mean, when you talk about training, that really is vast.
> *Musariri*: I also note that the British Council has been culturing contact between the University of Zimbabwe and British universities. How wide is the network in the UK with our university here?
> *Francis*: Well, it's very wide. I mean, it's *symptomatic of* the University of Zimbabwe which has such a high reputation that there are fifteen links between departments in the university here and equivalent departments in all sorts of institutions, universities, polytechnics in Britain. That is a huge number of links and reflects not only the closeness with which Zimbabwean and British educators have been working but, as I say, the level of the University of Zimbabwe.

Plainly, a form like *symptomatic of* is unlikely to be followed by a reference to the *high reputation* of the University of Zimbabwe. The rigidity of the semantic prosody can be established instantly, and the idiom *multitude of sins* seems to act as a precursor to the attitudes beneath this prosody. Here is the KWIC concordance for *symtomatic of* drawn from the 37 million word corpus. There were 44 citations, of which every second one is reproduced below. The reader is invited to consider what the concordance reveals about the possible attitude of Mr Francis.

```
1 these stages as symptomatic of our growth as a human spirit, t
2 s not in itself symptomatic of something deeply wrong in our o
3 and fathers are symptomatic of a parental paralysis which appe
4 is behaviour is symptomatic of other management inadequacies.
5 al difficulties symptomatic of little more than excess wealth
6 and planners is symptomatic of the need to compensate for inst
7  may be thought symptomatic of deeper and endemic tensions in
8 d particular is symptomatic of English Cultural Studies' prior
9 suspicions were symptomatic of a problem which was to bedevil
10  paranoia, one symptomatic of a system stagnating after being
11 ed in the war. Symptomatic of this was the fact that in Octob
12 ed, Jupiter is symptomatic of over-optimism, lack of forethou
13  of science is symptomatic of the general development of that
14  The forest is symptomatic of the quest which leads nowhere,
15 ugh in extremo symptomatic of the intended fate of trade unio
16 claim fully is symptomatic of the fact that even his youthful
17 Air Force were symptomatic of this organizational monomania w
18 intake, may be symptomatic of numerous disorders some of them
19 , which may be symptomatic of generalized internal bleeding m
20 ntury. This is symptomatic of your new broom, is it, Director
21  None alone is symptomatic of a clinical depression, but the
22 ou, Carter, as symptomatic of a vain or proud man who is not
```

This overwhelming evidence of a negative prosody in this concordance, considered in conjunction with the high level of linguistic and professional sophistication of the speaker, make the possibility of a slip of the tongue very unlikely. It begins to look as though a default hypothesis is possible.

It is plain that the speaker believes that the University of Zimbabwe badly needs assistance from Britain. Truly high standards at the University of Zimbabwe would render assistance in the form of more than a dozen funded links unnecessary. The speaker tries skilfully to avoid making the statement on Zimbabwe's national television that the University is dependent on Britain for assistance in the area of teaching and research. He tries instead to suggest that the links are in place because the University has high standards and a *high reputation.* However, even if this were the case at the level of policy, the choice of the form *symptomatic of* seems to be an indicator of the speaker's own attitude. The standards at the University of Zimbabwe, whatever they are in reality, are momentarily matched against the full paradigmatic range following the form *symptomatic of* in the language as a whole, in much the same way as Larkin's *utterly* precluded any unequivocally good outcome in the poem quoted at the beginning of this paper.

The full hypothesis may now be stated. Where encoders intend their remarks to be interpreted ironically, they 'write the device' in the form of an exception to an established semantic prosody. Conversely, where an utterance runs contrary to an established semantic prosody and it is clear from its context that it is not intended by the encoder to be interpreted ironically, we find that 'the device writes the encoder'.

4. Grading suasive texts: 'good' and 'bad' prosodies

There seem, *prima facie*, to be more 'bad' prosodies than 'good' ones, but the latter certainly exist and the principles on which they work are the same as those mentioned earlier. As the research stands at present, it looks as though the prosodies based on very frequent forms can bifurcate into 'good' and 'bad', using a grammatical principle like transitivity in order to do so. For example, where *build up* is used transitively, with a human subject, the form of the prosody is uniformly good. People *build up organisations*, *better understanding* and so on. Where things or forces, such as *cholesterol*, *toxins*, and *armaments build up* intransitively, of their own accord, they are uniformly bad. The forms offer us an exciting vignette of man either in control of his destiny or at the mercy of forces over which he has little or no control. This is an example, from the 37 million word corpus, of irony which is instantiated against a background of the 'good' prosody of *build up*, used transitively:

> And the reason they suffer is conditioning. You *build up* tension habits. Everybody builds up tension habits. And I mean everybody. Even people who think they're perfectly relaxed.

Equally, a form like *nurture* ought to be followed by desirable things worthy of the act of nurturing. Where this is not the case, we look for an ironic situation which we find in this citation:

> Once it had developed a structure that was seemingly secure, the structure would *nurture* antagonistic forces that would challenge and then destroy it. A new structure would then emerge, and then the process of conflict and destruction would begin again anew.

It is worth noting that the irony in both of these examples is not only textual but situational as well. That is, it points to a frequently observed situational irony which often characterises the human condition: that we often work against our own best interests. This is recoverable in the first citation above and appears in other prosodies such as the reflexive forms in *victim of*:

```
1 break it. Being the victim of her moods, she had so few that h
2 nd become less of a victim of her own feelings. The service re
3  into the offing, a victim of his own over emphatic persuasion
4 urself from being a victim of your past. You, and you alone, a
5  helpless, hopeless victim of your body and of the tradition b
6 t Thierfelder was a victim of his own flak and that there was
7 ray and box was the victim of his moods constructive today,des
8 wasn't ultimately a victim of his own charlatanism, Butler gav
9 ry fixation makes a victim of itself. If you pay attention to
```

5. Conclusion

By way of concluding this discussion, it is worth making two further observations about the behaviour of semantic prosodies.

First, it is clear that in many cases semantic prosodies 'hunt in packs' and potentiate and bolster one another in rather the same way that images are forced to cluster in poetry in order to prevent full 'intuitive' meaning from ebbing away into delexical meaning (Louw 1991). To return for a moment to the prosody of *bent on*, notice how other prosodically loaded forms crowd around the phrase in these two expanded citations, from the 37 million word corpus and *The Independent* respectively:

> I had only to look into the perambulators to realise that life, having invented the pattern, was *bent on* repeating it. There they sat reiterated suckling dynasts, regarding the world with lofty stares ...

> Michael Heseltine would not now be standing against his leader supported by a relatively small group of people *bent on* change and hopeful of drama. There were moments over the last 21 days when Mrs Thatcher might have been allowed to escape unchallenged.

One even finds that the assistance of metaphor can be enlisted both to prepare us for the advent of a semantic prosody and to maintain its intensity once it has appeared. Consider the following expanded concordance for the word *rampant*, from *The Times*:

> ... once more, with remarkable anticipation of a later wisdom it is to darkest Waterloo, we are in the *rampant* heartland of the two old problems. The twin bindweeds of homelessness and drink and drugs are deeply embedded, and have even started to destroy the fabric of the place.

A second point to be made here is that it should not be too readily assumed that semantic prosodies can be recovered with ease from corpora. Corpus linguistics has recently begun to reveal their existence, but it has certainly not begun to reveal their extent. It may well turn out to be the case that semantic prosodies are less accessible through human intuition than most other phenomena to do with language. As corpus linguistics reveals a greater and greater mismatch between the products of introspection about language and those of direct observation, it is easy to become blasé as we receive these new facts about language. A common problem is what might be called 'twenty-twenty hindsight': the tendency to claim that one 'felt' the presence of a form which was inaccessible to one's intuition until it was revealed through research. The only cure for twenty-twenty hindsight lies in constant exposure, of the most humbling kind, to real examples.

The following is a small test of the reader's intuitive perceptions of semantic prosodies, the 'answer' to which is to be found in the Appendix. Examination of the corpus shows that one of the frequent collocates to be found immediately to the left of the word *feeling* is *without*; in terms of frequency it follows only *a*, *the*, *of*, *was*, *and*, *are*, *this*, *you*, *that*, *be*, and *is*. Before turning to the Appendix, you, the reader, are invited to write down as many examples as you can invent which contain the two word structure *without feeling*, and then compare your examples with the concordance given there (drawn from the 37 million word corpus). If the concordance does not gratify your expectations, you might attempt to state why this is so.

It is plain that semantic prosodies will be of great assistance in the persuasion industry. Propaganda, advertising and promotional copy will now be gradable against the semantic prosodies of the whole language in the same way that the grading of literary texts against corpora will take the study of literary criticism into a new data-driven domain.

Semantic prosodies have, in large measure and for thousands of years, remained hidden from our perception and inaccessible to our intuition. They might never have been disclosed even through computational research and corpus linguistics had it not been for the particular demands, such as machine-driven lexicography from ever larger corpora, which John Sinclair's work has made on these disciplines. At present, corpora are just large enough to allow us to extract profiles of semantic prosodies from them. From these prosodies we may then extract the narrow band of irony or insincerity they contain. As corpora grow larger even instances of irony and insincerity will assume their own profiles. Semantic prosodies will be at the heart of 'radical stylistics' for many decades to come.

Author's Address:
English Department, University of Zimbabwe, Box MP 167, Harare, ZIMBABWE.

Notes

1. In this and subsequent examples and quotations, the emphasis on particular words is mine: these are the node words of the concordances in question.
2. This 37 million word corpus no longer exists as such. It has been incorporated into a much larger written corpus, which itself forms part of the Bank of English. The Bank of English is the name given to Cobuild's new corpus-building initiative. It is expanding rapidly, and currently comprises almost 200 million words of running text, offering a balanced representation of a wide variety of written and spoken language.
3. Only the first two of these are included in the concordance for *utterly* in section 1.1; this is because, for reasons of space, only one in three of the ilable citations could be listed.
4. Again, not all of these are included in the fuller concordance, as only one in three of the available citations could be listed.

References

Bréal, M. 1897. *Essai de Semantique*. Paris: Hachette. Translated as *Semantics: Studies in the Science of Meaning*. 1964. New York: Dover.

Firth, J. R. 1955. "Structural Linguistics". *Selected Papers of J. R. Firth 1952-1959* ed. by F. R. Palmer, 35-52. London: Longmans.

Freeman, D. C. 1970. *Linguistics and Literary Style*. New York: Holt, Rinehart & Winston.

Grice, H. P. 1978. "Further Notes on Logic and Conversation". *Syntax and Semantics* (Volume 9) ed. by P. Cole, 113-27. New York: Academic Press.

Larkin, P. 1964. *The Whitsun Weddings*. London: Faber & Faber.

Lodge, D. 1984. *Small World*. Harmondsworth: Penguin Books.

Louw, W. E. 1989. "Sub-routines in the Integration of Language and Literature". *Literature and the Learner: Methodological Approaches* ed. by R. A. Carter and R. Walker. British Council ELT Documents 130. 47-54. London: Modern English Publications.

__________. 1991. "Classroom Concordancing of Delexical Forms and the Case for Integrating Language and Literature". *Classroom Concordancing* ed. by T. Johns & P. King. English Language Research Journal 4. 151-78. University of Birmingham.

Mukařovský, J. 1932, 1964. "Standard Language and Poetic Language". *A Prague School Reader on Esthetics, Literary Structure, and Style* ed. by P. L. Garvin, 17-30. (Translated from Czech.) Washington D. C.: Georgetown University Press.

Sinclair, J. McH. 1966, 1970. "Taking a Poem to Pieces". *Linguistics and Literary Style* ed. by D. C. Freeman, 129-42. New York: Holt, Rinehart & Winston.

____________. 1987. *Looking Up*. London & Glasgow: Collins.

____________. 1991. *Corpus, Concordance, Collocation*. Oxford: Oxford University Press.

Ullmann, S. 1962. *Semantics: An Introduction to the Science of Meaning*. Oxford: Blackwell.

Appendix

Concordance of *without feeling*

```
ng these sessions without feeling a great deal of tension and without
he baby abandoned without feeling, all for the sake of an unfeminine c
ave extra energy, without feeling an image might be at stake. Leo risi
 and dead an area without feeling and sensation. There is a lack of mo
o do the exercise without feeling anxious. Depending on the severity o
 world in Troilus without feeling any threat to his or her own faith.
troll up its face without feeling any strain in his legs or any shorte
tched her gravely without feeling any call for alarm not, that is, unt
re is a dead area without feeling around the ankle, or the opposite. T
ntrol your weight without feeling at all deprived. You'll eliminate a
nd to accept this without feeling betrayed or disillusioned. When they
 accept a request without feeling criticized. Finally, keep in mind th
nnot visit it all without feeling decidedly unwell and even stolidly u
g you can look at without feeling depressed. Apart from that, at the d
le to accept this without feeling devalued and worthless. On the other
nt of pain relief without feeling dopey. The addictive properties of m
w to survive them without feeling either too guilty, distressed, or to
can talk out loud without feeling embarrassed. Scene 1. Imagine that y
til you can do so without feeling embarrassed or guilty about saying n
s of 50 years ago without feeling endlessly and personally betrayed. T
o about your life without feeling exhausted. The potential total benef
ern of bedwetting without feeling foolish or uncomfortable. The self-e
yourself out loud without feeling foolish. The fact is that you can, a
nts are relevant. Without feeling for historical continuity he assumed
cret of saying no without feeling guilty is to acknowledge certain bas
o refuse requests without feeling guilty. The right to have your own o
n't normally cook without feeling guilty. Since Andy Casey was leaving
ng all that money without feeling guilty myself. We all laughed then a
t difficult to do without feeling guilty or inadequate. There are many
gued, you should, without feeling guilty at all, allow yourself to res
a. He treated her without feeling. He had no gentleness in his charact
with such a woman without feeling her pain. The KGB instructors hadn't
y, his hands were without feeling, his arms racked. The first soldier
nd tied the laces without feeling his head lurch. His stomach was mere
foot on shipboard without feeling his stomach lurch, sink like a stone
ood night's sleep without feeling hung over the next morning. Lavender
tching others eat without feeling hunger pangs yourself. You will retu
dy's inner depots without feeling hungry. This happens already during
 be to date again without feeling intensely anxious. It would be a goo
is family's flat, without feeling it at all. And later, when he was a
t their sexuality without feeling it is a breach of contract, a betray
one can live with without feeling like a martyr. Then, as needed, fat
acity to be alone without feeling lonely. That capacity to bear separa
ortable situation without feeling nervous, and then try to reach that
to repeat himself without feeling offended. And progressively, it gave
oftly, mercifully without feeling or pain, into an eternity of silence
ects of mothering without feeling such a sense of failure. As Nell Mye
 able to say them without feeling tense. Second, pay some attention to
ugh this exercise without feeling tense or self-conscious about your e
 and comfortably, without feeling terribly sick or in pain. Another ex
 long as you want, without feeling that everything is lost. How you phr
ng as you can go, without feeling that it is painful. Each position ne
 Cosmic Community without feeling that we own the universe. We must fi
streets for hours without feeling that he is either visible or existen
my brother's name without feeling that dull sensation of loss. Miranda
```

```
ou my old friend, without feeling that I am disloyal to my commanding
ed over his death without feeling that her own inadequacy was somehow
er ends and means without feeling that the outcome is predetermined or
 nothing going on without feeling that it could be improved were they
ck. We can see S, without feeling the prick. But as soon as we feel th
d him to carry on without feeling the strain of responsibility to be t
y on their career without feeling the lack of closeness. They can get
anation for tears without feeling the need to explore their well-spri
l ability to talk without feeling the need to breathe. Yet now it was
more healthy diet without feeling they are missing out. Of course you
eir own mistakes, without feeling this makes them failures. Children a
re saying to you, without feeling threatened. Accept personal responsi
 are most capable without feeling threatened by one another. Parents a
ly on the new one without feeling unduly flustered. In fact their grea
 a sexual request without feeling uptight about it. They pay some atte
til you can do so without feeling uptight and until you find them at l
ng very bulky and without feeling very wet. Some have an adhesive stri
ht do. A romantic without feeling. Well, thank you. Actually, I'm not
bis at the moment without feeling you had got left behind. Rose had ne
```

Corpus Evidence of Language Change
The Case of the Intensifier

Alan Partington
University of Bologna, Italy

Abstract
This paper explores the meaning and function of intensification, both diachronically and synchronically. It argues that diachronic study is necessary to an understanding of how intensifiers have developed the meanings which they have today, and proposes that a number of frequent ones, like *very* and *utterly*, have undergone a shift from modal to intensifier. This is explained as part of a wider process of delexicalisation, whereby intensifiers have lost their independent meaning. It goes on to relate the notion of delexicalisation to Sinclair's concept of shared meaning, and to explore the effect of novel or unpredictable collocations. The final section examines the changes in syntactic behaviour which intensifiers have undergone over the centuries.

0. The status of diachronic linguistics

Ferdinand de Saussure (1916) uses the analogy of a game of chess to affirm the independence of the synchronic description of language from the diachronic. In order to describe the current state of play, it is not necessary to take into account the previous moves that have been made. The relationship between the pieces in the here-and-now is the vital consideration, not the way in which that relationship has been achieved. The same is true, Saussure argues, of language description. In order to have a picture of the structure of a language at any given moment, it is necessary to describe the relationships of its parts at that moment. How those parts came to be in those positions is strictly irrelevant to the description.

As a result of these ideas, historical linguistics, which had previously been considered the only respectably scientific branch of linguistics, went into steep decline. As Lyons (1981:54) says: "The principle of the priority of synchronic description, which is characteristic of most twentieth century linguistic theory, implies that historical considerations are irrelevant to the investigation of particular temporal states of a language".

No-one would wish to argue against Saussure by claiming that the synchronic description is totally dependent on the diachronic. However, to *describe* a linguistic system is not the same thing as to *understand* it. The synchronic approach tells linguists *how* the bits and pieces of a language fit together, but a diachronic approach can help us to understand *why*. It can help us to understand the forces and processes which put the pieces where they are. These historical processes are not totally arbitrary; they are not simply winds of change blowing from all directions. They reflect the communicative needs and desires which, over time, have shaped and continue to shape the nature of a language.

A central example of the kind of diachronic process I have in mind is provided by the linguistic function of *intensification*. Intensification is a direct indication of a speaker's desire to use and exploit the expression of hyperbole. In other words, s/he tells the listener that what is being said is sincerely vouched for. The importance of intensification in the communicative process is that it is a vehicle for impressing, praising, persuading, insulting, and generally influencing the listener's reception of the message. A diachronic examination of intensification should reveal what strategies speakers used in the past in order to achieve this goal, and suggest how these strategies came eventually to shape this part of the language. We see why language is as it is today. A complete understanding of a language, therefore, will require the admission of both synchronic and diachronic evidence.

In recent years, the synchronic description of language has been greatly facilitated by corpus linguistics. In particular, the type of corpus known as a sample corpus - a cross-section of language as written and spoken over a restricted period of time, such as the Brown and LOB corpora - is a tool for the synchronic study of language in use.

Diachronic study may also be facilitated by machine-readable corpora. The team of researchers working on the Birmingham-Cobuild corpus have been steadily collecting text for the past twenty years. As they continue to do so, it will become increasingly possible to compare texts dating from the earliest period of the project with those from later periods. In the meantime, it is already possible to compare a corpus of modern English with a corpus of English from the past. A number of such historical corpora are available in machine-readable form, for example, the Oxford Shakespeare corpus, the

Helsinki Middle English corpus, and the Century of Prose 1680-1780.

1. Intensifiers and hyperbole

This paper reports on a diachronic and synchronic study of the collocational behaviour of intensifiers, or, in Quirk et al's terminology, 'amplifying intensifiers (amplifiers)'. The data used for the diachronic part of the study was the corpus of historical information in the Oxford English Dictionary. The data for the investigation of the modern collocational behaviour of intensifiers was drawn from an 18 million word corpus of English texts collected by the Cobuild research group[1]. The methodology of the modern study was as follows. Each of the intensifiers was concordanced for investigation. On the basis of these concordances an ordered list of the common collocates for each intensifier was obtained using a program designed for this purpose. Another program, written by myself, was then used to compare the contents of any two such lists and thus provide a statistical impression of the overlap and differences between the collocational behaviour of the various items.

Lists of some of the lexical items classified as intensifiers can be found in Bolinger (1972:220,242), in Quirk et al (1985:590-1) and in Sinclair et al (1990:94-5). They include such items as:

absolutely	awfully	bitterly	completely
deeply	entirely	excruciatingly	heavily
highly	intensely	perfectly	quite
terribly	utterly	very	well and truly

A number of initial points need to be made concerning the behaviour of these items. First, they can modify verbs, as in *value highly*, and submodify adjectives and other adverbs, as in *absolutely crazy* and *perfectly well*. However, a number of intensifiers are restricted to one or other of these roles; for example *very* only premodifies adjectives and adverbs, while *very much* and *dearly* are found almost exclusively with verbs.

Second, all intensifiers are subject to restrictions on their ability to collocate with submodifiable items. Some can collocate with very few items indeed, as is the case of *stone*, which is restricted to the phrases *stone dead*, *stone cold* and *stone deaf*; others, like *very* and *quite*, have very wide collocational ranges.

A third point is that in theory intensifiers form an open set, in the sense that the creation of intensifiers is a productive process - new ones can be coined at any time. Bolinger (1972:23), possibly overstating the case, goes as far as to say: "Virtually any adverb modifying an adjective tends to have or develop an intensifying meaning". In other words, it is sufficient to employ an adverb in submodifying position for it to be interpreted as an intensifier.

During the course of the study, it quickly became clear that while the modern data can provide a description of the behaviour of this lexical group, this behaviour can be fully explained only by looking at intensification from a historical perspective. Intensification has of course been around ever since people first decided to emphasise or hyperbolise part of an utterance, but the actual mechanisms used to intensify are subject to rapid change over time. Bolinger (1982:18) tells us that the study of these words:

> is of more than intrinsic interest ... [They] afford a picture of fevered invention and competition it would be hard to come by elsewhere, for in their nature they are unsettled. They are the chief means of emphasis for speakers for whom all means of emphasis quickly become stale and need to be replaced.

In this sea of change, processes of expansion and contraction are occurring all the time:

> As each newcomer has appeared on the scene it has elbowed the others aside. The old favorites do not vanish but retreat to islands bounded by restrictions (for example *precious few*, but not *precious hot*), and the newcomer is never fully successful and extends its territory only so far.

The study of these two related phenomena - *linguistic fossilisation* (the restriction of lexical items to a small number of combinations, for example *stone dead*, *stone cold* and *stone deaf*) and *linguistic competition* (the jostling of lexical items to win the favour of speakers) - clearly requires the coupling of what might seem to be two rather odd bedfellows: pragmatics and diachronic linguistics. The historical study of intensification offers a fascinating view of some of the pragmatic and psychological mechanisms which have been at play over several centuries.

2. A historical process: the shift from modal to intensifier

In view of the phenomenon of linguistic competition, or jostling, mentioned above, it is interesting to explore the historical processes whereby certain words were 'successful' and became more or less permanent members of the class of intensifiers. One process which has produced an interesting set is what might be called the 'modal-to-intensifier shift'. A number of lexical items which today have an intensifying function began life with some modal semantic content, through which speakers comment on their assessment of the truth of the matter under discussion or vouch for the sincerity of their words. The most central of this group of words is the frequent *very*, which originally contained the sense *truly* or *genuinely*, as in the following citations from the O.E.D:

(1) The compyler here-of shuld translat *veray* so holy a story (1485: O.E.D: vol. XIX:569).[2]
(2) But for he was *verray* repentuant he was exciled for the fey (1387).

To *very* we must add the item *utterly*, which originally had the modal sense, according to the O.E.D, of "without reserve, sincerely, truly, plainly", as in:

(3) You wyll *utterly* saye unto me this proverbe (1539).

In this set we also find *entirely*, originally meaning "sincerely, earnestly":

(4) The griefe that by my selfe among many others, for his losse, is *entirely* conceived (1586).
(5) I *intirely* wish ... more wisdom to that Plantation (1647).

Absolutely has had a still more complex history, starting out with the semantic content "in a manner detached from other things, separately, independently", as in these citations:

(6) Yf he speke of him *absolutely*, without mencion of any speache before hadde with him (1532).
(7) Nothing is, nothing lives *absolutely*, but He; all other things, by participation from Him (1618).

It also had modal sense of "without doubt or condition ... certainly, positively" (now of course obsolete):

(8) Noon oughte to swear *absolutly* for a thinge but that by his owne eyen he be sure and certeyn that it is soo (Caxton 1489).

By the end of the sixteenth century the modal meaning is competing with a new exclusively intensifying sense:

(9) That they be very *absolutely* skilfull (1570).
(10) A Prince *absolutely* valorous and vertuous (1602).

It is not hard to understand the link between modality and intensification: it is a short step from averring truth to being emphatic about it, and in this case this step seems to be mirrored in a historical process. These items, which were once used by speakers to vouch for the truth of what they were saying, are today used to convey emphasis. The relationship between the two functions is underlined by the ability of a number of lexical items to shuffle across the border from one to another, even in their modern senses. Words like *really*, *truly*, and to some extent *certainly*, when used as sentence adverbs, maintain faith with their obvious modal origins:

(11) *Really*, I could hear her thinking.
(12) Aw, don't rub it in. Ah feel awfu'. I do, *really*.
(13) *Truly*, the British race is the finest on earth, Haig thought ...

Here, *really* has much the same meaning as *honestly*, and *truly* means *in truth*. However, in submodifying position, such items are far more likely to be perceived as intensifying:

(14) When the horsetail is *really* hot, wrap it up ...
(15) What a *truly* hateful man you are.
(16) Oh, by all means. How very kind, how *truly* thoughtful.

Bolinger gives examples of the intensifying use of the colloquial U.S. term *sure*:

(17) *Sure* she's pretty.
(18) She *sure* looks pretty.

Here *sure* has the modal meaning *in truth* or *surely*. This can be compared with the following citation, where *sure* is simply an intensifier:

(19) She's *sure* pretty!

This evidence shows that the modal-to-intensifier shift is a synchronic, as well as a diachronic, phenomenon.

3. The process of historical relexicalisation and delexicalisation

The phenomenon described above is an example of a still more general process active in the development of the language, namely, the process of the *relexicalisation* over time of lexical items. This can be defined as a progressive change in the meaning and function of an item. The process becomes of particular interest in cases like that of the modal-to-intensifier shift, in which it affects a whole set of words. Here we have seen the loss of modal content, with the lexical items concerned coming to function exclusively as intensifiers. I will now go on to examine a particular type of relexicalisation for which I will use a term borrowed from Sinclair (1992): *delexicalisation*. While Sinclair uses the terms *relexicalisation* and *delexicalisation* synchronically, to apply to phenomena which are present in the process of language production, I am applying them to a diachronic process. These different uses of the terms will be discussed further in section 4 below.

In its extreme form, delexicalisation can be defined as the reduction of the independent lexical content of a word, or group of words, so that it comes to fulfil a particular function but has no meaning apart from this to contribute to the phrase in which it occurs. For example, *very* adds little independent meaning: it is strictly an intensifier.

I will look specifically at two sets of words, which I call the *terribl*y group and the *highly* group, which have undergone the delexicalisation process with the result that their function has become almost entirely restricted to that of intensification.

Of great importance in terms of the present study is the existence of a correlation between the delexicalisation of an intensifier and its collocational behaviour. In fact, the more delexicalised an intensifier, the more widely it collocates: the greater the range and number of modifiers it combines with. *Very* is a central example here: it combines very widely indeed and is also the intensifier with the least independent lexical content. In fact the two phenomena - delexicalisation and width of collocation - are so closely linked that they are probably one and the same. Once a submodifier begins to collocate more and more widely, it automatically loses the independent lexical content it once had.

3.1 The 'terribly' group of words

One important group of intensifiers to have undergone delexicalisation is the group including *terribly* and *awfully*. The older senses of these are glossed similarly by the O.E.D. as "so as to excite terror or dread", but by the last century both came to collocate with items of a neutral or even a positive connotation:

(20) In the way of money-making, he is *awfully* clever (1859).

The collocates of these items in the modern corpus show that the process of delexicalisation is almost complete. In modern English *awfully* collocates as frequently with modifiers having positive connotations, like *good*, *nice*, *glad* and *well*, as with those with negative ones:

(21) ... larger than life, isn't she. Nearly mythic. And an *awfully* good person, too ...
(22) I'd be most *awfully* grateful if you could spare me a moment ...

The word *terribly* has gone a little less far along the road of delexicalisation, since its modern collocates show a slight degree of preference for negative items. Nevertheless, they also include *exciting*, *nice*, *funny* and *happy*. Examples of its positive collocates include:

(23) So I was *terribly* thrilled by this ...
(24) Yes, occupational therapy is a *terribly* good thing ...

Another word which might well be undergoing the same process at the moment is *dreadfully*. Judging by its collocational behaviour, in modern English this word still retains a negative semantic content. Yet, amongst its collocates in the corpus I found *dreadfully happy*, *dreadfully interesting*, and citations such as the following:

(25) They're *dreadfully* nice. You must meet them, Maria.
(26) Is that *dreadfully* handsome man the famous Hodders ...

3.2 The 'highly' group of words

Since delexicalisation is a continuous - and continuing - historical process, it is inevitable that items will be found at various points along the cline from full delexicalisation, where the item has a solely intensifying function, to more complete lexicalisation, where the item is usually used to convey meanings other than intensification. At various points along the cline, we have what we might call the spatio-physical words, of which I have investigated *highly*, *heavily* and *thoroughly*.

The history of the meaning and function of the first two of these can be divided into three main periods. In the earliest period, they generally conveyed a purely physical description of height and heaviness respectively. In the second period they underwent a movement from physical description to metaphorical statement. In the third, modern period, however, they have come to be used more frequently in premodifying position and their main function is to intensify: they tend to mean much the same as *a lot of* or *a large amount of*.

If we look for a moment at the development from the first to the second phases, it is interesting to note how each of these items moved into slightly different metaphorical territory. *Highly* became associated with emotions and feelings:

(27) When the Emperoure herd this, he was *hily* moved in all his bowels (1440).

However, the majority of the collocates of *highly* found in the modern corpus display little or no connection with either physical height or emotions. They are mainly fixed collocations (we might recall Bolinger's "islands") where *highly* simply means much the same as *very*, as in *highly disciplined*, *highly original* and so on. Citations include the following:

(28) ... royal jelly. It is a *highly* nutritious substance ...
(29) ... to pay discreet homage to a knowledgable and *highly* talented artist.

The word *heavily* was once associated with weight, but it senses were often extended metaphorically, to laboriousness or to sorrow and grief:

(30) Yf the servyse be sayd so *hauely* and dedely ... (1398).
(31) Hearing them so *heavily* lament ... (Spenser).

At a later period it conveyed severity and force, and finally it came to have the modern sense of simply *a large amount*:

(32) I have corrected so *heavily*, as almost to have rewritten it (Darwin 1859).

However, an examination of the modern collocates of *heavily* reveal a number of uses which reflect the 'fuller' semantic content that it had in the past. The collocations include *heavily clad* and *heavily laden*, both of which are to do with weight. There are also still occasional uses expressing laboriousness:

(33) ... when going through a *heavily* taxing task, such as examinations.
(34) ... by now it was ticked, crossed off, scored through, and *heavily* annotated.

Its collocates also include *populated, fortified*, *outnumbered* and several others which have to do with density and concentration.

These collocations suggest that the word *heavily* retains echoes of its earlier meanings more frequently than does *highly*: it has moved less far along the road to delexicalisation. This is also reflected in the fact that it is used more frequently as a verbal modifier than is *highly*. Nevertheless, the majority of the collocates of *heavily* are words which it simply intensifies:

(35) ... in which foreign capital was not *heavily* involved was coal-mining.
(36) The odds were *heavily* against the honest buyer and *heavily* in favour of the crook.

Thoroughly is another intensifier which once had a spatio-physical sense:

it used to have the sense of penetration, of going "through and through":

(37) Godard lokede on him *thoruth-like* (1300).
(38) I would have gone home wet *thoroughly* (1633).

This sense has been largely replaced by the modern, purely intensifying sense, where *thoroughly* means much the same as *completely*, as in:

(39) There are those who are *thoroughly* inaccessible ...

Nevertheless, many of the modern collocates of *thoroughly* seem to contain an echo of the older sense: many of them have to do with water or washing, recalling the quality of penetration:

(40) ... have feathers so constructed that they get *thoroughly* wet.
(41) Wash *thoroughly*, then rinse and dry.

A large number of other collocates are words relating to emotions, in phrases like *thoroughly disgruntled*, *thoroughly confused*, and *thoroughly annoyed.* Clearly, both water and emotions penetrate "through and through".

4. Delexicalisation and shared meaning

At this point I would like to compare two definitions of delexicalisation, the definition used in this paper, and that of Sinclair. As defined by Sinclair (1992:15), delexicalisation is closely bound up with the concept of shared meaning, which he outlines as follows:

> words in English do not normally constitute independent selections ... The item and the environment are ultimately not separable ...

Shared meaning is clearly at the heart of the principle of collocation. Very often, in the course of the on-line production of language, one word or group of words almost automatically 'calls up' another specific word or phrase, or at least, constrains the speaker to the choice of one of a limited set of possibilities.

If words are not chosen independently by speakers then they convey meaning only as part of the environment in which they are used: they are not meaningful as separate units. Individual words in language production therefore tend to be delexicalised. Delexicalisation, in Sinclair's synchronic definition, is the corollary of shared meaning.

The kind of historical delexicalisation described in this paper is not a wholly different phenomenon from that explored by Sinclair. Over the centuries, the items which have been described as delexicalised intensifiers have experienced a steady decline in their ability to function as independent lexical choices. Possibly the clearest example of this process is the behaviour of *absolutely*, which, as mentioned in section 2 above, used to mean "separately, independently" (O.E.D.). However, judging by the evidence from the modern corpus, the main function of the word today seems to be to modify terms already marked for strength or superlativity, such as *enchanting*, *delighted*, *shocking* and *appalling*, as well as the following:

(42) ... the way in which subjects are presented is *absolutely* crucial to the success of any system of education.

(43) ... a rule of life which is *absolutely* fundamental and inescapable.

The item *absolutely* seems to add little independent meaning to the adjectival group, apart, of course, from that conveyed by intensification. It functions purely in combination with the terms it modifies. More particularly, since it is against the rules of the language to say *marvellously marvellous* or *appallingly appalling*, the word *absolutely* steps into the premodifying position to give ++*marvellous* or ++*appalling*. In terms of the concept of shared meaning, we can see how the elements are interdependent and how meaning resides in the entire group, not in its individual elements.

The principle of shared meaning can operate with differing degrees of force and complexity. As Sinclair (1992:16) states: "The likelihood is of there being a continuum between occasional quite independent choices and choices which are so heavily dependent on each other that they cannot be separated, and so constitute in practice a single choice".

Those items which have not completed the delexicalisation process are more likely to be treated as independent units in the phrase, whereas those which are more fully delexicalised are much less likely to be used independently in modern English. We might take three typical but different cases as illustrations of this.

First, there are phrases such as *absolutely superb* and *highly disciplined*, which are predictable combinations and seem to constitute a single choice from a speaker's mental lexicon. The phrase functions as a unit and is an unmarked choice.

Second, there is the case where a delexicalised intensifier is used with an item whose collocational possibilities are relatively free. An example is the adjective *difficult*, which collocates in the modern corpus with a range of intensifiers including *awfully*, *terribly*, *extremely* and *exceptionally*. Here the choice is less predictable, but the combination of intensifier and adjective remains a single functional unit.

Third, there are phrases in which the choices are independent to a much greater extent and where both intensifier and adjective contribute separate meanings, as in:

(44) I had subjected a shallow and *horribly* articulate personality to a lifetime of ...

(45) Beyond them, higher still, *severely* rectangular, the stone portico ...

(46) The *thoroughly* conquered lady ...

(47) On the East, *violently* rigid, was the water, icy, scratched like a plate.

It is probable, nevertheless, that these 'nonce' collocations still have their effect on the basis of shared meaning, since they tend to function by exploiting, by turning upside-down, expected collocations. The effect the speaker wishes to produce is precisely that of relexicalisation (in a synchronic sense) and hence revitalisation, of the various elements of the phrase. The next section will deal with this phenomenon in more detail.

5. The meaning of novelty

So far in this paper I have dealt primarily with groups of delexicalised or semi-delexicalised intensifiers - 'official intensifiers', so to speak. We might now recall Bolinger's point that almost any adverb, placed in submodifying position, will be interpreted as having an intensifying function in addition to its descriptive meaning, as in *horribly articulate* and *excruciatingly gaudy*, both of which are found in the modern corpus.

Perhaps the most interesting aspect of this phenomenon is that the sheer novelty of a collocation is likely to make it more intensifying than a predictable one. Take these citations:

(48) Dennis could muster an *excruciatingly* sensible list of official data to disprove our arguments.

(49) It turned out to be as *excruciatingly* exquisite as she had hoped.

The choice of the intensifier *excruciatingly* makes the phrase much more emphatic than would be the case if a more neutral item like *perfectly* or *very*. It is the unusual combination of the intensifier and the adjective that creates the effect, rather than the choice of either one. The following example is also interesting from this point of view:

(50) They were *utterly* content in each other's company.

This collocation is unusual on two counts. First, *utterly* very frequently premodifies items with some negative semantic content (see Louw, this volume), and second, it often collocates with items which are already marked for superlativity: it occurs in phrases like *utterly exhausted*, *utterly ruined* and *utterly destroyed*. The novelty of the combination *utterly content* gives the phrase a much greater rhetorical force than a less marked combination, like *perfectly content* would have had. The concept of literary cliché is clearly bound up with predictability of collocation.

The use of unusual collocational patterns for rhetorical effect is very frequent in both literature and humour. A common device of this kind is the replacement of one item in a common collocation with its exact opposite:

(51) That night in Southern Australia brought its first snuffle of tidings of great horror (Anthony Burgess 1982:22).

This is an example of a process of relexicalisation whereby attention is focused on the independent lexical content of the individual words involved: meaning is not shared across the unit as a whole. The effect achieved is again that of revitalisation.

Relexicalisation can revitalise even collocations that have long ceased to be productive, that is, they have fossilised. The fossilisation of the phrases *stone dead*, *stone cold* and *stone deaf*, mentioned in section 1 above, is a case in point. Even such clear restrictions can be exploited, as is illustrated by the last two lines of Edward Thomas' poem *February Afternoon* (1981:91):

(52) And God still sits aloft in the array
That we have wrought him, stone deaf and *stone* blind.

6. Syntactic changes

Finally, I would like to discuss the syntactic changes in the behaviour of intensifiers which have taken place over the centuries. For example, it seems that it was once much easier to pile intensifiers one on top of the other in submodifying position, which can be interpreted as a symptom of the desire for hyperbole which is behind most uses of intensification. In the O.E.D. we find a number of examples of this, which usually, though not invariably, involve *very*:

(53) That they may be *very absolutely* skilfull (1570).

(54) We know them but not so *very thoroughly* as we might (1522).

Other citations include the phrases *exceedingly more rare*, *exceedingly too short* and *exceedingly well met*.

This is, however, only part of a more important characteristic of the words under examination. In general, they were able to appear in a greater number of syntactic environments than is acceptable today. I will take two of the central delexicalised intensifiers, *extremely* and *absolutely*, and compare their behaviour today with that they displayed in the past.

In the modern corpus, *extremely* occurs almost exclusively in premodifying position. The collocation program lists 56 significant collocates[3] for *extremely*. All 56 of these are adjectives, for example *active* and *busy*, premodified by *extremely*: not one is a verb. According to the O.E.D, however, *extremely* used to occur in a variety of different environments. For example, it used to occur both before and after the verb it modified, as well as before a prepositional phrase:

(55) I did *extremely* beat him (Pepys 1661).
(56) He was *extremely* racked, within half a finger breadth (1684).
(57) She ... look'd *extremely* at the opening door (Byron 1819).
(58) A sinecure which would fitt me *extremely* (Swift 1709).
(59) Two humours equall abounding together, *extremely* in superfluite (1562).

The word *absolutely* also occurs predominantly in modifying position in the modern corpus. However, like *extremely*, it was much more adventurous in the past. It also appeared both before and after the verb as well as in infinitive constructions, before a noun, and before a prepositional phrase:

(60) He was *absolutely* endeavouring to invent ... (1851).
(61) The spear was the symbol of property held *absolutely* (1875).
(62) To heare and *absolutely* to determine (Shakespeare 1597).
(63) To be *absolutely* a tyrant (Milton 1649).
(64) Three young ladies of the half-breed, *absolutely* with frocks on (1853).

Other intensifiers which seem today to have a more limited syntactic range than in the past include *very*, *utterly*, *perfectly*, *entirely* and *thoroughly*. The process whereby their syntactic behaviour became more limited is almost certainly linked with delexicalisation. As an adverb came to be used more and more in premodifying position, it slowly began to lose the meanings it had, except for that of pure intensification. This is in contrast to its use as a verbal adjunct, where it is typically an adverb of manner with independent meaning, as illustrated by the above examples of *extremely* and *absolutely*.

The above discussion is an illustration of the fact that meaning and syntax are interdependent. The lower the semantic content of an intensifier, the more restricted the syntactic environments in which it may occur. Or, put the other way round, the more restricted the syntactic flexibility of an item, the more reduced is its semantic potential.

7. Conclusion

In this paper I have tried to give some indication of the value of combining the study of synchronic and diachronic corpora in the investigation of intensification. It seems reasonable to hope that this method can be extended to other aspects of language. Still within the general area of intensification, one could look at adjectival intensifiers in such nominal groups as *a complete idiot* and *absolute rubbish*. It would be interesting to see whether the processes of historical delexicalisation and revitalisation apply to adjectives as well as to adverbs. One could also investigate the relationship between cognate intensifying adverbs and adjectives, for example *utter* and *utterly*, *terrible* and *terribly*, in order to find out what they have in common and how their relationship has changed over the years.

Other groups of submodifiers, like the set of items categorised as 'downtoners' by Quirk et al (1985:597-602), might also be examined through the comparison of corpora. Quirk et al subdivide this group into 'approximators' like *nearly*, 'compromisers' like *rather*, 'diminishers' like *only* and 'minimisers' like *hardly*. The modern corpus could be used to explore the meanings and functions of these words and to provide evidence which would enable us to decide whether such a division is justified by their collocational behaviour. A comparison with their use in a historical corpus might then tell us how and why they came to have the meanings they have today.

To draw this discussion to a conclusion, I would like to return briefly to intensifying adverbs, and cite my favourite example of intensifier-adjective collocation. It is part of a poem by W. H. Auden, entitled *Lullaby*, and is an intensely poetic illustration of the way in which normal patterns of collocation are open to the process of revitalisation discussed in section 5. In this case it is the intensifying adverb *entirely* which appears in an unusual environment:

(68) But in my arms till break of day,
Let the living creature lie,
Mortal, guilty, but to me,
The *entirely* beautiful.

Author's Address:
SSLMIT, Università degli Studi di Bologna, Corso della Repubblica, Forlí, ITALY.

Notes

1. The original 18 million corpus which was drawn on for this paper has now been superseded by the much larger corpora comprising the Bank of English. The Bank of English is the name given to the new corpus-building initiative at Cobuild: it consists of a collection of corpora currently totalling almost 200 million words. The 18 million word corpus is still available, however, in its original form.
2. References to page numbers in the O.E.D. are given only when the entry for a particular word spans more than one page. In all examples, whether from the O.E.D. or the modern corpus, the emphasis on the intensifier(s) is mine.
3. A significant collocate is any word which occurs with it more than twice in the 18 million word corpus.

References

Auden, W.H. 1966. *Collected Shorter Poems*. London: Faber.
Bolinger, D. 1982. *Degree Words*. The Hague: Mouton.
Burgess, A. 1983. *The End of the World News*. Harmondsworth: Penguin.
Louw, B. 1993. "Irony in the Text or Insincerity in the Writer? The Diagnostic Potential of Semantic Prosodies". This volume.
Lyons, J. 1981. *Language and Linguistics*. Cambridge: Cambridge University Press.
The Oxford English Dictionary. 1989. Oxford: Clarendon Press.
Quirk, R., S. Greenbaum, G. Leech & J. Svartvik. 1985. *A Comprehensive Grammar of the English Language*. London: Longman.
Sausssure, F. de. 1916. *Cours de Linguistique Generale*. Paris: Payot. Translated as *Course in General Linguistics*. 1959. New York: McGraw.
Sinclair, J. ed. 1990. *The Collins-Cobuild English Grammar*. London: Collins.
_______. 1992. "Trust the Text: the Implications are Daunting." *Advances in Systemic Linguistics: Recent Theory and Practice* ed. by M. Davies & L. Ravelli, 5-19. London: Pinter.
Thomas. E. 1981. *The Collected Poems of Edward Thomas*. Oxford: Oxford University Press.

Interpretative Nodes in Discourse

Actual and *Actually*

Elena Tognini-Bonelli
The University of Birmingham

Abstract

A corpus-based approach to language study enables us to describe the limitations of usage; this in turn makes it possible to describe the behaviour of certain lexical items which have a specific discourse function. The adjective *actual* and the adverb *actually* are a case in point; both have a structural role in organising discourse. Both frequently establish a correspondence between two elements, and at the same time signal a differentiation in their status.[1]

0. Introduction

This study examines the adjective *actual* and the adverb *actually* in an attempt to determine their functions in discourse. The data analysed comes from the Spoken Corpus, one of the corpora held in The Bank of English at Cobuild.[2] At the time of writing, the corpus contained 3.5 million words of natural spoken conversation. A sample of 200 instances was examined for both *actual* and *actually*. This was out of a total of 212 instances of *actual* and 1397 instances of *actually* in the corpus. Every instance was examined as the central node of a concordance entry; an overall context of 300 characters (five lines) was selected in order to determine and describe the features associated with the adjective and the adverb in question, and the specific functions which can be attributed to them.

1. The adjective *actual*

1.1 The functions of adjectives

Before embarking on a description of the specific role of this adjective, it would perhaps be useful to say something about the functions of adjectives in general.

When we look at the role of an adjective with respect to the noun group in which it occurs, we find that most grammars tend to imply - or state explicitly - that an adjective defines the noun either by adding a feature to it or by restricting it in some way. Quirk et al state that "adjectives, insofar as they attribute stable properties to the referents of nouns ... characterise the referent of the noun (directly)" (1985:74). Adjectives may be seen as restricting a noun group, that is "creating a narrower set of entities than that denoted by the common noun phrase alone"; alternatively they are seen as being used "non-restrictively" (Baker 1989:271), that is as not requiring us to form a smaller subset of the set denoted by the noun, and therefore to add some specific characteristic to it. In either case, they are usually taken to be separate choices from the noun, separate selections made against a range of other possible adjectival selections in order to 'characterise' the noun.

In contrast to this view, I would like to adopt the distinction posited by Sinclair (1992) between the *focusing* and the *selective* functions of adjectives. In its selective function, an adjective is chosen independently from the noun. When we talk, for instance, of 'the Italian Department', the adjective *Italian* answers the question 'which department?' and classifies the noun accordingly. The selective function presupposes a selection from a range of other possible adjectives also available for describing the noun; it therefore characterises the noun by classifying it in some way. This is the traditional way of viewing the function of adjectives. An adjective with a focusing function, on the other hand, is 'co-selected' with the noun and, as such, is closely linked to the nominal choice rather than being a choice in it own right. The "necessary correlate of co-selection" (ibid:16) is the phenomenon of delexicalisation. Sinclair gives *physical assault* and *scientific experiment* as examples where the adjective typically has a focusing function and is hence delexicalised: the meaning of the adjective seems to repeat part of the meaning already built into the noun, rather than add to or restrict the noun in any way.

In a corpus-based study of the adjectival modifier *real* (Tognini-Bonelli 1993), I have shown that this adjective may have either a selective or a focusing function; this of course entails a change in meaning. The selective function represents the marked and less frequent choice; it is strongly associated with a particular environment, typically the preceding definite article or the possessive. In this case, *real* has what I call an 'inferential meaning':

(1) The British Government believes *the real reason for the ban* is an attempt to cushion French farmers from competitively priced British products ... (BBC Corpus)

Here, *the real reason for the ban* is selected against 'another alleged reason' which is implicitly referred to as 'not correct'. The existence of another element to make up the contrastive set and the value judgement associated with it are both inferential, since they are not explicitly stated in the text.

The focusing function of *real*, which covers 80% of the instances in my data, involves a delexicalisation of the adjective to the point where it simply intensifies the noun. Consider for example:

(2) Mrs. Thatcher said the only way to tackle inflation was by high interest rates even though they were causing *real difficulties* for some people ... (BBC Corpus)

In this example, *real* reinforces the meaning already realised by the noun, focusing on a particular aspect of it. Through a process of delexicalisation, it comes to emphasise the 'typicality' of the noun, i.e. the consensus-based view that the meaning of it is what we all expect it to be. Here, *real difficulties* can be glossed as 'genuine difficulties'.

I believe that it can be revealing to introduce the study of *actual* by comparing it first with *real*, because in a way the two adjectives are very similar. There is, for instance, an interesting circularity in their dictionary definitions: *actual* is defined in terms of what is 'real' and *real* in terms of what is 'actually' the case. *Actually*, as a related form, is defined as referring to "the reality of a situation" (*Collins Cobuild English Language Dictionary*, Sinclair 1987). This circularity may, on the one hand, point to a certain similarity in the meanings of the two adjectives and on the other, to the difficulty of defining items which belong more to the world of discourse than to the world around us. Having noted the similarities, however, I would like now to move on to consider the differences.

In the light of the distinction outlined above, we note that *actual*, unlike *real*, is always found in a focusing function, with the specific role of adding weight to and backing the meaning of the noun without restricting it or classifying it in any way. I will deal shortly with the specific type of prominence set up by *actual*. For the moment, in order to clarify the issue, let us consider two examples, one of *real* and the other of *actual*, where the two adjectives have similar colligational and collocational patterning: they are both preceded by the definite article and followed by the noun *problems*:

(3) ... the summit here last week concluded that deep emission cuts are both possible and urgently needed. So *the real problems* still lie ahead ... (BBC Corpus)

(4) ... women's magazines that matter get grottier and grottier if Claire will

forgive my saying so. *The actual problems* that people are allowed to ask advice about have become disgusting beyond belief ... (Spoken Corpus)

In example 3, *real* has a selective role and establishes a contrast at discourse level in a similar way to a logical connector (for example *however* or *but*). Here, *the real problems* are set against 'other problems' which are implicitly labelled as less relevant or important. The meaning of *actual* in example 4, on the other hand, can be glossed as 'in themselves', 'as such', or 'very' when it is a post-determiner as in *the very man I've been looking for*. In this case, *actual* does not modify the meaning of *problems* by selecting this type against other types of problems; rather it focuses on a particular aspect of *problems*, the problems encountered, and relates this to a specific set of circumstances.

As I will be arguing in the following section, *actual* establishes a kind of parallelism between two elements. However, this cannot quite be labelled as contrastive, as is the case with *real*. This leads us on to the specific environment and functions associated with the adjective *actual*.

1.2 The specific features associated with 'actual'

If we consider the syntactic patterning around this adjective, we notice that *actual* is preceded, in 99% of the instances, by the definite article. This is an interesting feature if we compare it again with the profile of *real*, where a preceding definite article is associated only with 15% of the instances. This points to the fact that each word can be seen to have its own grammatical profile, as Francis (this volume) argues. Therefore, even two adjectives that are, *prima facie*, similar, appear consistently associated with different collocational and/or colligational patternings. Francis (1991) also points out that "when adjectives are used in this way [focusing] they are inextricably linked to their head noun and must precede it; in other words when adjectives have this function they are always attributive"; indeed, *actual*, which always has a focusing function, is invariably found in attributive position. In only 1% of the instances is it followed by another modifying adjective, thereby acting as a coordinated modifier.

As noted in the preceding section, *actual* does not select one attribute of the noun among or against possible others. At the level of the noun group, its function is to focus on the meaning already realised by the noun and thus to add weight to it. It is only when we start considering its role in discourse that we notice certain features that differentiate the role of *actual* from other adjectival modifiers.

1.2.1 The correlation between two elements

One of the features that is immediately apparent when looking at the data is that, very frequently, *actual* is found marking a correspondence between two elements: the noun group that includes *actual* appears to be one of two elements in a set:

(5) ... and *the three lanes are marked Oxford Oxford Watford* you see and eventually just before *the actual junction* where there are traffic lights ...

(6) ... the National Anthem and talking about our - *Britain ruling the waves* if you can't produce *the actual ships* ...

In example 5, the correlation is between two noun groups: *the three lanes* and *the actual junction*. The set established between *Britain ruling the waves* and *the actual ships* in example 6 embraces larger units, comprising a semi-fixed phrase as well as noun groups. The correspondence operates at a semantic level and therefore between syntactically different elements.

We find *actual* signalling a more explicit type of correspondence when the two elements are marked by a logical connector:

(7) I don't want to read a book that's about *psycho-analysis* but I think *the actual presentation of Freud himself* is amazingly successful ...

(8) ... has been at the level of *general policies* statements rather than *actual policies* ...

Here, the presence of *but* and *rather than* reinforces and makes explicit the correspondence between the two elements, making the relationship one of contrast. In this particular environment, a logical connector always signals a contrastive relationship between the two elements. I think it is worth making a point, at this stage, about the type of correlation established and the contrastive function that may be associated with it. *Actual* does not establish a correlation *between two different elements* as is the case with *real* (Tognini-Bonelli 1993). Rather, it presents the same semantic unit from two different perspectives. If there is a contrast expressed, as in examples 7 and 8 above, then it is a contrast in the environments and not between two entities, since in this case there is only one entity which is common to both.

The two types of correlation - what I call the 'positive correlation', as in examples 5 and 6, and the 'contrastive correlation' as in examples 7 and 8, account for 50% of the total instances (25% each of the groups). The marked presence of this correlation, which is often contrastive, in the environment of *actual* might account for the development of a semantic prosody (Louw, this volume). This may lead a reader or hearer to perceive parallelism and contrast 'between the lines' when encountering *actual*, even when these are not explicitly present. This is what Sinclair (personal communication) calls "the paradox of language", by which he means that when something is sufficiently expected, that is strongly 'co-selected', it does not need to occur in order to be perceived. Sinclair gives the example of *happen*, which is consistently co-selected with things that are 'not very nice' and, although not entailing 'not very nice', it has come to prospect it. This is a very interesting area indeed; it is however beyond the scope of this paper to explore it in more detail. Here, having noted the existence of the two types of correlation discussed above (namely positive and contrastive), I would like to go on to examine the specific features of this correlation, and to do so independently from the explicit element of contrast which is sometimes present in the context. I will therefore proceed to examine in what way the two elements are related to one another, and how they are differentiated.

Now, if we consider the type of correlation marked by *actual* in general, we notice first of all that it is, in 99% of the cases, retrospective: the element containing *actual* performs an act of anaphoric reference to a preceding element. In this respect, *actual* plays the role of a cohesive device. From a semantic point of view, the two elements belong to the same field; in example 5 *the three lanes* are definitely related to *the actual junction*, and in example 6 *ships* play a major role if Britain is to *rule the waves*. Similarly, in example 4 *the actual problems ... disgusting beyond belief* are related to other problems which, though not explicitly present within the textual span examined for this study, are a matter of common knowledge in the world of agony columns in women's magazines.

The correlation established between the *actual* element and the other element can be regarded as a general vs. specific (or broad vs. narrow) view of the same object. The *actual* element gives us the narrower, more particular view of the object in question. Taking example 5, for instance, *the three lanes marked Oxford Oxford Watford* outline the broad view of what the junction is: a junction starts as soon as the first indication of it appears, for example a notice saying it is beginning, lane instructions, or lane signs. In a way this is the abstract schema of a junction. There is here an appeal to shared knowledge, to general consensus: in the world as we know it lanes which are sign-posted are likely to be so because of coming junctions. The narrow view of the same object, in this instance, is the junction itself, the place at the heart of it where the motorist passes a point of no return, *the actual junction*. The main issue here, therefore, is the angle from which the entity is viewed.

Actual, then, is a device which allows us to present different viewpoints on the same object, separating things that one would have thought were not separate. It invokes shared knowledge for the more general perspective and offers a specific - and in a way more privileged - viewpoint, more relevant to the circumstances, something which is assumed to be non-shared and has to do with individual experience. Consider for example:

(7) I found that dialogue lacked conviction, again going more for *pattern* and not *as actual people really speak* ...

(8) ... and have some sort of *allegiance* to one church or another. But as far as *actual practice* is concerned ...

(9) From a survey published today, it appears that the British *grumble* a lot, but few *make actual complaints* ...

In these instances, it is clear that the *actual* element comes with a tinge of unexpectedness. Having mentioned *pattern* in dialogue, the further distinction signalled by *as actual people really speak* is unexpected and introduces a new, unforeseen dimension. Similarly, we all know about *allegiance to one church or another*, but in order to talk of *actual practice* one has to have first-hand knowledge or experience of it. Again the difference between *grumble* and *actual complaints* is one between 'in theory' and 'in practice', where you have to be 'there' to be able to assess the latter. Everybody knows what happens in theory, but what is the case in practice is assumed to be non-shared.

Actual, to sum up, appears to bring about a retrospective differentiation, implying greater specific knowledge on the part of the speaker. The differentiation is between two perspectives on the same entity. On the one hand there is the *we all*, on the other the *I* of the discourse: the *we all* element represents shared knowledge, the *I* of the discourse offers the new, slightly unexpected, non-shared angle. A greater familiarity with the world is assumed by the *I* of the discourse, implied behind the *actual* element.

1.2.2 Other uses of 'actual'

The function of establishing a correlation with a preceding element, which is very prominent with the adjective *actual*, is not always as explicit as we have observed in the instances discussed above where the 'other' element can be located in the co-text. Sometimes, the correspondence is just implied by the presence of *actual* and the hearer/reader is left to identify the other element for him/herself.

The phrase *in actual fact*, for instance, which is a prominent use of the adjective in my data (15%), establishes a correspondence which goes beyond the boundaries between utterances and embraces larger stretches of discourse. Because *actual* forms part of an adjunct, and does not establish a parallelism as immediate as the one found between noun groups, the identification of the two elements of the correlation is often inexplicit. Here again, though, we find features similar to the ones established by the non-idiomatic use of the adjective. Consider for instance:

> (10) ... amount of leisure and we're saying, "Oh dear, what are we going to do with all this?" {JL} *In actual fact* we don't have all that much more leisure in the sense of actual time that people ...
>
> (11) Well where does one draw the line? *In actual fact* it's not just the Old England Tennis Club, there are eight, I think, tournaments ...
>
> (12) ... recently in Sierra Leone, would be a meeting of finance ministers. *In actual fact*, the twenty-seven ministers were ministers of culture ...

Here, the idiomatic expression *in actual fact* is a thematic adjunct which has a similar function to an 'orienting theme' (Mauranen, this volume) in written discourse, in that it provides "the necessary interpretative framework for establishing the connection [with the preceding utterance]". This bridging function is again what I see as the established correlation between two elements, although the two elements are not explicitly marked. It is interesting to note that, whereas the adjective establishes the correspondence between two noun groups or even larger units, but is still confined within the boundaries of an utterance, the adjunct relates two elements in two different utterances. Again the correspondence is between the general and the specific, the 'given' and the 'new'. What is redefined is shared knowledge, and the ground is prepared for an unexpected correction.

Another use where the adjective does not explicitly establish a correlation with a preceding element in the text, but where the correlation is still felt implicitly, is when it is co-selected with one element in a lexical set comprising nouns of discourse reference. Typical collocates of *actual* with this specific function are nouns such as *cause, example, basis, explanation, question, reason, line of argument, value* and *issue*:

> (13) I'm afraid I really can't comment on *the actual basis* of this particular story, but talking about what effect on ...
>
> (14) ... experiencing a strike on the railways. One thing must be said, it seems to me, is that *the actual cause* of this argument, particularly between ASLEF and the British Rail ...

Although *actual* does not establish a correlation with a preceding stretch of discourse in these instances, it still makes an implicit differentiation between a shared view and a privileged, individual interpretation. In example 13, *the actual basis of this particular story* is set against the background of what might be generally believed by public opinion to be typical in the circumstances. The result is that it is felt as slightly unexpected, but something that the speaker believes 'might turn out to be so' after all. Although the reporting structure *I can't comment* negates the offer of the privileged (*actual*) point of view, it still implicitly differentiates the status of the two possible interpretations.

The correlation may sometimes not be present at all; the adjective then is left to its local focusing function within the noun group. No correlation is specifically involved, for example, when *actual* is co-selected with nouns belonging to a lexical set which includes units of measurement and technical words. Typical collocates of *actual* with this function are words like *value, voltage, amplitude, depth, entity, drop, figure, number, output, rate* and *ratio*:

(15) ... what we call background and what do we call foreground, here *the actual figure itself* is liable to ambiguous interpretation ...

(16) ... it's very hard on Mr Average, in fact for the single person, *the actual net tax* now, with the increase of National Insurance ...

Here, the meaning of *actual* can be glossed as emphasising the precision of the measurement, the specificity of the instance. The sheer focusing function of the adjective emphasises the typicality of the noun and the validity of the observation without making any reference to preceding elements. This technical function of *actual* accounts for 15% of the data.

1.2.3 The discourse angle

The type of parallelism so frequently established by *actual* could be considered to belong to the domain of stylistics. I would like to argue here that this particular adjective has a discourse role; this differs from a stylistic or a simply cohesive role because it is structural in nature. As observed in sections 1.2.1 and 1.2.2 above, this role may vary in scope. It may operate below the level of the utterance, as in examples 5-9 above, where the two elements of the correlation are both present within a single utterance and *actual* is the signal to the reader or speaker to go back and identify the 'other' element. On the other hand, it may also operate above the level of the utterance: this is specifically the case of the expression *in actual fact* in examples 10-12. Moreover, *actual*, when co-selected with discourse words, is seen to establish a correspondence which is at the inferential level, since the 'other' element of the correlation is not present and has to be inferred.[3]

In order to understand and describe the structural role of *actual*, I am adopting the dynamic model of discourse outlined by Sinclair (1981, 1985, 1993). This model is able to account for shifts in text with reference to the state-of-the-text at the moment of their occurrence: "at any one time a text exists only as the portion of it that is being processed, interpreted and understood" (NERC 1993).[4] Discourse, whether spoken or written, is seen as constantly unfolding from one point or 'posture' to another in the achievement of a purpose. The basic units considered in this model are the sentence, "or segments of a sentence which appear to act as independent portions" (ibid) in written discourse, and the move in spoken discourse; these are taken to represent the state of the discourse at any point.

Coherence is progressively structured by (i) a process of default 'encapsulation', where each utterance makes reference to the preceding one and takes over from it the status of 'state-of-the-text', thereby organising the text that has already occurred (Sinclair 1993:7), and at the same time by (ii) acts of prospection where "each utterance sets the scene for the next" (Sinclair 1985:15), thereby organising in advance the text that has yet to occur.

The progressive segmentation of text into postures is fairly self-evident when we consider the orthographic layout of the sentence in written discourse or turn-taking in conversation. Other devices, however, contribute to postural changes. From the retrospective point of view, for instance, the 'internal acts of reference' (Sinclair 1985:10-12), that is logical acts (for example *and*, *however*, *but*, *in fact*), deictic acts (for example *this*, *that*) and ellipsis, are indications of argument, as are other acts of self reference where an utterance is explicitly referred to in the discourse (as in *that's a very interesting question*). All these implement and sometimes underline postural shifts. Sinclair goes on to point out that "each time an author or narrator takes up a noticeably different attitude to his subject matter, there is a *prima facie* case for a change in posture" (ibid:22). This is where I would like to consider the role of *actual* in discourse. In the light of the discussion above I see this adjective as determining a postural change. It is important to note here that *actual* has an organisational role in discourse in that it signals the mental structure of the interpretation and the subsequent organisation of text accordingly. As I have shown, it functions by establishing a correlation with a preceding stretch of discourse: the two elements of this correlation are taken to be equivalent, or semantically related to each other. At the same time, *actual* marks the 'demotion' of one element of the correlation to the advantage of the other: the differentiation between the two elements is one between a broader, consensus-based view and a more specific, unexpected and implicitly privileged interpretation over the same entity. This organisational process of demotion and promotion is different from what Sinclair (1993) sees as the simple 'point-to-point' cohesion, which does not have a structural role as such. We can now say that *actual* qualifies as a device of postural change on two grounds: (i) it performs an act of self-reference, i.e. it refers to a preceding unit of discourse and, by virtue of the correlation established, redefines it, and (ii) it offers a different interpretative angle on the subject matter.

In this respect, *actual* can be seen as an 'anaphoric adjective' with a very similar role to the anaphoric nouns that Francis (1986) describes: it has a metadiscursive organisational function. At the same time, it adds "something new to the argument in terms of attitudinal meaning" (ibid:68). I equate 'attitudinal' with 'interpretative' as they both define the ways in which discourse is provisionally categorised by one speaker.

2. The adverb *actually*

Before going on to describe the role of the adverb *actually* and discuss its functions, it is worth noting that this adverb has a very high frequency in spoken discourse. The frequency is indeed remarkable if we compare the total of 212 instances of *actual* in the Spoken Corpus with the total of 1397 instances of *actually* in the same corpus. This frequency of occurrence is inextricably bound up with the strategic role that the adverb has in discourse. As I will try to show in the next sections, *actually*, by virtue of the correlation established with a preceding stretch of discourse, is probably the most common device available to a speaker who wishes to make his/her own perspective stand out with respect to the general, and more common, consensus-based view, or to other preceding textual claims or events. Because of this, the use of *actually* is also a very common way of implicitly acknowledging what has gone on before, that is paying lip-service either to another participant's contribution or to one's own stated position before going on to contradict it or correct it in some way.

2.1 The correlation between two elements

The discourse angle that I have adopted for the analysis of *actual* is also at the basis of my understanding of the role of *actually*. As in the case of *actual*, this adverb can be seen as a signal for a change of posture in text, repositioning the point of view according to which a certain entity is presented. Again, the correlation that we observe between two 'different' elements at the textual level is more like an interpretative distinction made within a single entity. Like *actual*, then, *actually* has the function of dividing and differentiating where there is, at first sight, unity. This function is quite clear when we look at how, within a single utterance, a unit such as a subordinate clause can be 'split' by the postural change signalled by *actually*:

(17) ... children who are *actually* learning subjects at school ...
(18) ... the direction in which the snake has *actually* travelled ...
(19) ... I mean words that are *actually* spelt as they sound ...

In example 17 the relative clause *who are* [...] *learning subjects* classifies the head noun *children.* The introduction of *actually* brings about a postural change, and with it an implicit differentiation between *children (at school)* on the one hand and those who are *actually learning subjects* on the other. Implicitly, we are made aware that children may go to school, but they may not necessarily learn subjects. *Actually* here has a dual focus whereby the 'actual learning activity' is foregrounded and comes out emphasised in its own right as the most relevant element in the correlation. In example 18, a similar mechanism is at work. The noun *direction* already sets the scene for a verb of movement to follow. *Actually* is a signal that what is to come is not to be taken as shared, does not go along with the set scene. Again, the result of this differentiation is an emphasis on the fact that *the snake actually travelled.* Similarly, in example 19 we may take it that in English words are often not spelt as they sound. *Actually* breaks this assumption: the postural change acts as a choice node where the text could take another direction: the element of choice is felt in that we understand that *words* may or may not be *spelt as they sound.* As a result of the possibility of choice the element marked by *actually* is emphasised, as in examples 17 and 18, or felt as unexpected, as in example 19.

2.1.1 Repositioning the interpretative angle

I will now go on to look at the way in which this adverb, by virtue of the correlation between two elements outlined above, can take on the tactical discourse function of changing the interpretative angle with respect to the state-of-the-text. Consider:

(20) ... what we're going to do is to *denationalise* it, *give it actually back to the people* ...

(21) ... well especially *when I am with cockneys* yes this is it I think *being actually in the environment* and hearing it around you ...

We have here a case of paraphrase, which, in these examples, is marked explicitly, in example 20 by the two co-ordinated verbal groups and in example 21 by the lexical paraphrase *yes this is it* which emphasises the transition from an element in adjunct position to the same element in subject position. This is what I called 'positive correlation' when discussing *actual* in section 1.2.1 above. We observe that, as was the case with *actual*, the status of the two elements is differentiated: in the process of rephrasing, what has changed is the point of view on the subject matter, that is the position of the speaker with respect to the text. In example 20, while *denationalise* is a fairly neutral and objective word, to represent the same process as *[to] give it actually back to the people* is not just a paraphrase; it brings to the fore the political standpoint of the speaker. In example 21, the transition is from an entity in the real world (*when I am with cockneys*), in the form of a finite verb clause (which of course must have a subject), to a discourse entity (*being actually in the environment*) realised by a non-finite verb clause (which does not need a subject). This is what Davies (1988) sees as a question of the visibility of the writer with respect to the text. Although the first element is the more specific one in terms of individual experience (the participant in the action is 'visible'), it is repackaged under a more appropriate label in order to attend to the current discourse purpose. The transition is from the world to the discourse, from the generally relevant to the more specifically relevant. The entity which is marked by *actually* - and isolated from the rest of the utterance by the postural change - is presented as more appropriate to carry the status of 'state-of-the-text'.

Actually is frequently found introducing or backing a report or a preface. In this case, it acts as a "personal point of view" preface (Stubbs 1983), indicating a "break in the perceived topical coherence of the discourse" (ibid):

(22) ... for young people to leave home when they go to University? {SL} I think, it's very important *actually* I wouldn't want to stay at home while I was at university. ...

(23) ... it seems to me, given that one gets older, one can find all sorts of good reasons for morals, *actually I think in many ways* the so-called permissiveness has made people a great deal happier ...

In these instances, re-positioning the interpretative angle emphasises the reporting or the prefacing structure.

We note that, in example 22, the reporting structure highlighted by *ac-*

tually introduces a new move: the postural change therefore coincides with turn-taking. This is where we can observe the process of encapsulation. As mentioned in section 1.2.3 above this is the default hypothesis in a dynamic model of discourse, operating at the level of sentences or moves; it "reclassifies a previous sentence (or move) by 'demoting' it into an element of the structure of the new sentence" (Sinclair 1993:8). Here, the process of encapsulation is used tactically in order to give more prominence to the reporting structure by acknowledging the preceding move and building it into the present utterance. This strategy is very common in conversation, which is understandable if we consider that participants often build another participant's contribution into their own as a way of showing awareness of the interactive process, thus making their contribution less abrupt and less face-threatening. In example 23, *actually*, although not encapsulating a preceding move, still serves the purpose of achieving a reformulation of the preceding utterance and preparing the ground for a statement of opinion.

2.1.2 Unexpectedness

The correlation between two elements is probably most obvious when *actually* has the function of introducing an instance of re-phrasing (as in examples 20 and 21) because the two elements are presented as unmistakeably parallel. The correlation is also quite noticeable, as we have observed, when an opinion or a new angle on the subject matter is introduced (as in examples 22 and 23), where paying lip-service to what has gone on before is codified almost as a social norm. In other instances, the differentiation between the two elements, and the consequent reassessment of the status of one, are more strongly marked:

(24) You may just be disturbing his peace {TA} I see {A} so you can be *committing damage* without actually *breaking down the gates* ...

(25) ... and just *the problem of heavy fire doors* in some buildings actually presents *a problem to handicapped students* ...

In these two examples we observe the process by which an element takes priority over another. The two elements are, again, related: in example 24 *committing damage* represents the more general view, expected to cover things such as *breaking down the gates*. In this instance, the correlation is contrastive because of the presence of *without*. In example 25, *the problem of heavy fire doors* predicts *a problem to handicapped students*. As I noted in the case of *actual*, though, the second element is offered as specific and unexpected, and is presented as something which might be of interest because it is not shared.

The position of the speaker is involved in this differentiation and a specific point which is closer to his/her heart - or to his/her discourse purpose - is being foregrounded.

The presence of both elements in the co-text of the utterance is not always necessary. When only the element marked by *actually* is present, another element, which is related to it, will be inferred:

(27) ... but I do think, since nothing is clearer to me than the fact that *Mr Paisley actually wants to govern Ireland*, that he will discover ...

(28) Are you going to move out? {A} Yeah well, I may, I don't know, well I've got this, erm, job, *which is alright actually*. The pay is not bad for what I do.

In example 27, the inferred element is that there is reason to believe that Mr. Paisley doesn't *really* mean to govern Ireland; it is with reference to this unstated assumption that the element introduced by *actually* is perceived as unexpected. Similarly, in example 28, the *job, which is alright actually* is interpreted with reference to the fact that, given the circumstances, it might not be *alright*. The fact that the *job is alright actually* is again perceived as non-shared.

Often we note that the element marked by *actually* has a bad connotation: *actually* seems to keep company with some problem or difficulty. This is evident in examples 24 and 25, where the unexpected seems to merge with the undesirable. The dividing line between unexpectedness and unpleasantness is a tenuous one. The 'problem' in example 24, for instance, may not necessarily be a problem only in the real world: *breaking down the gates* might indeed get you in trouble with the police, but the fact that you can *commit damage* [..] even if you do not *break down the gates* may be a 'discourse problem' forcing you to reassess the meaning of the word *damage* in this context. This is where *actually* is clearly seen to have a discourse role entailing the re-assessment of a preceding element.

In other instances, however, the unexpectedness marked by *actually* has more to do with the unpleasant and the undesirable in the real world, thus acquiring a bad 'semantic prosody' (Louw, this volume):

(29) Did you *actually* hit him? ...
(30) ... would *actually* have him killed ...
(31) He is *actually* afraid of his own children ...

The feature of unexpectedness here is implicitly relying on the assumption of given norms; again, there is an inferential correlation with another element, the 'shared view' on the matter, the fact that you do not usually go about hitting people, having someone killed, or even worse, being afraid of your own children.

Although *actually* anticipates the unexpected - which frequently coincides with the unpleasant - if it is found marking what is generally and obviously taken as shared and normal it seems to trigger irony:

(32) We are *actually entitled not to die* ...

(33) They [lawyers] *actually enjoy* an interesting case like this.

(34) The modern economists' income-analysis (to the surprise of some economists) *actually seems to work* ...

What is at stake here is again a phenomenon that Louw (this volume) describes as related to semantic prosodies: in this case the semantic prosody prospected by *actually* is one of unexpectedness and, often, unpleasantness. Louw describes a special effect related to this phenomenon, whereby once a specific prosody is established consistently "it admits the possibility for irony". In the instances above, *actually* is associated with things that are taken to be either obvious or normal. The clash between prospected unexpectedness and delivered normality triggers the irony.

1.2.3 Self-correction, mitigation and challenge

In the sections above, I have noted the repositioning of the interpretative angle and the introduction of something as unexpected as two of the most common functions associated with *actually*. Indeed, the two functions overlap in certain areas, for instance when the interpretative angle marked by *actually* is presented as new and therefore unforeseen.

Related to these two functions is a third one, where the textual shift and the consequent re-positioning of the state-of-the-text may be due to embarrassment. This may be connected to a sensitive issue:

(35) I didn't realise that you two lived so close *actually* that you and Bob .. and I yes well I didn't realise till just now .. oh damn it ...

Similar embarrassment can be present, for example, when a speaker is about to do or say something which might be perceived as insulting or in some way threatening to another speaker:

(36) {A} Thanks. Is mine with, with milk or without? {B} Sorry oh .. without milk {A} No, that's fine *actually*, as long as I have sugar I don't mind ...

(37) {A} suppose I for example or Andrew go to Cuba {B} no {A} and

{TA} I would like to go there *actually* {B} not just now ...

In example 36, even the simple question (*with milk or without?*) could be perceived as face-threatening, that is understood as hinting at the fact that no choice has been offered. Participant {A} is quick to correct the impression s/he may have given by bringing a less threatening topic, in this case *sugar*, to the fore. Similarly, in example 37 there is a potentially face-threatening situation where {TA} ignores the fact that the plan to *go to Cuba* has already been dismissed and insists that he *would like to go there*. *Actually*, in these two instances, has the function of toning down the potential embarrassment of the situation by ensuring a smooth transition where there could, possibly, be a break.

This use of *actually* is also very common when a speaker is about to contradict him/herself, or to introduce a point that might go against him/her. *Actually*, then, masks a contradiction as an act of rephrasing:

(38) ... oh plenty of scope in teaching, do they want anyone to teach German in the University *actually* I don't know German ... and English and French aha and Linguistics ...

(39) ... I mean, I was taken to see Alexei Sale - you know the New Wave comedian. *Actually* I think he is very old-fashioned in some ways ...

In example 32, although the contradiction is not explicit, it is subsumed by the question *do they want*.., the answer to which would be 'I can'. Similarly, in example 33, the fact that *the New Wave comedian* is felt to be *very old-fashioned* is again a contradiction in terms. *Actually*, by marking the shift between one element and the other, ensures a smooth transition between the two. As a result, the contradiction is felt as almost a planned differentiation in the discourse and does not affect coherence.

3. Conclusion

Corpus studies have brought about a major shift in the relationship between data and theory. We are now able to derive generalisations from the cumulative effect of single instances of usage. This has allowed an important change in emphasis in language theory, namely from a linguistics of *langue* to a linguistics of *parole*. Patterns of usage which were previously not available to the sharpest of intuitions are now surfacing and becoming noticeable in the light of the available evidence. A set of alphabetically-ordered concordances, for instance, which is the standard tool of the corpus linguist, merges the two dimensions of language, the instance and the codified pattern, in a single set of axes. The vertical axis yields the patterns, the 'repeated events' related to the node word, in other words its 'grammar'; the horizontal axis, on the other hand, represents the specific instance, in its own individual and unrepeatable context. If Saussure was justified in postulating *langue* because of the unattainability of *parole*, this is no longer justifiable in the light of the data.

Among the aspects of language description which have lagged behind, ignored by traditional introspective linguistics, is the study of lexis; this was taken to belong to the domain of parole and was consequently set aside, if not dismissed, as contingent on the specific situation. Cohesive patterns were traced, but the strict interdependence between lexis and grammar (Francis, this volume), which is obvious when one draws one's observations from a corpus, had no place in traditional grammars. A similar fate was consigned to pragmatics, which has become the ragbag for anything that cannot be shown to be rule-governed and is, therefore, assumed not to have strict textual realisations.

This paper is to be seen in the wider perspective of a linguistics of parole where a bottom-up approach allows generalisations to be drawn out of instances of actual usage. In it, I have argued the case of two words, an adjective and an adverb, which are usually assumed to belong to the realm of lexis, but which have been shown consistently to have a structural role at the level of discourse.

I have traced the way in which these two words have come to lose their meaning in the outside world, a meaning which is unmistakeably connected with 'reality', and have become, to a certain extent, empty words. This process of delexicalisation has coincided in both cases with the acquisition of an organisational as well as an evaluative role, so that both the adjective and the adverb have taken on the function of structuring discourse, on the one hand, and of introducing a new interpretative angle, on the other. The result of this double role has been identified as the maintenance of discourse coherence, achieved by bridging two separate elements in the text.

I have argued that the evaluative nature of the differentiation means that what is structured and monitored is the interpretation of a given entity. Ironically, the general assumption according to which we use shared knowledge in order to interpret a text is reversed here: *actual* and *actually* are two examples where textual evidence is used to structure and interpret the world. They both act as pointers to what is assumed to be common knowledge, implicitly referring to a set of norms which are presented as known.[5]

Author's address:
Department of Italian, School of Modern Languages, The University of Birmingham, Edgbaston, Birmingham, B15 2TT. U.K.

Notes

1. This paper has benefitted from the valuable comments of Mr. W.E. Louw of the University of Zimbabwe. I would like to take this opportunity to thank him for being the first to stimulate my interest in Linguistics, in the course of my M.A., under his supervision, many years ago.
2. The Bank of English is the name given to the new corpus-building initiative at Cobuild. It is a collection of current English speech and writing now being expanded to over 200 million words of running text, offering a balanced representation of a wide variety of spoken and written language.
3. This inferential function, when determined from a corpus-driven perspective, poses a problem which can only be referred to in passing in this study. The problem is that if inference is codified and therefore predictable, it ceases to be an inference. A corpus allows us to describe the limitation of usage, and often yields evidence concerning the predictability of inference. There is a need therefore for a redefinition of inference in the light of what the corpus can provide.
4. This quotation comes from one of John Sinclair's contributions (WP8, 1993) to the final report to NERC (the Network of European Reference Corpora). This is an EEC-sponsored feasibility study for the establishment of a European Corpus Network.
5. Writing a paper can be a painful job at times. This one was fun, although I say this with hindsight. There was, however, an excruciatingly painful side to it: that was having to present my data and my interpretation of it without *actually* using the words *actual* and *actually*! If I had not been convinced of the vital roles of these two words in structuring and evaluating discourse I would certainly now be convinced!

References

Alexandrova, O. 1987. *English Syntax: Collocation, Colligation and Discourse*. Moscow: University of Moscow Press.

Baker, C. L. 1989. *English Syntax*. Cambridge, Massachusetts and London: The MIT Press.

Davies, F. 1988. "Reading Between the Lines: Thematic Choice as a Device for Presenting Writer Viewpoint in Academic Discourse". *The ESPecialist 9. 1-2*.

Francis, G. 1986. *Anaphoric Nouns*. Discourse Analysis Monograph No 11. English Language Research, University of Birmingham.

________. 1991. "The Functions of Adjectives". Unpublished manuscript. Birmingham: Cobuild.

________. 1993. "A Corpus-driven Approach to Grammar: Principles, Methods and Examples". This volume.

Louw, W.E. 1993. "Irony in the Text or Insincerity in the Writer? The Diagnostic Potential of Semantic Prosodies". This volume.

Mauranen, A. 1993. "Theme and Prospection in Written Discourse". This volume.

Quirk, R., S. Greenbaum, G. Leech & J. Svartvik. 1985. *A Comprehensive Grammar of the English Language*. London and New York: Longman.

Saussure, F. de. 1916. *Cours de Linguistique Générale* ed. by T. de Mauro 1972. Paris: Payot.

Sinclair, J. M. 1985. "On the Integration of Linguistic Description". *Handbook of Discourse Analysis* ed. by T. A. van Dijk, 2. 13-28. London: Academic Press.

__________, ed. 1987. *Collins Cobuild English Language Dictionary*. London: Collins.

__________. 1992. "Trust the Text". *Advances in Systemic Linguistics* (Proceedings of the Systemic conference 1990, Sterling) ed. by M. Davies & L. Ravelli, 5-19. London and New York: Pinter Publishers.

__________. 1993. "Written Discourse Structure". *Techniques of Description: Spoken and Written Discourse* ed. by J. Sinclair, M. Hoey & G. Fox, 6-31. London and New York: Routledge.

Stubbs, M. 1983. *Discourse Analysis. The Sociolinguistic Analysis of Natural Language*. Oxford: Basil Blackwell.

Tognini-Bonelli, E. 1993. "From a Reliable Source: Uses and Functions of the Adjective *Real*". Dialogue Analysis IV, Papers from the 4th Conference, Basel 1992, 429-36. Tübingen: Max Niemeyer Verlag.

Who Can Make *Nice* A Better Word Than *Pretty*?
Collocation, Translation, and Psycholinguistics

Kirsten Malmkjær
University of Cambridge

Abstract
Corpus linguistics has brought into sharp relief the question of the relationship between speakers' intuitions about meaning and use, and the relative frequency of occurrence of linguistic items. Here, this question is addressed using (i) data from a study in translational stylistics and (ii) data from an experiment in the psycholinguistics of late bilingualism. Both sets of data suggest that translators might produce higher degrees of equivalence if they are translating *out of* their mother tongues than if they are translating *into* their mother tongues. This suggestion is substantiated with reference to theories of first and second language acquisition, and it is argued that these same theories also offer an answer to the question raised by corpus linguistics. Finally, it is proposed that co-operation between corpus linguists, psycholinguists and translation theorists is likely to speed up the journey towards satisfactory explanatory hypotheses in each discipline.

0. Introduction

0.1 The relationship between intuition and data

The availability of large, machine-readable corpora of linguistic data means that it is possible to answer almost any question we care to ask about word and pattern frequencies and about co-textual constraints on their occurrence. On the other hand, corpus linguistics has brought into sharp relief at least one question which appears unanswerable from within the discipline itself, namely the question of the relationship between speakers' intuitions about the basic meaning and most common use of a lexical item, and the relative frequency of occurrence of the item with that basic meaning and use. The question is raised by Sinclair and Renouf (1988:151-2) who point out that while

> textual evidence shows us that the first and second most frequent uses of *see* are those found in the familiar phrases *you see* and *I see* ... Most native speakers ... would probably say ... that it is the seeing through one's eyes which is surely the major use/meaning associated with *see*.

The problem resurfaces in Sinclair (1991:39) where it is stated that "actual usage plays a very minor role in one's consciousness of language". This statement makes it plain why the problem is unresolvable from within corpus linguistics itself: the corpus provides evidence of actual usage. But that usage is not mirrored in speakers' consciousness of language. Therefore, consciousness of language cannot be derived solely from exposure to actual usage, and actual usage evidence cannot suffice to explain consciousness of language.

Sinclair and Renouf (1988) themselves suggest where a further element of explanation might be found when they refer to Lyons' (1977:247) psycho- or neurolinguistic notion of 'salience'. This may be a plausible explanation for the mismatch between intuitions and use in the case of lemmas like *see*, since humans are biologically endowed with sight, a faculty which probably has just the kind of psychological salience which Lakoff and Johnson (1980), for instance, use in their explanations of metaphorical language use. Because much information is gained through the faculty of sight, it is quite plausible to see the expressions *I see* and *you see* as sight-based metaphors for understanding.

Psycho-neurolinguistic theory might also explain a finding which appears at first sight to contradict Sinclair's (1991) statement mentioned above. Dixon (1982:11) says that in all cases he has investigated, "intuitive judgements as to basic-part-of-speech membership coincided with analytic decisions [made on the basis of a word's morphology and grammatical behaviour]". It is not inconceivable that the reason all languages seem to have categories like noun and verb is that humans are predisposed to notice things and their behaviour.

However, the lexis of any language contains countless lemmas which are very far removed from any basic conceptual categories that might be proposed for humans, and something other than universal syntactic parameters has to explain why these are set differently for different languages. Furthermore, Aitchison (in press) shows (i) that the native language of even highly proficient speakers of another language influences their prototypicality judgements in the foreign language, and (ii) that children's prototypicality judgements differ from those of adults but resemble them more closely the older the child. It is difficult to see how anything other than amount of exposure to language in use could explain these phenomena.

It is clear that exposure to language in use must play a part in the development of intuitions, as no other plausible source for many intuitions presents itself, and there are good, general psycholinguistic theories of how this use-data is treated by the language learning child (see section 4). However, as adult intuitions turn out not to mirror the use-evidence from which they are derived, the source of the intuitions needs careful consideration if psycholinguists are to arrive at a clearer understanding of the process through which they are developed. It is impossible to come to understand the steps through which the output of a developmental process is arrived at, unless one is clear about the nature of the basic input. So to arrive at more detailed hypotheses, scholars concerned with language processing and development need to take corpora of actual use into account. In other words, the question about the relationship between intuitions and data which corpus linguistics has raised is unlikely to receive a satisfactory answer if the data are ignored.

0.2 Translation studies as a data resource for other disciplines

In this paper I want to address the question raised in section 0.1 above using data from a study in translational stylistics. By 'translational stylistics' I mean stylistic analysis of translations, involving anything from one translation to any number of Source Texts (STs) and any number of translations of them.

Translational stylistics has many uses. Translators and teachers can look to existing translations to see what they have to offer (Lefevere 1981; Toury 1981). Comparisons of different translations of a text can highlight areas of ambiguity in the ST. Such areas of ambiguity are often highly significant, but they can be harder to spot in monolingual stylistics. Comparisons can also highlight differences between languages, genres and cultures, which often have serious implications for comparative linguistic, literary and cultural studies. Differences in translators' styles, and general difficulties in translating (for example technical terms, metaphors, similes, rhyme, distantly vs. closely related languages, etc.) are clearly important for the discipline itself. In conjunction with protocol studies (in which translators are asked to comment on their thought processes while they are translating, or immediately afterwards) the examination of different translators' products can highlight the effects of their preferred translation strategies (Toury 1986; Lörscher 1991; Tirkkonnen-Condit 1991). All of this has implications for both the training of translators, and translation in language teaching and testing.

In my own study, summarised in section 2 below, two corpora were examined, though not, unfortunately, in machine-readable form. One corpus consisted of H.C. Andersen's stories and novels in the original Danish. The other consisted of a number of different translations of some of the stories into

English. These, and the Danish STs, were searched manually for collocates for a number of adjectives. These were selected on the basis of the recurrence in the stories of one particular theme, and a feeling that a particular pattern of modifier + noun use played a central role in highlighting the theme in the STs, but that this highlighting was far less effective in the Target Texts (TTs) which did not reproduce the ST's patterning.

I begin, however, with one sentence from the ST corpus and five translations of it from the TT corpus (see section 1 below). This sentence is interesting in view of the question raised in section 0.1, because the translators seem to have experienced a conflict of intuitions about two aspects of it: (i) the meanings of two adjectives used in the sentence, one of which was among the nodes in the collocation search; (ii) the semantics of the construction in which the two adjectives are used.

Only one of the translators has resolved this probable conflict in a way which accords with the close textual analyses summarised in section 2 below. It is possible that he was able to do this because his intuitions were in some sense clearer than those of the other translators. His translations of this and other stories proved, moreover, to mirror more closely than any of the other translators' the use in the STs of the adjective nodes in the collocational study. In order to explain why this should be the case, I refer to an experiment whose findings are, in turn, explained by just those psycholinguistic theories of what speakers do with usage data that I mentioned at the end of section 0.1.

This suggests that corpus-based comparative translational stylistic studies might be useful in the search for the answer to my focal question, since they can provide important additional evidence for the psycholinguistic theories that address it. During the quest for independent status for their discipline, translation scholars have drawn freely on insights provided by other disciplines (Bassnett-McGuire 1991:xi), but it is my belief that it would enhance the status of translation studies if we could reverse the direction of the flow of ideas by beginning to make serious claims about the way in which translation studies can, in turn, contribute to other disciplines.

In this way, then, three areas of study are drawn together in this paper: corpus linguistics; translation studies; and psycholinguistics. The underlying theme is the question posed by corpus linguistics, of the relationship between textual evidence and speakers' intuitions. The motivation for the paper is a finding in a study of a corpus of STs and a corpus of their TTs which also raises this issue. An embryonic answer is sought in psycholinguistic theory but I suggest that the data of corpus linguistics and translation studies has to be fed into that theory before a fully developed account can emerge.

1. A case of conflicting intuitions

1.1 The data

In H.C. Andersen's story *Svinedrengen* (*The swineboy*), the following clause complex occurs:

> *"Den er mere end nydelig" sagde Keiseren, "den er pæn!"*
> ("It is more than - " said the emperor, "it is - !")

By *it*, the emperor is referring to a rose. In English translations (see Appendix for sources) this clause complex is translated as follows:

(1) 'It's more than nice!' said the Emperor. 'It's lovely!'
(2) The clause complex is not included.
(3) "It is more than pretty", said the Emperor, "it's beautiful."
(4) "It is more than pretty; it is nice," remarked the emperor.
(5) "It's more than pretty," said the Emperor, "It's handsome."

1.2 Intuitions about the sentence

The semantics of an 'it is more than adj x, it is adj y'-construction suggests (i) that 'x' and 'y' should refer to qualities of the same basic 'semantic type' (Dixon 1982:1-7) and (ii) that 'y' should indicate that the item referred to by *it* possesses more of the quality, or possesses it to a greater degree, than 'x' would suggest. A mini-survey involving native speakers provided the following x and y substitutes in the frame:

it is more than	*adj x*,	it is	*adj y*
	cold		freezing
	bad		awful
	good		wonderful
	funny		hilarious
	bad		appalling

The Bank of English[1] corpora further provide:

content	happy
disheartening	contagious
good	perfect
pretty	beautiful
radical	revolutionary
uneasy	frightened

If we add to this evidence the evidence provided by the four translations which include the frame, we get:

it is more than	*adj x,*	it is *adj y*
translation 1	nice	lovely
translation 3	pretty	beautiful
translation 4	pretty	nice
translation 5	pretty	handsome

It is fairly clear that the choice of adjectives in translation 4 differs from all of the other choices in moving from the more to the less 'emphatic' or 'expressive'. However, in this, translation 4, by Erik Haugaard, is the closest to the original ST clause complex, according to some intuitions about the two Danish adjective-lemmas *nydelig* and *pæn.*

It is interesting to observe, then, that the intuitions about the construction apparently held by the translators who produced translations (1), (3) and (5) appear to have over-ridden their understanding of the meaning of the adjectives in ST.[2] An 'abnormal' usage in ST has been replaced with a 'normal' usage in TT.

1.3 The sentence in the context of textual analysis

Haugaard's chosen translational equivalent for the text part under scrutiny corresponds to the equivalent suggested by Knowles and Malmkjær (1991: 206), who did not then know the translation by Haugaard. In that paper, and in Knowles and Malmkjær (1989; forthcoming) and Malmkjær (forthcoming) the choice of "it is more than pretty ... it is nice" was backed by reference to close textual analyses of, among other textual features, patterns of collocation in Andersen's stories for the lemma *nydelig*; a summary is given in section 2.

Haugaard's translations of other fairytales to which we referred in those papers reproduce more closely than any other translations we have come across what we refer to in the 1991 paper as Andersen's key terms. In section 3, I suggest a possible reason for this. In section 4, I suggest that my hypothesis finds support in the psycholinguistics of bilingualism.

2. Collocational translational stylistics: an example

2.1 A story theme

In a set of earlier papers, Knowles and Malmkjær (1989; 1991; forthcoming) and Malmkjær (forthcoming) argue that an important theme in the writing of H.C. Andersen is the opposition between artificiality, which tends to be associated with social conventions, and genuineness, which tends to be associated with natural phenomena.

Thus, in *The princess on the pea*, a prince sets out to find a princess who has to be genuine, but is unable to determine whether any of the many princesses he comes across satisfies this criterion. The one he eventually selects presents herself to him looking quite unlike a conventional princess, but she is sensitive to natural phenomena and is therefore judged genuine. In *The swineboy*, the princess is able to appreciate only artificial items, and is hence judged unworthy of the prince. In *The steadfast tin soldier*, the soldier is genuine in a way that the paper dancer is not: when they are burnt, he remains as a small tin heart, whereas all that is left of her is the large spangle, now blackened by the fire, which had adorned her dress. In *The emperor's new clothes*, only a small child, unworried by social convention and by what others might think, dares admit what everyone sees, that the emperor is naked: the fine clothes the swindlers have convinced everyone else that they have been making are not only artificial, but are actually non-existent. In the novel *The Two Baronesses* a person of low birth is shown worthy of baroness status because of her character.

Numerous other stories and novels display similar themes. The opposition is clearly given at the story level of the texts, but it is powerfully supported at the linguistic level too, largely through collocational patterns which are created by associating adjectives with nouns in a systematic way (see section 2.3 below).

2.2 *The translation of adjectives*

Adjectives are notoriously difficult to translate, even though they may realise semantic universals (Dixon 1982), and even when both languages in question have the word class. Although it is possible to isolate and gloss certain 'basic' adjectival meanings which pattern across languages in similar ways to basic colour terminology (Berlin and Kay 1969; Dixon 1982:4-7), it is rarely the case that terms in the two languages correspond to each other on a one-to-one basis, especially if the two languages have large numbers of descriptive adjectives, as English and Danish do. So even if it is possible to say, for each language, that a set of adjectives have a loosely pejorative quality and another set a loosely approving quality, the further semantic detail indicated by individual lexical items tends to be unevenly distributed and to overlap between items.

The adjectives that interest us here are dealt with as follows in Cassell's Standard Danish-English/English-Danish Dictionary (I have translated some explanations from Danish into English; not all details are included):

nydelig adj pretty, nice (of person also), attractive
pæn adj nice (e.g. face, dress)
pretty adj *køn, pæn, net, nydelig*
nice adj *pæn, net, rar, flink, tiltalende, elskværdig, lækker, dejlig, fin*

By looking up all the terms used as equivalents in the examples above, and their equivalents, and theirs, and so on, one could construct a complicated network of terms and their interrelations, showing how they overlap and differ. There is obviously neither one fixed use nor one fixed meaning for each term in each language, and there is no one-to-one correspondence between the terms across the languages.

Cassell's is not a bad dictionary; it reflects usage pretty well for both languages. It shows that all four terms, the two English and the two Danish adjectives, are used as terms of approval. (It also gives examples of ironic usages which may be pejorative, of course, but that does not affect the argument I am about to present.) An innocent translator could easily come to the conclusion that the exact choice of term is not very important, and this is probably true for many adjectives in many texts, though not for these adjectives in H.C. Andersen's stories.

2.3 'Nydelig' and its co-textual context

In these stories, *nydelig* takes on a pejorative sense through its constant collocation with nouns that refer to artificial objects and through patterns of opposition set up between the collocations '*nydelig* + artificial object referent' and the collocations 'approval adj + noun referring to a natural phenomenon'. The texts make it clear which phenomena and things are natural and which are artificial. In this section, I provide a summary of the lexical patterns entered into by the lemmas *nydelig* and *pæn*[3] in three of Andersen's stories, taking as my starting point the sentence from *The swineboy*, mentioned in section 1.1 above:

"Den er mere end nydelig" sagde Keiseren, "den er pæn!"

The context in which this sentence occurs is the following. A prince sends a princess two gifts: a nightingale and a rose. The princess initially likes the gifts, but when she discovers they are real, not made, she rejects them and their giver:[4]

(1) "Oh, how *nydeligt* it is *made*!" said all the ladies-in-waiting [of the rose].
(2) "*It* is more than *nydelig*!" said the emperor, "*it* is *pæn*!"
(3) But the princess felt it and almost started crying.
(4) "Fie Papa!" she said, "*it* is not *kunstig*, *it* is *virkelig*!"
(5) "Fie!" said all the courtiers, "*it* is *virkelig*!"
(6) "Now, let us just see what there is in the other case, before we get angry!" suggested the emperor, and then the nightingale was taken out...
(7) "I certainly hope I am not given to understand that *it* is *virkelig*!" said the princess.
(8) "Yes, *it* is a *virkelig* bird!" said those who had brought it.
(9) "Well, then let that bird fly," said the princess...

The prince then covers his face with dirt and goes to seek employment in the palace. He is given the job of swineboy, and has to live in a little hut down by the pigsty. Here he sits and makes artificial, silly play-things which really appeal to the princess, who buys them off him with kisses:

(10) ... but all day long he [the prince] sat and worked, and by evening he had *made* a *nydelig* little pot, there were bells on it and as soon as the pot boiled they rang so delightfully and played the old tune: "*Ach, Du lieber Augustin, Alles ist gone, gone, gone*!" but the *Allerkunstigste* was that when one put one's finger in the steam from the pot, one was immediately able to smell what food was being cooked in every chimney in town; you see, this was certainly something different from a rose.

In (1), *nydelig* is used by the ladies-in-waiting, who clearly think that the rose is an artifact. It collocates with the term for *made*. In (2), it is said that the rose is more than *nydelig*, namely *pæn*. According to common intuitions about the use of this construction (see section 1.2 above), this makes the sentence somewhat odd. However, if my understanding of the terms in question as indicating, respectively, artificiality (*nydelig*) and naturalness (*pæn*) is correct, the sentence is simultaneously odd and perfectly sensible, in exactly the way that "it is more than artificial, it is real" is odd, in that the construction is normally used with adjectives referring to things of the same basic semantic type, but sensible, given the author's preference for the real over the artificial.

In (4), the princess refers to the rose with the term *it*, and bemoans the fact that rather than being *kunstig*, the rose is *virkelig*. The first meaning the dictionary provides for *kunstig* is *artificial*. The first meaning it gives for *virkelig* is *real* (though see Malmkjær, forthcoming, for reasons in favour of translating it as *genuine* in Andersen's stories; the dictionary gives *virkelig* as the last of four equivalents for *genuine*). So at this point, the rose has been described as *pæn* by the emperor, and as *virkelig* by the princess, and *virkelig* has been opposed to *kunstig*. *Nydelig* has been associated with the verb *make*. By (7) - (8) we can classify the nightingale with the rose in terms of being *virkelig*:

Set 1	Set 2
virkelig (rose; nightingale)	*nydelig* (not rose)
pæn (rose)	*made* (not rose)
	kunstig (not rose)

In (10), the pot the prince *had made* is described by the authorial voice as "a *nydelig* little pot", and one of its qualities is said to be the *Allerkunstigste*. *Kunstig*, as we have seen, means 'artificial'; *kunst* means 'art'; the *-ste* ending indicates superlative, and *aller* means 'most ... of all'. The nearest translation for this term is 'most artful of all'. The pot is, in addition, said to be certainly "something different from a rose" which clearly sets the two things in opposition. In fact, a comparison of the descriptions of the power of the rose and the power of the pot also reinforces this opposition. We have seen the pot's qualities described in (10) above. The comparable passage for the rose tells us that the rose comes from a rose tree which grows on the prince's father's grave. Every five years only, this tree bears a single rose so sweetly scented that it makes one forget all one's sorrows and worries. The semantic parallelisms/oppositions are between (i) the steam from the pot and the smell of the rose, and (ii) the gaining of trivial information and the forgetting of serious worries and sorrows. So now our two sets look like this:

Set 1	Set 2
virkelig (rose; nightingale)	*nydelig* (pot)
pæn (rose)	*made* (pot; not rose)
	kunstig (not rose)
	artful (pot)

The lemma *virkelig* had already been used repeatedly, and, I would want to argue, as a key term, by Andersen in the story *The princess on the pea*. This is a story of 347 words; the lemma *prindsesse* occurs 12 times in it, three times in immediate collocation with *virkelig* and four times in immediate collocation with the near-synonym *rigtig* (real). The text makes it clear that a *virkelig/rigtig* princess is desirable (at least to the prince), and that such a person is different from a princess with whom "something is wrong" or who shows evidence of "something that was not quite *rigtigt*". So we can add to our two sets:

virkelig (rose; nightingale; desired princess)	*nydelig* (pot)
rigtig (desired princess)	*not quite rigtig* (other princesses)
	something wrong (other princesses)
pæn (rose)	*made* (pot; not rose)
	kunstig (not rose)
	artful (pot)

Nydelig is used frequently in *The steadfast tin soldier* as a modifier of (the surroundings of) a character, the dancer, who contrasts with another character, the tin soldier, in interesting ways. The story begins by introducing the tin soldier as one of twenty five brothers who had all been *born* of an old tin spoon and who resembled each other "to the life", except that one, the one of the story's title, had only one leg. But he is the *noteworthy* one. In contrast, the dancer is presented as follows:

> On the table where they [the soldiers] were arranged there were many other toys; but what was most *eye-catching* was a *nydeligt* castle of *paper*. Through the small windows you could see straight inside the halls. Outside were small trees around a small mirror that was supposed to look like a lake ... It was all *nydeligt*, but the *nydeligste* was a small *maiden* who stood in the middle of the open castle door, *she* was also *cut out* of paper, but she wore a skirt of the clearest linen and a small, narrow, blue ribbon over her shoulder like a sash! in the middle of it all was a shining sequin, just as big as her face.

The *nydelig* dancer is *cut out* of (fragile) paper, and she and her surroundings are *eye-catching*. We can add these terms to the two sets:

Set 1	Set 2
virkelig (rose; nightingale; desired princess)	*nydelig* (pot; dancer)
rigtig (desired princess)	*not quite rigtig* (other princesses)
	something wrong (other princesses)
pæn (rose)	*made* (pot; not rose)
	kunstig (not rose)
	artful (pot)
born (soldier)	*cut out* (dancer
noteworthy (soldier)	*eye-catching* (dancer)

And we could go on and on adding terms and amplifying the oppositions by analysing more and more stories. What should be clear is that Andersen has pushed *nydelig* across from the set of terms for the real (approval) to the set of terms for the artificial (disapproval); and if it is not, then reading the stories will help. What reading the stories will not do, though, if it is done in English, is show the patterning set out above. This is because translators tend not to be consistent *across* stories in the translational equivalents they give for the Danish terms, even if they are consistent to a degree within stories. But the most consistent is Haugaard. In the following section I should like to suggest why this might be the case.

3. Two types of translator

3.1 Translators' relationships with their languages

Haugaard does not have a professional interest in linguistics or in theories of collocation, but he is, to the best of my knowledge, the only translator cited here whose first learnt, or acquired, language is Danish. Not having English as a first language ought, according to received wisdom, to be a disadvantage in translating into English, as one is supposed to translate into one's mother tongue (see Newmark 1981:180). In Haugaard's case, this rule might in fact be waived (as it has been by the publishers of his translation from Danish, Chinese and Japanese), since English is his language of habitual use. Having language L as one's 'language of habitual use' is accepted as an alternative qualification to having L as one's 'mother tongue' for translating into L by, for example, the Institute of Translation and Interpreting (ITI).

According to some hypotheses about first and second language acquisition/learning, the acquisition of a first language is different from the learning of subsequent languages and there are differences between the products of the learning/acquisition processes (see section 4 below). It is possible that the differences are such that a bilingual who has acquired a first language before another language was introduced, and has reached adulthood before switching to the other language as the language of habitual use (although instruction in that language may have been received during childhood) will remain a more sensitive receptor of the first language than of any language subsequently learnt. I want to pursue the fairly commonsense hypothesis that, in the case of experienced translators, this might imply that a 'Source Language (SL) mother tongue but Target Language (TL) habitual use'-translator will be especially sensitive to the nuances of ST as well as being sufficiently proficient in TL to be able to reproduce them as fully in TT as TL allows.

Of course, the other translators may not have wished to remain close to

ST, but, if so, then this might again be partly because SL is not their first language. Perhaps the root cause for the common preference for translators who work into the mother tongue is that a kind of natural affinity is for them a powerful inducement, in addition to the conscious effort they obviously also expend, to their production of natural TTs; they may feel less regret in letting go of SL characteristics in the interest of TL naturalness. The special quality of 'SL mother tongue but TT habitual use'-translators might then be that, in addition to being sufficiently proficient in L2 to be able to produce it fluently and naturally, they are, whether consciously or not, particularly geared towards retaining as many SL and ST characteristics as possible. This may make them particularly well-qualified to produce translations of texts in which mode of expression is inextricable from propositional content.

3.2 Intuition, exposure and psycholinguistics

The psycholinguistic theories and data referred to in section 4 below can account, at least in part, for the development of what is commonly referred to as native speaker 'intuitions' or 'judgements'. Native language proficiency obviously comes with increasing exposure to the language being acquired. This long-term exposure is collocational in nature (the acquiring infant does not meet words in isolation from one another, except in some circumstances, and certainly not for long). Yet, as we have seen, there is often a conflict between native speaker intuitions and textual evidence (see section 0.1). Clearly, psycholinguistic accounts need to take this into consideration, and in section 5 I suggest ways in which the example of the 'SL mother tongue but TL habitual use'-translator, and the methodology of translation in general, might be used in providing this broader view of what is involved in language acquisition.

4. The psycholinguistics of late bilingualism

4.1 Two views about the relationship between L1 and L2 acquisition

Although the risk of oversimplification looms large, it is possible to claim that there are two basic views of the relationship between L1 and L2 acquisition. According to one view, the two processes are essentially identical (Dulay and Burt 1974; Krashen 1982; 1985; Dulay et al 1982; Felix 1981; Pfaff 1984; Ellis 1984). Coppieters (1987:546) summarises the reasoning behind this view as follows:

> since native and non-native acquisition of a language apparently follow much the same developmental path ... both forms of language acquisition must reflect the operation of a single learning mechanism ... Krashen states (1985:3) that "deep down, the 'mental organ' for language ... produces one basic product, a human language, in one fundamental way."

It is the other view, according to which L1 and L2 acquisition differ, that I want to explore. Both this view, and my hypothesis about translators, have found evidential support in a study by Coppieters (1987) on differences in native and native-like speakers' command of a language.

4.2 Differences between native and native-like speakers

Coppieters (1987:547-49) discusses in detail the difficulties involved in discovering whether there are differences between native and near-native underlying competence in a language. In his own study, he attempted to minimise these by (i) eliciting intuitions concerning not only well-formedness of sentences but also their interpretation and (ii) using an interview format for eliciting the data :

> the data ... were elicited in the course of simple conversations [taped and lasting around 50 minutes each] which started from the list of [107] sentences which I had elaborated. Beyond well-formedness and interpretation, speakers were encouraged to express whatever reactions they had to the sentences presented, and were always reinforced in their intuitions.

Coppieters' near-native speakers (NNSs) of French were people for whom (ibid: 545) "L2 has become the functional equivalent of a mother tongue - people who, except possibly for their accent, can no longer be reliably distinguished from native speakers by the 'mistakes' they make, or by the inappropriateness of their language". They were "L2 learners who, after years of exposure and practice, function correctly, naturally, and automatically", but who had acquired French as adults (ibid:551): "i.e., although many had obtained some school training in French during their adolescent years, none had been placed in circumstances requiring them to use French for normal communicative purposes before the age of 18". My category 'SL mother tongue but TL habitual use'-translators would form a subset of just such a group, but would need, in addition, to be proficient, practising, and preferably professional or semi-professional translators, since translation ability appears not to develop automatically along with language proficiency (Toury 1986).

The differences Coppieters uncovered did not surface in the speakers' use of the language, but only in the interviews on comprehension. So a NNS's *production* is not affected, but their *interpretation* may be. Therefore, an 'SL mother tongue but TL habitual use'-translator should be able to *produce* fluent TTs while being a better *interpreter* of ST than a TL mother tongue translator.

Coppieters' study shows that native and near-native speakers of French differed markedly in their "interpretations of sentences involving basic grammatical contrasts such as the two past tenses (imparfait and passé composé), the 3rd person pronouns *il* and *ce*, and the placement of the adjective before and after the noun" (ibid:544). While both groups had evolved grammars capable of choosing between *il* and *ce*, for example, and of interpreting the contrast between them, their comments on the sentences in which they had to choose between the terms revealed that they used different criteria for choosing, even in those cases where their choices were the same (555-6). The results illustrate (ibid:544):

> the relative independence of the two levels of language: on the one hand, language use, and on the other hand, underlying grammar as reflected by speaker's intuitions ... Near-native speakers diverge less from native speakers in formal features, such as those currently covered by studies in Universal Grammar, than in 'functional' or 'cognitive' aspects of grammar.

Notice that this helps explain why Dixon (1982:11) reports that intuitive judgements about part-of-speech seem reliable whereas Sinclair and Renouf (1988:151-2) and Sinclair (1991:39) suggest that intuitions about use are not: use is functional while part-of-speech categories are formal.

4.3 Coppieter's study and the difference between two types of translator

The fact that non-native speakers diverge from native speakers in functional features supports my hypothesis about the differences between TL and SL translators. An SL native language translator is more likely to share the author's interpretations of ST linguistic choices than a TL native language translator: s/he will be more sensitive to their significance, and will therefore, obviously, be more likely to reproduce them in TT (it being less likely that a translator will reproduce a feature s/he has not noticed, particularly the kind of foregrounded patterning discussed in section 2.3 above).

Coppieters' NNS subjects were given sentences out of context, whereas my translators worked with sentences in context. He suggests that the NNSs might have been able to interpret the sentences in the same way as the NS group had they been in context. Again, this difference between the two groups supports my hypothesis: Haugaard is able to reproduce ST key terms as key terms in TT because although each of these terms has several potential translational equivalents, he is consistent in the actual equivalents he provides for them. This requires an ability to consider individual terms simultaneously as *isolated* from their context (otherwise they cannot be held constant in their patterns) and *within* the context (otherwise it would be impossible to decide which terms function as key terms in the first place). It is the ability to deal with terms in relative isolation from context that seems less developed in the TL translators whose tendency is to translate the whole contexts without apparently having been able to isolate the key terms.

4.4 The psycholinguistic explanation

The psycholinguistic theories concerning language acquisition which might explain these differences include Karmiloff-Smith's (1986; reported in Coppieters 1987:569) according to which children initially develop an internal representation of language which includes a great deal of information about context for each term or expression's form/function pairing. This initial system is unanalysed and often contains several forms where the adult system would only have one form (for example, past tense verb forms may be stored separately from non-past forms). The final, adult, analysed, complex, abstract system only develops gradually and slowly. Coppieters argues that what native speakers have and non-native speakers lack is, then, just this history of long, slow, gradual development which may have left memory traces that are available if required for interpretive purposes:

> No matter how close to child language acquisition one conceives adult SLA to be, it would seem rather implausible that the initial unanalysed stages of child language acquisition would be duplicated by the adult learner. The form of the L2 is new to an adult; but the content, by and large, is not.

4.5 The explanation for the discrepancy between intuitions and use

It seems likely that when speakers are asked to state their intuitions about individual lexical items and patterns, they draw on their fully analysed, but highly abstract, representation of linguistic knowledge. This abstract system does not, of course, prevent speakers from understanding those actual usages that differ from their intuitions. That is, speakers do not lose that sensitivity to the input which enabled them to abstract, but when asked for intuitions about items out of context they have only the abstract system to rely on.

It cannot be the case, however, that this abstract system stores items totally devoid of context, because if it did speakers would not be able to respond to questions about meaning/use. The response has to carry with it one particular context. Otherwise the only possible answer to a question like: *What do you think 'see' means*? would be *See*. So the abstraction is not carried out through a simple stripping away from the forms of all functions. Rather, one of the form/function pairings of the original input gradually comes to be selected for the abstract representation. The question of why *that* form/function pairing, rather than any other, is selected still has to be answered. While it is possible to explain the choice for *see* via the notion of biological or psychological salience, this is a less plausible explanation for items like *certain*, for example. Speakers' intuitions about *certain* appear to be that it is used when one feels that something is correct or true, even though a more frequent use is 'particular' (Sinclair 1990). How might the process of abstraction to the less frequent form/function pairing be explained for items like *certain*?

5. Towards an explication of speakers' gradual manipulation of the evidence

During language acquisition, unanalysed knowledge becomes analysed and reanalysed through successive stages, and "the development of language proficiency ... includes as one of its essential components the continual analysis of linguistic knowledge to create representations based on formal or symbolic structure" (Bialystok 1991:116).

One possible way of tapping child intuitions and modes of processing input might involve the use of translations, oral or written, made by bilingual children of various ages and by bilingual adults, both late bilinguals and those who have become bilingual before the age of five, say. These could be compared with monolingual children's, and monolingual adults' performance on cloze tests involving the same terms, and, importantly, with corpora of usage in both languages. The bilinguals would differ from the monolinguals in having had two sets of input, and any systematic variation between the groups' output could be significant in the quest for explanation of the problem with which we set out in section 0.

6. Conclusion

It seems to me likely that comparisons of translations made by TL native speakers and translations made by SL native speakers - if they turn out to be systematically different as my own investigation of translations of H.C. Andersen's stories into English suggest - can add to the debate about whether or not L1 and L2 acquisition proceed differently and result in different products. Translations are particularly interesting because translators have to be able to tap both their analysed knowledge store and the history of its development in their attempts to balance their intuitions about the meanings of terms and about how the terms have to be used in particular contexts.

I have argued for the hypothesis that the systematic differences I have found between Haugaard's translations and those made by native speakers of TL can be explained by reference to differences in the products of the two processes of L1 and L2 acquisition. If this is the case, then comparisons of translations made by the two kinds of translator might help clarify the nature of the differences in products, and perhaps also, by implication, the nature of the differences in the processes. Machine-readable corpora of the kinds of text I have referred to would greatly advance research of this kind.

Author's address:
University of Cambridge, Research Centre for English and Applied Linguistics, Keynes House, Trumpington St., Cambridge, CB2 1QA, U.K.

Notes

1. The Bank of English is the name given to the new corpus-building initiative at Cobuild, Birmingham. It currently comprises almost 200 million words of running text.
2. I am, of course, assuming that the translators had near-native competence in Danish.
3. Adjectives inflect in Danish for gender (neuter or common) and number (singular or plural) of the modified noun, as well as for comparative and superlative.
4. For ease of reading, I have translated all but the forms of the lemmas that interest me. Nodes and collocates are emphasised. The term *Allerkunstigste* is not in immediate collocation with a significant term, but the specification of what the *Allerkunstigste* is, in the following clause, aids interpretation.

References

Aitchison, Jean. Forthcoming. "Understanding Words". *Language and Understanding* ed. by Gillian Brown, Kirsten Malmkjær, Alastair Pollitt & John N. Williams. Oxford: Oxford University Press.

Bassnett-McGuire, Susan. 1980, 1991. *Translation Studies*. London & New York: Routledge.

Berlin, B. & P. Kay. 1969. *Basic Colour Terms: Their Universality and Evolution*. Berkeley & Los Angeles: University of California Press.

Bialystok, Ellen. 1991. "Metalinguistic Dimensions of Bilingual Language Proficiency". *Language Processing in Bilingual Children* ed. by Ellen Bialystok, 113-40. Cambridge: Cambridge University Press.

Cassell's Standard Danish Dictionary. 1984, 1986. Holt, Rinehart & Winston.

Coppieters, René. 1987. "Competence Differences Between Native and Near-native Speakers". *Language* 63,3. 544-73.

Dixon, R. M. W. 1982. *Where Have All the Adjectives Gone? And other Essays in Semantics and Syntax*. Berlin, New York & Amsterdam: Mouton Publishers.

Dulay, H. & M. Burt. 1974. "Natural Sequence in Child Second Language Acquisition". *Language Learning* 24. 37-53.

_________________ & Krashen, S. D. 1982. *Language Two*. New York: Oxford University Press.

Ellis, R. 1984. *Classroom Second Language Development*. Oxford: PergamonPress.

Felix, S. W. 1981. "The Effect of Formal Instruction on Second Language Acquisition". *Language Learning* 31. 87-112.

Karmiloff-Smith, Annette. 1986. "From Meta-process to Conscious Access: Evidence from Children's Metalinguistic and Repair Data". *Cognition* 23. 95-147.

Knowles, M. & K. Malmkjær. 1989. "Translating Ideology: Language, Power and the World of the Tin Soldier" in *Language and Ideology* ed. by M. Knowles & K. Malmkjær. *English Language Research Journal* (University of Birmingham) 3. 205-41

_________________________. 1991. "Key Terms in H.C. Andersen's Fairytales and their Translations into English". *Babel: International Journal of Translation* 37, 4. 203-12.

_________________________. Forthcoming. "The Lexicalisation of Magic: The Self-conscious Fairytale". To appear in *Ilha do Desterro*.

Krashen, S. D. 1982. *Principles and Practice in Second Language Acquisition*. Oxford: Pergamon Press.

___________. 1985. *The Input Hypothesis*. London: Longman.

Lakoff, George & Mark Johnson. 1980. *Metaphors We Live By*. Chicago & London: The University of Chicago Press.

Lefevere, André. 1981. "Programmatic Second Thoughts on 'Literary' and 'Translation' or: Where do We Go from Here". *Poetics Today* 2,4. 39-50.

Lörscher, Wolfgang. 1991. *Translation Performance, Translation Process and Translation Strategies: A Psycholinguistic Investigation*. Tübingen: Gunter Narr Verlag.

Lyons, John. 1977. *Semantics*: Volume 1. Cambridge: Cambridge University Press.

Malmkjær, Kirsten. Forthcoming. "Translating the Real Story: Dancing with Multiple Partners". To appear in *Linguistic Approaches to Literature* ed. by Jonathan A. Payne & John Wheatley. English Language Research Journal: University of Birmingham.

Newmark, Peter. 1981. *Approaches to Translation*. Oxford: Pergamon Press.

Pfaff, C. 1984. "On Input and Residual L1 Transfer Effects in Turkish and Greek Children's German". *Second Language: A Crosslinguistic Perspective* ed. by R. Andersen, 271-98. Rowley, MA.: Newbury House.

Sinclair, J. McH. 1990. Seminar presented to the Research Centre for English and Applied Linguistics, the University of Cambridge: 16 May 1990.

_____________. 1991. *Corpus, Concordance, Collocation*. Oxford: Oxford University Press.

__________ & Antoinette Renouf. 1988. "A Lexical Syllabus for Language Learning". *Vocabulary and Language Teaching* ed. by Ronald Carter & Michael McCarthy, 140-60. London & New York: Longman.

Tirkkonnen-Condit, S. ed. 1991. *Empirical Research in Translation and Intercultural Studies*. Tübingen: Gunter Narr Verlag.

Toury, Gideon. 1980, 1981. "Translated Literature: System, Norms, Performance: Toward a TT-oriented Approach to Literary Translation". *Poetics Today* 2, 4. 9-27.

__________. 1986. "Monitoring Discourse Transfer: A Test-case for a Developmental Model of Translation". *Interlingual and Intercultural Communication: Discourse and Cognition in Translation and Second Language Acquisition Studies* ed. by Juliane House & Shoshana Blum-Kulka, 79-94. Tübingen: Gunter Narr Verlag.

Appendix: Sources for the translations

1. *Hans Anderson's Fairy Tales: A Selection*. Translated from the Danish by L.W. Kingsland. London: Oxford University Press. 1959.
2. *Hans Christian Andersen: Fairy Tales*. Translator not named. London: Hamlyn. 1959.
3. *Hans Christian Andersen: Fairy Tales*. Translated by Marie-Louise Peuleve. Odense: Skandinavisk Bogforlag. No date.
4. *Hans Andersen: His Classic Fairy Tales*. Translated by Erik. Haugaard. London: Lynx. 1988.
5. *Hans Christian Andersen: Eighty Fairy Tales*. Translated by R.P. Keigwin. Odense: Skandinavisk Bogforlag. 1976.

Corpus Linguistics and Translation Studies
Implications and Applications

Mona Baker
Cobuild, Birmingham

Abstract
The rise of corpus linguistics has serious implications for any discipline in which language plays a major role. This paper explores the impact that the availability of corpora is likely to have on the study of translation as an empirical phenomenon. It argues that the techniques and methodology developed in the field of corpus linguistics will have a direct impact on the emerging discipline of translation studies, particularly with respect to its theoretical and descriptive branches. The nature of this impact is discussed in some detail and brief reference is made to some of the applications of corpus techniques in the applied branch of the discipline.

0. Introduction

A great deal of our experience of and knowledge about other cultures is mediated through various forms of translation, including written translations, sub-titling, dubbing, and various types of interpreting activities. The most obvious case in point is perhaps literature. Most of us know writers such as Ibsen, Dostoyevsky and Borges only through translated versions of their works. But our reliance on translation does not stop here. Our understanding of political issues, of art, and of various other areas which are central to our lives is no less dependent on translation than our understanding of world literature.

Given that translated texts play such an important role in shaping our experience of life and our view of the world, it is difficult to understand why translation has traditionally been viewed as a second-rate activity, not worthy of serious academic enquiry, and why translated texts have been regarded as no more than second-hand and distorted versions of 'real' texts. If they are to be studied at all, these second-hand texts are traditionally analysed with the sole purpose of proving that they inevitably fall short of reproducing all the glory of the original. A striking proof of the low status accorded to translated texts comes from the young but by now well-established field of corpus linguistics. A recent survey commissioned by the Network of European Reference Corpora, an EEC-funded project, shows that many corpus builders in Europe specifically exclude translated text from their corpora.[1] This is presumably done on the grounds that translated texts are not representative and that they might distort our view of the 'real' language under investigation. It is perhaps justifiable to exclude translated texts which are produced by non-native speakers of the language in question, but what justification can there be for excluding translations produced by native speakers, other than that translated texts per se are thought to be somehow inferior or contrived? Biased as it may be, this traditional view of translation implies, in itself, an acknowledgement of the fact that translational behaviour is different from other types of linguistic behaviour, quite irrespective of the translator's mastery of the target language.

The starting point of this paper is that translated texts record genuine communicative events and as such are neither inferior nor superior to other communicative events in any language. They are however different, and the nature of this difference needs to be explored and recorded. Moreover, translation should be taken seriously by related disciplines such as linguistics, literary theory and cultural and communication studies, not least because these disciplines can benefit from the results of research carried out in the field of translation. At the same time, as a phenomenon which pervades almost every aspect of our lives and shapes our understanding of the world, the study of translation can hardly be relegated to the periphery of other disciplines and sub-disciplines, those listed above being no exception. What is needed is an academic discipline which takes the phenomenon of translation as its main object of study. For many scholars, this discipline now exists. Some refer to it as the 'science of translation', others as 'translatology', but the most common term used today is 'translation studies'.

Eco (1976:7) distinguishes between a discipline and a field of study. The first has "its own method and *a precise object*" (my emphasis). The second has "a repertoire of interests that is not as yet completely unified". It could be argued that translation studies is still largely a 'field of study' in Eco's terms. The vast majority of research carried out in this, shall we say emerging discipline, is still concerned exclusively with the relationship between specific source and target texts, rather than with the nature of translated text as such. This relationship is generally investigated using notions such as equivalence, correspondence, and shifts of translation, which betray a preoccupation with practical issues such as the training of translators. More important, the central role that these notions assume in the literature points to a general failure on the part of the theoretical branch of the discipline to define its object of study and to account for it. Instead of exploring features of translated texts as our object of study, we are still trying either to justify them or dismiss them by reference to their originals.

It is my belief that the time is now ripe for a major redefinition of the scope and aims of translation studies, and that we are about to witness a turning point in the history of the discipline. I would like to argue that this turning point will come as a direct consequence of access to large corpora of both original and translated texts, and of the development of specific methods and tools for interrogating such corpora in ways which are appropriate to the needs of translation scholars. Large corpora will provide theorists of translation with a unique opportunity to observe the object of their study and to explore what it is that makes it different from other objects of study, such as language in general or indeed any other kind of cultural interaction. It will also allow us to explore, on a larger scale than was ever possible before, the principles that govern translational behaviour and the constraints under which it operates. Therein lie the two goals of any theoretical enquiry: to define its object of study and to account for it.

Section 1 below offers an overview of the emerging discipline of translation studies and explains why translation scholars are now in a position to use the insights gained from corpus linguistics, and some of the techniques developed by it, to take translation across the threshold of 'field of study' and into the realm of fully-fledged disciplines.

1. Translation studies: the state of the art

1.1 Central issues: the status of the source text and the notion of equivalence

Until very recently, two assumptions dominated all discussions of translation and were never questioned in the literature. The first is that of the primacy of the source text, entailing a requirement for accuracy and faithfulness on the part of the translator. The second is a consequence of the first and is embodied in the notion of equivalence which has been the central concern of all discourse on translation since time immemorial. Translations should strive to be as equivalent to their originals as possible, with equivalence being understood mainly as a semantic or formal category. The implied aim of all studies on translation was never to establish what translation itself is, as a phenomenon, but rather to determine what an ideal translation, as an instance, should strive to be in order to minimise its inevitable distortion of the message, the spirit, and the elegance of the original.

The essentialist question of how equivalence per se might be established in the course of translation has gradually been tempered by experience and by an explosion in the amount and range of texts which have come to be translated in a variety of ways on a regular basis. Hence, we now have a massive amount of literature which attempts to classify the notion of equivalence in a multitude of ways, and the question is no longer how equivalence might be achieved but, increasingly, what kind of equivalence can be achieved and in what contexts. This in itself is a noticeable improvement on the traditionally static view of equivalence, but it still assumes the primacy of the source text and it still implies that a translation is merely a text striving to meet the standards of another text.

1.2 Developments which support a move towards corpus-based research

The attempt to extend and classify the notion of equivalence has brought with it a need to explore not only the source text as the model to be adhered to but also the target language, and the specific target language text type, in order to give meaning to such categories as stylistic equivalence and functional equivalence. If the idea is not simply to reproduce the formal structures of the source text but also to give some thought, and sometimes priority, to how similar meanings and functions are typically expressed in the target language, then the need to study authentic instances of similar discourse in the two languages becomes obvious.

There have been other developments which have played a more direct role in preparing the ground for corpus work. One such development is the decline of what we might call the semantic view of the relationship between source and target texts. For a long time, discourse on translation was dominated by the idea that meaning, or messages, exist as such and can, indeed should, be transferred from source to target texts in much the same way as one might transfer wine from one glass to another. The traditional dichotomy of translating word-for-word or sense-for-sense is a product of this view of meaning. At about the same time that the notion of equivalence began to be reassessed, or perhaps a little earlier, new ideas began to develop about the nature of meaning in translation. Firth (1968:91) was among the first to suggest that, difficult though as it may appear, an approach which connects structures and systems of language to structures and systems in the context of situation (as opposed to structures and systems of thought) is more manageable and "more easily related to problems of translation". Similarly, Haas (1968:104) stresses that, in practice, correspondence in meaning amounts to correspondence in use and asserts that "unless we can succeed in thus explaining translation, the mystery of bare and neutral fact will continue to haunt us". Two expressions are equivalent in meaning if and only if "there is a correspondence between their uses" (ibid). The importance of this change in orientation, from a conceptual to a situational perspective and from meaning to usage, is that it supports the push towards descriptive studies in general and corpus-based studies in particular. Conceptual and semantic studies (in the traditional sense) can be based on introspection. Studies which take the context into consideration, and even more so, studies which attempt to investigate usage, are, by definition, only feasible if access is available to real data, and, in the case of usage, to substantial amounts of it.

Apart from the decline of the semantic view of translation, another, and very exciting, development has been the emergence of approaches which undermine both the status of the source text vis-à-vis the translated text and the value of the very notion of equivalence, particularly if seen as a static relationship between the source and target texts. The move away from source texts and equivalence is instrumental in preparing the ground for corpus work because it enables the discipline to shed its longstanding obsession with the idea of studying individual instances in isolation (one translation compared to one source text at a time) and creates a requirement which can find fulfilment in corpus work, namely the study of large numbers of texts of the same type. This is precisely where corpus work comes into its own.

1.2.1 New perspectives: polysystem theory

In the late seventies, Even-Zohar, a Tel-Aviv scholar, began to develop a theory of literature as a polysystem, that is as a hierarchical and dynamic conglomerate of systems rather than a disparate and static collection of texts. A given literary polysystem is seen as part of a larger cultural polysystem, itself consisting of various polysystems besides literature, for example politics and religion. These polysystems are structured differently in different cultures.

Polysystem theory has far-reaching implications for the status of translated literature in general and for the status of the source text vis-à-vis the target text in particular. First, the theory assumes a high level of interdependence among the various systems and sub-systems which underlie a given polysystem, as well as among the polysystems of literature in various cultures. This means that, for instance, "literature for children would not be considered a phenomenon sui generis, but related to literature for adults" and, similarly, "translated literature would not be disconnected from original literature" (Even-Zohar 1979:13). As a consequence, the status of translated literature is elevated to the point where it becomes worthy of investigation as a system in its own right, interacting with its co-systems and with the literary polysystems of other cultures. By recognising translated literature as a system in its own right, polysystemists shifted the attention away from individual literary translations as the object of literary studies to the study of a large body of translated literature in order to establish its systemic features.

Second, one of the main properties of the polysystem is that there is constant struggle among its various strata, with individual elements and systems either being driven from the centre to the periphery or pushing their way towards the centre and possibly occupying it for a period of time (ibid:14). This constant state of flux suggests that no literary system or sub-system is restricted to the periphery by virtue of any inherent limitations on its value. Thus, the approach stresses that translated literature may, and sometimes does, occupy a central position in the polysystem and is therefore capable of providing canonised models for the whole polysystem. Moreover, given that polysystem theory recognises that intra- and inter-relations exist within both systems and polysystems, leading to various types of interference and transfer of elements, models, canons, and so on, it becomes obvious that "semiliterary texts, translated literature, children's literature - all those strata neglected in current literary studies - are indispensable objects of study for an adequate understanding of how and why transfers occur within systems as well as among them" (ibid: 25). And finally, polysystemists reject the popular view of translation as a derivative activity and stress instead that literary translation is "a creatively controlled process of acculturation in that translators can take an original text and adapt it to a certain dominant poetics or ideology in the receiving culture" (Heylen 1993:21).

This view of literature as a conglomerate of systems, as well as the growing interest in transfer and interference across systems, has gradually undermined the status of the source text in translation studies. Since the early eighties, Toury, another Tel-Aviv scholar, has been stressing that a translation belongs to one textual system only, namely the target system, and the source text has gradually been assuming the role of a stimulus or source of information rather than the starting point for analysis. Questions regarding how a translated text came into being or what type of relationship it has with a given source text are becoming secondary to its classification as part of the target textual system. As Toury puts it in a more recent publication (1985:19):

> It is clear that, from the standpoint of the source text and source system, translations have hardly any significance at all, even if everybody in the source culture 'knows' of their factual existence ... Not only have they left the source system behind, but they are in no position to affect its linguistic and textual rules and norms, its textual history, or the source text as such. On the other hand, they may well influence the recipient culture and language, if only because every translation is initially perceived as a target-language utterance.

It is worth noting that similar, though not quite so radical, assessments of the status of the source text have also emerged among groups of scholars not specifically concerned with literary translation. For example, Vermeer (1983: 90)[2] suggests that the function of the translated text is determined by the interests and expectations of its recipients and not by the function of the source text. The SL text is a source of information and, like other sources of information, it may be exploited in a variety of ways to meet the expectations of an envisaged audience.

1.2.2 From equivalence to norms

From the late seventies onwards, the source-oriented notion of equivalence has been gradually replaced by notions which clearly take the target system and culture as a starting point. Some of these notions have evolved within theories designed to account for translation within a commercial environment. They include, for example, Vermeer's notion of coherence, defined as the agreement of a text with its situation (Vermeer 1983), and Sager's definition of equivalence as a function of the specifications that accompany a request for translation (Sager 1993). The most important, however, has been the notion of norms, introduced by Toury (1978, 1980).

Toury has developed a tripartite model in which norms represent an intermediate level between competence and performance. If we think of competence as an inventory of all the options that are available to translators in a given context, and performance as the subset of options which are actually selected by translators from this inventory, then norms are a further subset of these options. They are options which are regularly taken up by translators at a given time and in a given socio-cultural situation. In this sense, the notion of norms is very similar to that of typicality, a notion which has emerged from recent work on corpus-based lexicography and which contrasts sharply with the standard, absolute dualisms in linguistics: competence and performance, langue and parole.

Norms, then, are a category of descriptive analysis. They can be identified only by reference to a corpus of source and target texts, the scrutiny of which would allow us to record strategies of translation which are repeatedly opted for, in preference to other available strategies, in a given culture or textual system. The concept of norms tips the balance not only in favour of the target text (as opposed to the traditional obsession with the source text), but, more important, it assumes that the primary object of analysis in translation studies is not an individual translation but a coherent corpus of translated texts. Norms do not emerge from a source text or a body of source texts. Equally, they do not emerge from the target system nor from a general collection of target texts. They are a product of a tradition of translating in specific ways, a tradition which can only be observed and elaborated through the analysis of a representative body of translated texts in a given language or culture. They can therefore be seen not just as a descriptive category but also as providing a functional, socio-historical basis for the structure of the discipline (Lambert 1985:34).

1.2.3 The rise of descriptive translation studies

Since the seventies, several scholars have begun to express dissatisfaction with the heavy reliance on introspective methods in translation studies. Holmes (1988:101) makes the point most clearly:

> Many of the weaknesses and naiveties of contemporary translation theories are a result of the fact that the theories were, by and large, developed deductively, without recourse to actual translated texts-in-function, or at best to a very restricted corpus introduced for illustration rather than for verification or falsification.

Newman (1980:64) similarly suggests that the way out of the dilemma posed by the notions of equivalence and translatability is to look at actual instances of translation and to determine, on the basis of those instances, "the kind of generalities that might form the basis of a theory of competence or systematic description". It is however Toury who has done more to elaborate the concept of descriptive translation studies than anyone else in the discipline.

For Toury, it is vital for translation studies to develop a descriptive branch if it is ever to become an autonomous discipline. Without this, translators will continue to rely on other disciplines such as linguistics to provide them with theoretical frameworks and the means to test their hypotheses. Descriptive Translation Studies, or DTS for short, is not reducible to a collection of case studies or comparative analyses of source and target texts. It is that branch of the discipline which must provide a sound methodology and explicit research procedures to enable the findings of individual descriptive studies to be expressed in terms of generalisations about translational behaviour. Its agenda consists, primarily, of investigating what translation is "under any defined set of circumstances ... and WHY it is realized the way it is" (Toury 1991a:186). One of its main objectives is to render the findings of individual studies intersubjective and to make the studies themselves "repeatable, either for the same or for another corpus" (Toury 1980:81).

It is perhaps worth noting at this point that although the words *corpus* and *corpora* are beginning to figure prominently in the literature on translation, they do not refer to the same kind of corpora that we tend to talk about in linguistics. Corpora in translation studies have so far been very modest affairs. Their size is not generally expressed in terms of number of words but of number of texts, and they are searched manually. For example, Vanderauwera (1985) is a study of "50 or so novels" translated from Dutch into English in "roughly the period 1960-1980" (ibid:1-2). This is a very small corpus, and yet the experience of searching it manually leads Vanderauwera to suggest that "serious and systematic research into translated texts is a laborious and tiresome business" (ibid:6). Toury himself seems torn between the need to set an ambitious program for DTS and the recognition that "the larger and/or more heterogeneous the corpus, the greater the difficulties one is likely to encounter while performing the process of extraction and generalization" (1980:66-7). In an earlier publication, Toury (1978:96) argues for a distributional study of norms based on statistical techniques but concludes that

> ... as yet we are in no position to point to strict statistical methods for dealing with translational norms, or even to supply sampling rules for actual research (which, because of human limitations, has nearly always been applied to samples only, and will probably go on being carried out in much the same way). At this stage we must be content with our intuitions ... and use them as keys for selecting a corpus and for hitting upon ideas.

One of John Sinclair's major achievements for linguistics has been his success, through the collection of computerised corpora and the development of a relevant research methodology, in providing ways of overcoming our human limitations and minimising our reliance on intuition. His work can provide solutions for precisely the kind of problems that translation scholars are still struggling with today.

2. Corpus work in translation studies: the potential

There is no doubt that the availability of corpora and of corpus-driven methodology will soon provide valuable insights in the applied branch of translation studies, and that the impact of corpus-based research will be felt there long before it begins to trickle into the theoretical and descriptive branches of the discipline. Sinclair (1992:395) touches very briefly, and strictly from the point of view of a linguist, on one obvious application:

> The new corpus resources are expected to have a profound effect on the translations of the future. Attempts at machine translation have consistently demonstrated to linguists that they do not know enough about the languages concerned to effect an acceptable translation. In principle, the corpora can provide the information.

In the above statement, which is one of the very few Sinclair has made on translation, the concern is merely with improving the performance of translators and of machine translation systems in terms of approximating to the structures and natural patterns of a given language or languages. This same concern underlies most of the expressions of interest in corpus studies which are beginning to take shape in the literature.[3] It is of course a legitimate concern and one which will be shared widely by scholars within and outside translation studies, theorists and practitioners alike. I would, however, like to think that the 'profound effect' which Sinclair refers to will not be understood merely in terms of knowing enough about the languages concerned to approximate to their patterns. After all, once we are in a position to describe and account for our object of study, namely translation, we might find that approximating to the patterns of the target language, or any language for that matter, is not necessarily as feasible as we seem to assume, and that it is not the only factor at play in shaping translational behaviour. Several scholars, for example Toury (1991b:50) and Even-Zohar (1979:77) have already noted that the very activity of translating, the need to communicate in translated utterances, operates as a major constraint on translational behaviour and gives rise to patterns which are specific to translated texts. Thus Even-Zohar (ibid) stresses that "we can observe in translation patterns which are inexplicable in terms of any of the repertoires involved", that is patterns which are not the result of interference from the source or target language. Examples of these patterns are discussed as *universal features of translation* in section 2.1 below. The profound effect that corpora will have on translation studies, in my view, will be a consequence of their enabling us to identify features of translated text which will help us understand what translation is and how it works. The practical question of how to improve our translations will find more reliable and realistic answers once the phenomenon of translation itself is explained in its own terms.

Practical applications aside, what kind of queries can access to computerised corpora help us resolve in our effort to explicate the phenomenon of translation? Given that this question, to my knowledge, has not been addressed before, what follows has to be seen as a very tentative list of suggestions which can provide a starting-point for corpus-based investigations in the discipline but which do not, by any means, address the full potential of corpora in translation studies.

2.1 Universal features of translation

The most important task that awaits the application of corpus techniques in translation studies, it seems to me, is the elucidation of the nature of translated text as a mediated communicative event. In order to do this, it will be necessary to develop tools that will enable us to identify universal features of translation, that is features which typically occur in translated text rather than original utterances and which are not the result of interference from specific linguistic systems.

It might be useful at this point to give a few examples of the type of translation universals I have in mind. Based on small-scale studies and casual observation, a number of scholars have noted features which seem, intuitively, to be linked to the nature of the translation process itself rather than to the confrontation of specific linguistic systems. These include:

(i) A marked rise in the level of explicitness compared to specific source texts and to original texts in general (see for instance Blum-Kulka 1986:21). In Baker (1992), I discussed several examples of translations which build extensive background information into the target text. In one case (*Autumn of Fury: the Assassination of Sadat* by Mohamed Heikal; 1983:3), a simple clause - *The example of Truman was always present in my mind* - is rendered into Arabic as follows:

> In my mind there was always the example of the American President Harry Truman, who succeeded Franklin Roosevelt towards the end of World War II. At that time - and after Roosevelt - Truman seemed a rather nondescript and unknown character who could not lead the great human struggle in World War II to its desired and inevitable end. But Truman - faced with the

> challenge of practical experience - grew and matured and became one of the most prominent American presidents in modern times. I imagined that the same thing could happen to Sadat.

Toury (1991b:51) accepts that 'explicitation'[4] is a feature of all kinds of mediated events, including interaction in a foreign language, but wonders whether there are any differences in the level and nature of explicitation by, for instance, language learners vs. translators, professional vs. non-professional translators, or in oral vs. written translation. The techniques currently available in corpus linguistics can in principle cope with such tasks as the measurement of expansion in two or more corpora. Moreover, specific conventions and software tools are now available for recording and retrieving automatically such things as information about a writer or speaker, for example his/her first language, nationality, and gender.

(ii) A tendency towards disambiguation and simplification. For example, Vanderauwera (1985:97-8) notes that, in her corpus of English translations of Dutch novels, potentially ambiguous pronouns are replaced by forms which allow more precise identification, and difficult syntax is made easier. Similarly, she reports that "where quotation marks fail to distinguish a person's speech or thought in the source text, they are almost invariably restored in the target text" (ibid:94). Again, corpus linguistics now has the necessary tools to measure such things as simple vs. complex syntax, and the automatic retrieval of pronominal as opposed to precise forms of identification, as well as elements of punctuation such as quotation marks, is a fairly simple operation in terms of current corpus techniques.

(iii) A strong preference for conventional 'grammaticality'. In interpreting, that is oral translation, this manifests itself in an overriding tendency to round off unfinished sentences, 'grammaticise' ungrammatical utterances and omit such things as false starts and self-corrections, even those which are clearly intentional in a courtroom context (Shlesinger 1991:150). Vanderauwera (1985:93) records a similar tendency towards general textual conventionality in a corpus of English translations of Dutch novels.

(iv) A tendency to avoid repetitions which occur in source texts, either by omitting them or rewording them (Shlesinger 1991, Toury 1991a). Toury (ibid:188) reports this feature as "one of the most persistent, unbending norms in translation in all languages studied so far".

(v) A general tendency to exaggerate features of the target language. For example, binomials composed of synonyms or near-synonyms, which are a common feature of Hebrew writing, tend to occur more frequently in translated than in original Hebrew texts and to replace non-binomials in source

texts (Toury 1980:130). Vanderauwera (1985:11) suggests that translations overrepresent features of their host environment in order to make up for the fact that they were not originally meant to function in that environment.
(vi) Point (v) above not withstanding, it has been shown that the process of mediation often results in a specific type of distribution of certain features in translated texts vis-à-vis source texts and original texts in the target language. For example, Shamaa (1978:168-71) reports that common words such as *say* and *day* occur with a significantly higher frequency in English texts translated from Arabic than they do in original English texts. At the same time, their frequency of occurrence in the translated English texts is still considerably lower than the frequency of the equivalent Arabic items in the source texts. Shamaa suggests that, although subtle and elusive, this unusual distribution of features contributes to the identification of a text as a translation and "leave[s] a vague impression of being culturally exotic" (ibid:172). It is a symptom of what is sometimes referred to as "the third code" (Frawley 1984:168), which is a result of the confrontation of the source and target codes and which distinguishes a translation from both source texts and original target texts at the same time.[5] Another example of the operation of the 'third code' is provided by Blum-Kulka (1986:33) who suspects that research is likely to reveal "that cohesive patterns in TL texts are neither TL nor SL norms oriented, but form a system of their own". Research into the nature of the third code in translation might also provide answers to queries about the nature of 'pseudotranslations', that is texts which are regarded as translations but for which no genuine source texts exist.

It will clearly take some time and ingenuity to develop a corpus methodology for capturing universal features of the type discussed above. But, assuming that we have a corpus of texts translated into, say, English from a variety of languages, we might attempt to isolate patterns which
(i) occur across the corpus, irrespective of whether the source texts are French, Hebrew or Chinese,
(ii) do not occur, or do not occur to the same degree/with the same frequency, in original English texts.
We then need to compare our results with the results of similar research carried out on corpora of translated texts in, let us say, French and German. As a starting point, this will help us identify certain patterns or tendencies which occur in the three corpora and which, therefore, are good candidates for universal features of translation. Further research on similar corpora in other languages will either confirm or disprove our hypotheses.

2.2 Translational norms operating in a given socio-cultural context

Universal features such as those discussed in 2.1 above can be seen as a product of constraints which are inherent in the translation process itself, and this accounts for the fact that they are universal (or at least we assume they are, pending further research). They do not vary across cultures. Other features have been observed to occur consistently in certain types of translation within a particular socio-cultural and historical context. These are a product of norms of translation which represent another type of constraint on translational behaviour (see section 1.2.2 above). Like universal features of translation, textual exponents of translational norms can be efficiently and reliably investigated if access to computerised corpora is made available to scholars in the field.

Toury (1978) distinguishes between various types of norms which operate at different stages of the translation process. The techniques of corpus linguistics are particularly suited to the identification of what Toury calls 'operational norms': norms which "affect the matrix of the text, that is, the modes of distributing linguistic material [matricial norms] ..., and the actual verbal formulation of the text [textual norms]" (ibid:87). Matricial norms cover the occurrence of omissions, additions, substitutions and transpositions in translated texts. For instance, Toury (1980:127) reports that in translations of prose fiction from English and German into Hebrew during the years 1930-45, there was a tendency to omit substantial subtitles (that is subsidiary rather than main titles of literary works), and that this contrasts with a tendency to add subtitles of a certain type in literary translation into French in the early eighteenth century.

Textual norms have been discussed more widely in the literature, and examples include the following:

(i) Toury (1980:129) reports, in the same corpus of Hebrew translations of prose fiction, a high level of dependence on a repertory of fixed collocations derived from canonised religious texts. The same corpus also reveals that special importance was attached to direct speech: pieces of dialogue were regularly turned into independent paragraphs, indirect speech was replaced by direct speech, and phrases which indicated a move from narration to dialogue were omitted.

(ii) In a corpus of 425 mystery books translated into Hebrew since the early sixties, Toury (ibid:104) identifies a strong norm which he expresses as "The title should never be too complex, witty or sophisticated". This norm manifests itself in two ways. First, sophisticated titles (in terms of lexis) are replaced by simple titles which contain one of a stock of items that include the Hebrew equivalents of 'mystery', 'murder', 'blood', 'death', and so on. Thus, *The Case of the Ice-Cold Hands* becomes "The Mystery of the Murder in the Motel". Second, the Hebrew titles follow a limited range of simple syntactic patterns, notably 'the x of y', as in the above example, and 'the x's y' as in "The Black Candle's Mystery".

(iii) In a corpus of Dutch novels translated into English, Vanderauwera (1985:93) notes that foreign words and dialogue which occur in the source texts are either replaced in the translation by target language items (for example Malay *klamboe* is replaced by *mosquito curtain*) or glossed in the target language.

(iv) Based on a study of a limited corpus of translations of modern, non-literary English texts in a variety of languages, I have suggested (Baker 1992:36) that Japanese seems far more tolerant of the use of loan words in translation than, for instance, Arabic and French. The norms that govern translational behaviour in the three languages are noticeably different in this regard.

2.3 Other issues for corpus research

Apart from universal and norm-oriented features of translational behaviour, access to computerised corpora should enable us to explore a number of other theoretical issues which are difficult to deal with on the basis of small-scale studies.

First, there is the question of the intermediate stages of translation, or how the final product evolves over a period of time. Recent developments in the working practices of professional translators will soon make this type of research feasible. For instance, the Institute of Translation and Interpreting of Great Britain is currently considering a proposal to adapt a variant of the British Standard BS5750[6] for professional translators working in a variety of contexts: as staff translators, on their own or through agencies or co-operatives. If adopted, this standard will require translators to follow certain routines in documenting the work they receive and despatch, whether in manuscript or disk form, and to maintain clearly labelled and dated versions of each translation. Access to this type of text in electronic form can be used to explore the process of translation through a retrospective analysis of successive versions of the product. New software tools, including tools for fuzzy matching (that is identifying stretches of text which are similar but not identical), are currently being developed by corpus linguists. Svanholm (1992) reports that IBM already offer their translators fuzzy matching software as a basic tool for revising and updating existing translations of large documents.

Second, central questions which have been on the agenda for decades can be resolved more efficiently and reliably through the investigation of large computerised corpora. These questions include the size and nature of the unit of translation, the type of equivalence which is achieved in practice and the level at which it is achieved. Again, new software tools are now available for the investigation of parallel corpora, that is corpora of source texts and their translations (see for example Marinai et al 1991, Brown & Cocke 1988, Church & Gale 1991). New and sophisticated methodologies are also evolving for investigating the nature and limits of equivalence on the basis of comparable corpora, that is corpora of comparable original texts in several languages. Sinclair has pioneered the investigation of comparable corpora through a project on multilingual lexicography (Sinclair 1991).

3. Conclusion

I have argued in this paper that translation studies has reached a stage in its development as a discipline when it is both ready for and needs the techniques and methodology of corpus linguistics in order to make a major leap from prescriptive to descriptive statements, from methodologising to proper theorising, and from individual and fragmented pieces of research to powerful generalisations. Once this is achieved, the distinction between the theoretical and applied branches of the discipline will become clearer and more convincing.

There is now an urgent need to explore the potential for using large computerised corpora in translation studies. It seems to me that most of the components for realising this potential are in place. The emphasis has shifted from meaning to usage, and the notion of equivalence is gradually giving way to that of norms. The status of the source text has been undermined and we have managed to make the leap from source-text-bound rules and imperatives to descriptive categories. There is increasing interest in features of translated texts per se and we are beginning to develop a descriptive branch of the discipline with well-defined objectives and an explicit program. What we need is a research methodology and a set of tools that can help us put this program into action. A suitable methodology and a set of very powerful and adaptable tools are now available from corpus linguistics.

Author's Address:
Cobuild, Westmere, 50 Edgbaston Park Road, Birmingham. B15 2RX. U.K.

Notes

1. INL Working Paper 92-11. J. G. Kruyt & E. Putter, Corpus Design Criteria: report submitted to the European Commission by the Instituut voor Nederlandse Lexicologie, Leiden, as a contribution to a European enquiry into corpus design criteria.
2. I do not have access to the original German text quoted in the references, but have relied on an unpublished translation by Professor J. C. Sager, UMIST, of short extracts from various German sources.
3. For example, Lindquist (1984) is a corpus-based study of adverbials which aims not at explicating translational behaviour but at providing a 'translation grammar' which can guide translators' choices.
4. This is the term used in the literature.
5. In some cases, when an unusual distribution of features is clearly a result of the translator's inexperience or lack of competence in the target language, this phenomenon is referred to as 'translationese'.
6. BS5750 is a British standard for quality assurance. It specifies a set of controls which have to be implemented in order for an organisation to claim that it is 'quality assured'. It is therefore a form of company certification, guaranteeing that the final product marketed by a given company meets the national standard.

References

Baker, M. 1992. *In Other Words: A Coursebook on Translation*. London & New York: Routledge.

Blum-Kulka, S. 1986. "Shifts of Cohesion and Coherence in Translation". *Interlingual and Intercultural Communication: Discourse and Cognition in Translation and Second Language Acquisition Studies* ed. by J. House & S. Blum-Kulka, 17-35. Tübingen: Gunter Narr.

Brown, P. & J. Cocke. 1988. "A Statistical Approach to Language Translation". *Proceedings of the 12th International Conference on Computational Linguistics: COLING '88* ed. by D. Vargha, 71-6. Budapest.

Church, K. & W. Gale. 1991. "Concordances for Parallel Text". *Using Corpora: Proceedings of the Seventh Annual Conference of the UW Centre for the New OED & Text Research.* St. Catherine's College: Oxford.

Eco, U. 1976. *A Theory of Semiotics*. Bloomington & London: Indiana University Press.

Even-Zohar, I. 1978. "The Position of Translated Literature within the Literary Polysystem." *Literature and Translation* ed. by J. S. Holmes, J. Lambert & R. van den Broeck 117-27. Leuven:ACCO.

__________. 1979, 1990. "Polysystem Theory." *Poetics Today (Special Issue on Polysystem Studies by Itamar Even-Zohar)* 11, 1. 9-26.

Firth, J. R. 1956, 1968. "Linguistics and Translation" *Selected Papers of J. R. Firth 1952-59* ed. by F. R. Palmer, 84-95. London: Longman. (first read at Birbeck College, London in 1956).

Frawley, W. 1984. "Prolegomenon to a Theory of Translation". *Translation: Literary, Linguistic and Philosophical Perspectives* ed. by W. Frawley, 159-75. London & Toronto: Associated University Presses.

Haas, W. 1968. "The Theory of Translation" *The Theory of Meaning* ed. by G. H. R. Parkinson , 86-108. Oxford: Oxford University Press.

Heylen, R. 1993. *Translation, Poetics, and the Stage: Six French Hamlets*. London & New York: Routledge.

Holmes, J. S. 1988. *Translated! Papers on Literary Translation and Translation Studies*. Amsterdam: Rodopi.

Lambert, J. 1991. "Shifts, Oppositions and Goals in Translation Studies: Towards a Genealogy of Concepts." *Translation Studies: The State of the Art* ed. by K. M. van Leuven-Zwart & T. Naaijkens, 25-37. Amsterdam: Rodopi.

Marinai, E., C. Peters & E. Picchi. 1991. "Bilingual Reference Corpora: A System for Parallel Text Retrieval". Unpublished manuscript. Istituto di Linguistica Computazionale: Pisa, Italy.

Lindquist, H. 1984. "The Use of Corpus-based Studies in the Preparation of Handbooks for Translators". *Translation Theory and its Implementation in the Teaching of Translating and Interpreting* ed. by W. Wilss & G. Thome, 260-70. Tübingen: Gunter Narr.

Newman, A. 1980. *Mapping Translation Equivalence*. Leuven: ACCO.

Sager, J. C. 1993. *An Extended Communicative Theory of Translation*. Unpublished Manuscript, University of Manchester Institute of Science and Technology.

Shamaa, N. 1978. *A Linguistic Analysis of Some Problems of Arabic to English Translation*. D. Phil. Thesis, Oxford University.

Shlesinger, M. 1991. "Interpreter Latitude vs. Due Process. Simultaneous and Consecutive Interpretation in Multilingual Trials". *Empirical Research in Translation and Intercultural Studies* ed. by Sonja Tirkkonen-Condit, 147-55. Tübingen: Gunter Narr.

Sinclair, J. M. 1991. *Council of Europe Multilingual Lexicography Project*. Unpublished report submitted to the Council of Europe under contract no. 57/89.

___________.1992. "The Automatic Analysis of Corpora". *Directions in Corpus Linguistics: Proceedings of Nobel Symposium 82, Stockholm 4-8 August 1991* ed. by J. Svartvik 379-97. Berlin & New York: Mouton de Gruyter.

Toury, G. 1978. "The Nature and Role of Norms in Literary Translation." *Literature and Translation* ed. by J. S. Holmes, J. Lambert & R. van den Broeck 83-100. Leuven: ACCO.

________. 1980. *In Search of a Theory of Translation*. Tel Aviv: The Porter Institute for Poetics and Semiotics.

________. 1985. "A Rationale for Descriptive Translation Studies". *The Manipulation of Literature: Studies in Literary Translation* ed. by T. Hermans, 16-41. London & Sydney: Croom Helm.

________. 1991a. "What are Descriptive Studies into Translation Likely to Yield apart from Isolated Descriptions." *Translation Studies: The State of the Art* ed. by K. M. van Leuven-Zwart & T. Naaijkens 179-92. Amsterdam: Rodopi.

_______. 1991b. "Experimentation in Translation Studies: Achievements, Prospects and Some Pitfalls". *Empirical Research in Translation and Intercultural Studies* ed. by Sonja Tirkkonen-Condit, 45-66. Tübingen: Gunter Narr.

Vanderauwera, R. 1985. *Dutch Novels Translated into English: The Transformation of a "Minority" Literature*. Amsterdam: Rodopi.

Vermeer, H. J. 1983. *Aufsätze zur Translationstheorie*. Heidelberg: Groos.

3. Text and Technology: Computational Tools

A Prototype Boundary Marker

David Coniam
Chinese University of Hong Kong

Abstract

This paper describes the development of a software tool which is designed to parse text on the basis of the orthographic space between words. The task of this tool is to determine what type of syntactic boundary is marked by each space: a word boundary, a group boundary or a clause boundary. This parser is not envisaged as a stand-alone tool; it forms part of a complementary suite of programs discussed in Sinclair 1992.

0. Introduction

Sinclair (1992) lays the foundation for the next generation of analytic software in corpus linguistics and sets general guidelines for software standards. These guidelines include, among other requirements, that (a) analysis should be restricted to what the machine can do without human intervention, (b) analysis should be done in real time, and (c) operations should be designed to cope with unlimited quantities of text material (ibid:396). These requirements, in addition to the fact that "parsing is a very complex business", lead Sinclair to suggest the following (ibid:383-4):

> It may be expedient, as we move to absorb the evidence of long texts, to lower the priority of an all-purpose comprehensive parser and consider in its place a number of simple operations. Each one would contribute to an eventual parse strategy, but in the meantime each can give some valuable information for research and for limited applications.

This paper describes the development of a partial parser, conceived and designed as part of a suite of software tools described briefly by Sinclair in the above publication. The partial parser in question is a boundary marker, a syntactic tool which attempts to interpret the space between words. Its original brief was to assign a word, group or clause boundary between any two words. Like the other software tools under development, the current program has a small-scale, very finite and unsophisticated task to do, in an artificial intelligence sense of semantic and real-world knowledge. It is first-order Markovian in that the majority of its decisions are made on the basis of the current word and one word either side of the current word. On this limited basis, it attempts to deliver as much information as possible about word boundaries.

Research in the boundary marking project commenced in late 1990. The current paper presents an overview of its progress to date, along with potential refinements.

1. Natural language parsing: the state of the art

Gazdar & Pullum state that "there can be no parsing without a grammar" (1985:274). Grammar, or rules in some form, is the cornerstone of most natural language processing systems. Some systems are syntactically-oriented and some are semantically-oriented. Those based on syntactic rules include, for example, Augmented Transition Networks: LUNAR (Woods 1970); Phrase Structure Grammars: Linguistic String Project (Sager 1981); Transformational Grammar: Petrick's Transformational Parser (Petrick 1973); quasi-TG: PARSIFAL (Marcus 1980); Chart Rewrite Parser: Functional Unification Grammar (Kay 1982). Semantically-oriented systems include, for example, Conceptual Analysers: ELI (Riesbeck 1978); BORIS (Dyer 1983). The main feature that all these systems have in common is a tremendous number of rules that are specific to their formalism. There is one system that appears to have fewer formal syntactic or semantic rules: the Word Expert Parser (Small 1983). This, however, appears to have an extremely large knowledge database from which its interpretations of generally very limited input are formed.

The argument has further tended to be that parsing which integrates syntactic and semantic analyses represents a more sensible use of resources and is closer to the manner in which humans process language (see for instance Charniak 1983). This argument rests on the fact that, in spite of early TG claims, we do not appear first to compute a syntactic representation of the input and then pass it to a semantic and pragmatic processor for interpretation against the real world. As far as computer processing of language is concerned, it may well be feasible to integrate syntactic and semantic analyses for short input sentences. Where the input is a lengthy text, however, such to-ing and fro-ing may not be easy to achieve, since considerable interaction would need to take place between the components of a natural language processing system.

The parsing of natural language has therefore been generally viewed as an extremely complex undertaking, and parsers, as outlined above, have been very large and monolithic affairs which have attempted to encompass the many aspects of linguistic and non-linguistic knowledge in one platform. A good case in point is the often-quoted SHRDLU (Winograd 1972).[1] With the advent of corpora, researchers have come to realise that monolithic machines, which have traditionally performed in-depth analysis of rather short input sentences, are extremely unsuitable for the parsing of unrestricted text. New tools, using different strategies, are now beginning to appear. Examples include CLAWS (Garside et al 1987); Fidditch (Marcus & Santorini 1992), and Constraint Grammar (Karlsson 1990).

The suite envisaged by Sinclair and mentioned briefly in section 0 above is also a set of tools which have been conceptualised along different lines from pre-corpora parsers. The premise is that it is desirable to have a set of small-scale software tools, each having a finite task to do, and each passing its processed data on to other tools in the suite for refinement and interpretation (Sinclair 1992). These tools should be self-sufficient, robust, and flexible. The fact that they are stand-alone tools offers the added advantage that each can be readily amended or refined without affecting the operation of the other tools in the suite.

The idea that each partial parser can have a limited and finite task to do suggests that parsing can be achieved by a system which does not incorporate innumerable rules. One objective of the project described in this paper is, therefore, to investigate the extent to which Gazdar and Pullum's dictum, as quoted above, can be circumvented. The question, in other words, is: how much can be achieved in the analysis of language without rules, or with as few rules as possible?

2. The boundary marker: system overview

The original prototype of the boundary marking program, Version 1, is outlined below in some detail. Certain design features have been substantially modified or superseded over the past few months, but the component program, the structure of the data files, and the way those files interact have laid the foundations for a successful implementation of the boundary marking concept. The current implementation of the system differs in some respects from its predecessor, but the essential design remains the same.

The system consists of a set of data files and the main operating program.

The data files consist of: a set of text files, a dataset of word-class information, the lexicon of known words and an affix dataset.

2.1 Text files

The texts to be analysed are short horoscopes from the British newspaper *The Daily Mirror*. The average length per text is 60 words. Here is an example.

> A bright new future could be opening up. But your natural caution may make you hesitate a shade too long. So strike while the iron is hot - for you may not get such an opportunity again. Support comes your way through a close friend or neighbour.

2.2 The dataset of word-class information

This dataset is based on the list of special classes from the Introduction to *The Collins COBUILD English Language Dictionary* (Sinclair et al 1987). The task of this part of the system is to record very specific information about word combinatory possibilities. The dataset therefore consists of valency information on word classes, that is, the word classes with which a particular word class can coexist.

The word-class categories are arranged as far as possible into major groupings, with each category assigned its own unique code, namely a mnemonic four-digit number. For example, predicative adjectives all begin with 11 and attributive adjectives with 12. Qualitative predicative adjectives are assigned the code 1150 and classifying predicative adjectives the code 1120. In addition, each category has 14 data entry slots assigned to it as follows:

(i) three slots for the category of word which can precede a given category. For example, *big*, as a qualitative attributive adjective, can be preceded by an adverb, a determiner, or a verb.
(ii) three slots for the gap each preceding category expects. Thus, in the above example of *big*, a preceding adverb would expect a word gap, and so would a preceding determiner, but a preceding verb would expect a group gap.
(iii) three slots for the category of word that can follow a given category.
(iv) three slots for the gap each following category expects.
(v) one slot for whether a category can be clause initial, with *Y* and *N* indicating *yes* and *no* respectively.
(vi) one slot for whether a category can be clause final.

The original set of special classes in the *Collins COBUILD English Lan-*

guage Dictionary ran to 94 entries. These were reduced to 82 categories for the purposes of Version 1 of the program and systematised in various ways:[2]

(i) certain words were given entries of their own, for example, *of* and *to*.
(ii) categories of a similar nature, such as qualitative predicative and classifying predicative adjectives, were grouped together at one level: they shared the first two digits of the mnemonic code assigned to them (*11*50 and *11*20 respectively in this case).
(iii) certain categories were, on the other hand, further subdivided. The category of coordinating conjunction was split into 'coordinating-conjunction-*but*' and 'coordinating-conjunction-*and*', on the grounds that *but* is always preceded by a clause gap whereas the boundaries of *and* are more varied.

The above grouping together and subdividing of categories was done on two grounds. The grouping together of categories supports a simple fall-back position by having a superordinate category in case an exact match is not found. In the case of adjectives, for instance, the fall-back position is that of 'general adjective'. Suppose the system is attempting to analyse *an expensive woollen sweater* and does not find an exact match between *expensive,* a qualitative adjective, and *woollen,* a classifying adjective. The fall-back position is provided by a match between the qualitative adjective *expensive* and the general adjective *woollen.* The further subdividing of categories was done on the grounds that the more exact a category is, the more precise will be the match, once a match is found. Consequently, the type of boundary involved is identified more accurately.

2.3 The lexicon of known words

This consists of the 100 most common words in English as detailed in Sinclair (1991:143), plus a number of the commonest elements of punctuation, and the words that appear in the sample texts for analysis. Each word is allowed up to three separate slots, one for each different category that the word can belong to. For example, *the* can only be a determiner, and hence has only one slot assigned to it; but *as* can be a subordinating conjunction, an adverb or a preposition, and hence has three separate slots. The category slots for each word are arranged in the same order as that given in the *Collins COBUILD English Language Dictionary*. Generally speaking, the most frequent senses/word classes of a particular word are given first, followed by less frequent ones. The entries for the various words are set up as follows:

(i) syntactic valency information on each category that the word can belong to, as outlined in 2.2 above.

(ii) three slots for preceding close collocates, plus three slots for the likely syntactic boundary between each preceding collocate and the word in question.

(iii) three slots for close-following collocates, again with three slots for the likely syntactic boundary between each collocate and the word in question.

The specification of close collocates was based on inspecting 500 KWIC concordance lines from the Bank of English[3] for each word.

2.4 The affix dataset

The system has two simple datasets which consist of the 30 commonest prefixes and 62 commonest suffixes in English. These provide a fall-back position for when the system meets a word which is not in its lexicon. The datasets are set up according to the type of word class that a particular affix is likely to be, together with the likely gap that there might be between the word and a preceding or following word class. For example, *-er* can be an adjectival or nominal suffix, *-ise* is a verbal suffix, *re-* a verbal prefix, and so on. As an adjectival suffix, *-er* can be preceded by an adverb (with a word gap), an adjective (a word gap), or a determiner (also a word gap). It can be followed by a noun (with a word gap), an adjective (a word gap), or a verb (with a group gap). The contents of the datasets are arranged so as to avoid, as far as possible, any duplication of information. For instance, adjectives are ordered so that the system would read *-ical* before *-ial* before *-al*; otherwise with an adjective like *numerical* the system would get loaded with information for both *-ical* and *-al*.

3. The boundary marker: program operation

When a given text is input to the boundary marker, the system proceeds by examining each word against its lexicon of known words (see section 2.3 above). If the word is found in the lexicon, the system moves to fill information-holding arrays for the current word and the current-plus-one word. These arrays are filled with whatever detail the lexicon has on the category class that the current and current-plus-one words might belong to, together with details on close-collocating preceding and following words.

If the word is not found on first examination, it is sent to a routine which strips off common endings such as *-s*, *-ing*, and *-ed*. After some rudimentary tidying-up of the base, which might include, for example, reducing double consonants to single ones or changing '*e*' and '*i*' endings to '*y*', the stripped word is checked against the lexicon once more. If the system still fails to find it, some elementary disambiguation of possible word classes is attempted. For example, an attempt might be made to decide whether an *-ing* ending is adjectival, verbal or nominal. If no match is found, the final check is to examine the word against the affix datasets. If an affix match is found, and that affix has more than one word class possibility (*re-*, for example, can be nominal or verbal), elementary word-class disambiguation is then attempted again.

3.1 Assigning boundaries

Once a word is found, the next stage is to assign the boundaries between it and those words which precede and follow it. The success of this part of the program depends on whether or not an actual match is found between the data being parsed and the grammatical and collocational information held in the lexicon.

The most direct match is achieved with a close-collocating preceding or following word. As explained in 2.3 above, the lexicon holds information not only on the collocates of each word but also on the boundary that is deemed to be normal between each two collocates. If such a close-collocating match is found, the boundary held in the word's entry in the lexicon is set as the current gap. Note that close collocates can be morphological as well as lexical elements. For example, *be* and *go* have the structural element *-ing* as a following close collocate.

A match can also be achieved on the basis of syntactic categories. As outlined in 2.1, each syntactic category is assigned a four-digit number, and categories which appear to share similar boundary characteristics are grouped together so that the first two digits display the common element in a larger grouping. An entry such as 1203, 'classifying attributive adjective', would have 1305, 'adverb plus adjective/adverb', as one of the categories that it is likely to be preceded by. The sequence ... *very new* ... would therefore result in the following analysis:

- *very* has 1305 as one if its category types; category 1305 has 1203 as one of its 'followed by' categories.
- *new* has 1203 as one of its category types; category 1203 has 1305 as one of its 'preceded by' categories.

A four-digit match between the two categories 1305 and 1203 is therefore found and the appropriate boundary set. If, however, no exact four-digit match is found between two words, the category numbers are re-examined on the basis of their first two digits. In other words, a match is sought against the superordinate member of the category.

In some instances, a number of matches (generally two-digits) may be found. For example, the gap between *open* and *water* in the sequence *open water* can be interpreted in at least three ways:

(i) VERB -VERB (clause gap with a value of 3)
If the gates *open water* the plants.

(ii) VERB - NOUN (group gap with a value of 2)
Open water canisters now.

(iii) ADJECTIVE - NOUN (word gap with a value of 1)
Open water canisters leak.

If the parser is unable to disambiguate the three contexts, it simply adds up the values of the gaps between the two words (in this case six) and divides the result by the number of instances (three) to achieve an average of two, i.e. a group gap. This is a rather unsatisfactory matter which is currently receiving attention.

If a sequence of three known words offers no information in terms of a two-digit match against the current word, the following and preceding gap expectations are averaged out. If the sequence involves two unknown words, no guess is made and the boundary is defined as '??'.

3.2 Program output

The program output is displayed as symbols between words:

(i) An asterisk ,*, signifies a word boundary.
(ii) Square brackets, [ ], signify a group boundary.
(iii) Curly brackets, { }, signify a clause boundary.

Here is an example of a text analysed by Version 1 of the boundary marker.

{[A * close * friend][may * choose][the * wrong * time][to * let * slip] [a * secret]}{[.]}{[Relationships][in * the * family][are][a * little * tense][, ?? but ?? you][know][the * reason][why][and][should * be * able * to * cope]}{[.]}{[In * a * career][matter][,][however][, ?? you] [may * have * to * go][back][to * square * one]}{[.]}{[You] [will * not * be][happy][about * this]}{[.]}

4. Evaluation of version 1 and development of version 2 of the program

Version 1 of the boundary marker parses texts with a reasonable degree of accuracy on the whole. If the text to be analysed contained only words in the program's lexicon, the resulting analysis would be approximately 70% acceptable in 'precision' terms, that is the ratio of received correct readings to all received readings.[4] This is achieved without access to any grammatical rules on the part of the machine. The system looks for a match between two words and, on the basis of best fit, assigns a boundary. It does not deduce what the word class of each word is or should be, even though this is the main piece of information that it tacitly draws upon, nor does it bear in mind its past decisions concerning previous word classes or previous boundaries when it moves to the next word and examines the gaps between the two succeeding words. Each boundary is assigned independently of its predecessor(s). There is a noticeable cline of success from word to clause level. The program is fairly successful in assigning word boundaries (80% accurate), not so successful in assigning group boundaries (66% accurate), and least successful of all in assigning clause boundaries (only 60% accurate). This means that, for example, setting a clause gap before *as* where it is operating as a subordinating conjunction is considerably more problematic for the system than setting a group gap before *as* where it is operating as a preposition.

With postmodifying phrases and clauses, the system experiences considerable difficulty. It knows for example that a common collocate after *man* is *who*, and that it is appropriate to set a group gap between the two. After *who* however, the program continues as though it were still operating at the same level as before, not being aware that its level of operation should be rankshifted down one. This is understandable inasmuch as the system is first-order Markovian and only looks at one word either side of the current word. Nevertheless, rankshifted groups and clauses need to be given further consideration.

The rest of this section looks in more detail at areas which proved problematic in the execution of Version 1 of the system and outlines some of the amendments that have been incorporated in Version 2.

4.1 The collocation dataset

The common collocation dataset posed a number of problems. The gaps were set by the intuition of the researcher, and while *to* before *learn* should always result in a word-gap boundary (to * learn), the same is not always true of, for example, *again* followed by *and*. The latter may signal a word boundary as in (again * and * again) or a clause boundary as in (he [asked] [again] {[and]

[left]}). A proportion of the information in the collocation dataset is also necessarily redundant. Knowing that *will* is a close-preceding collocate of *learn* duplicates information that the program will draw on anyway; namely that a modal may be followed by a main verb. The conclusion would appear to be that the collocation dataset may only be useful for idioms or fixed syntagms such as *again and again, only* + pronoun, or *human being*.

Between versions 1 and 2, the number of preceding and following close collocates has been increased from three to six slots. While in certain cases this has resulted in an increase in accuracy, it generally also increased system overheads. For the sake of expediency and speed, it may well be worth considering reducing the number of collocate slots to three again, or at least removing redundant or duplicated information, since the additional information on collocates does not appear to result in any improvement in accuracy.

4.2 The category dataset

The category dataset is the system's main brain in terms of gap assignment and, to a lesser extent, disambiguation of word class type. The original file consisted of 82 categories and permitted each word to be preceded or followed by up to three categories. This set turned out to be wasteful in that many of the combinations were never accessed. Moreover, many of the categories themselves were never used because they did not relate to any of the words in the lexicon held by the system. The categories 'adjective after noun' and 'verb ergative' for example were never used.[5]

The information held in the category dataset proved to be overspecified in terms of the finer distinctions made within each category. A word which belonged to a number of different categories proved easier to parse than one which had several entries within the same category. For example, *like* has four different entries: as a preposition, a subordinating conjunction, a classifying attributive adjective and a transitive verb. *Man,* on the other hand, has three different entries under the category 'noun': countable noun, uncountable noun and singular noun. The program has considerably more success in deciding whether *like* is a preposition or a verb than in deciding whether *man* is a countable or singular noun. In the latter case, the program has to use the expedient solution of reducing all occurrences of *man* to the superordinate category 'noun'.

The list of verbs held by the system provides another example of unutilised information. As the system is first-order Markovian and only looks one word either side of the current word, with verbs entered as V+O+A or V+O+O, only the first element after the 'V' is accessed. At the moment, 'V' is the only item of information which is actually made use of. This does not however mean that information on objects and adjuncts will not be utilised in the future, especially when the size of the lexicon is increased.

In Version 2, the category dataset has been pruned back to 25 entries. These consist of the main superordinates from Version 1 plus a number of new entries. The new entries include special words such as *of* and *to*, comma, clause-end-punctuation, non-finite verb, and a division of the category 'pronoun' into 'subject pronoun' and 'object pronoun'. These entries have been added because it was felt that they would improve the performance of the system in terms of assigning group and clause boundaries. They are organised in the form of a 25 x 25 matrix, with the exact boundary between any two elements specified.

4.3 The word dataset

The word dataset has been reorganised in three ways. Firstly, the entries have been reorganised to fit in with the remodelling of the category dataset. As a consequence, the three entries for the word *man* (countable noun, uncountable noun and singular noun) have been replaced by the one entry 'general noun'. Secondly, as a consequence of some of the issues discussed in 4.1, a number of the close-collocating entries have been deleted, on the grounds that they supplied information which was either conflicting at times, or which duplicated existing information. And thirdly, the number of possible entry slots in the dataset has been increased from three to four since three slots do not seem to give adequate coverage to certain words. Not all the senses of *like* and *close*, for example, had been captured in the previous dataset.

4.4 The introduction of new boundaries

The original brief of this system was simply to mark word, group and clause boundaries. In the light of current experience, this would appear to underestimate the facts of the language, with boundaries between certain words an area which requires further investigation.

One new gap that has already been introduced is the 'negative' gap, which in principle is similar yet different from a 'word' gap. The negative is often regarded as an adverb, and this would suggest that it introduces a group gap. This however is a simplistic view since *not* clearly forms part of the group in which it resides, and hence should not interfere with the building of the verb group.

Further boundaries that are currently being investigated include those between a head noun and a postmodifying group, and between a head noun and a postmodifying clause.

4.5 The introduction of rules

It had originally been hypothesised that it might be possible to assign boundaries without recourse to rules. The only 'rules' would involve identifying the permissible gaps between any two words on the basis of word class and collocational information. This hypothesis has been strengthened to a certain extent with the category dataset having been pruned back as described in 4.2. However, a number of situation-specific rules and heuristics have since been developed. Some of these, such as word sense disambiguation and lemmatisation, will almost certainly be superseded by other parsing modules in the suite of software tools proposed by Sinclair (1992). At this moment, the boundary marker has to undertake these tasks itself. A number of examples of the types of rules that the system is currently utilising are given below.

4.5.1 Verb group rules

The boundary marker now differentiates between auxiliary/modal verbs, main verbs and second/non-finite verbs. It makes use of rules which instruct it, for instance, to parse two main verbs as separate verbal groups if an intervening infinitival *to* is encountered, as in "At work, things *start to go* your way". This helps in assigning word or group boundaries between immediate constituents of a verbal group or groups. It may also prove useful later in assigning clause boundaries when the system needs to know how many finite verbs there are in a sentence.

4.5.2 Pronoun rules

Subject and object pronouns are now treated differently and this has proved helpful in distinguishing between group and clause boundaries.

4.5.3 Punctuation rules

Punctuation gives a considerable amount of information which, oddly enough, parsers generally ignore. Items of punctuation have now been included as individual categories. There are two sub-categories: one for elements which are always clause-final, for example the full-stop, semi-colon and question mark, and another for the comma.

All elements of punctuation, apart from the comma, mark clause boundaries. On reaching an element of punctuation, then, the system checks its current status as to how many finite-verb groups have been built, with only one finite-verb group permitted per main clause (postmodifying clauses to head nouns are excepted). If the system encounters more than one, it backtracks to check the main verb/conjunction situation and attempts to build a missed clause gap.

The comma has been given a different treatment, on the grounds that it may at times mark a clause boundary, while at others may only signal a group or even smaller boundary.

4.5.4 Phrasal verbs

The system has a rudimentary phrasal verb look-up table. This has been compiled by reference to the *Collins Cobuild Dictionary of Phrasal Verbs* (Sinclair 1989) and by further inspecting concordance lines for the verbs in question in the Bank of English. On encountering a word which can be either a preposition or adverb, the system checks the phrasal verb dataset to see if the combination is a legal phrasal verb. If a positive match is found, the word is given the appropriate preposition/adverb tag.

4.5.5 Postmodifying groups

Sinclair (1972:132ff) identifies a number of postmodifying group types. This is a slightly amended list:

nominal group	the old man
pronominal group	she / just him
prepositional group	in a red hat
non-finite verbal group	dreaming / interviewed
finite verbal group	was dreaming / intercedes
adjectival group	very expensive indeed
semi-adverb group	nearby
full-adverb group	usually

These group types provide a starting point for the program. The system also has a limited number of rules that it now applies in its effort to disambiguate group types. For example:

IF	(first_item = preposition)
AND	(last_item = noun OR last-item = pronoun)
THEN	group_type = prepositional group

On this basis, *in a much quieter mood* and *with friends* would be identified as prepositional groups.

4.5.6 Postmodifying clauses

The structure of a rankshifted clause within a postmodifying clause needs to be maintained while still demonstrating the cohesiveness of the nominal group. For example, in "Once again others you have trusted in the past won't always give the right advice", the system should ideally identify *you have trusted in the past* as a rankshifted clause postmodifying *others*. This is currently achieved in the following way. When a noun or pronoun is followed by a relative *who*, *which* or *that*, or by a pronoun or a determiner, the program starts building a postmodifying clause. It continues to analyse the words and build groups within this sub-clause. The process continues until a finite verbal group is built. Once this has been encountered, the preceding sub-clause is appended to the previous noun or pronoun as a postmodifying clause.

4.5.7 Coordination

For a program which essentially only looks one word either side of the current word, coordination presents considerable problems. At the current stage of development, the treatment of coordination is still rather rudimentary, but the system does utilise a number of basic rules.

On meeting a major coordinator, namely *and* or *or*, the system adopts a wait-and-see attitude. If the words preceding and following the coordinator are of the same word class, they are coordinated as two words with a word gap and the coordination symbol & between them, as in the following example:

> Don't forget to find time to relax with *friends or family.*
> friends or family
> [friends & or & family]

If the element either side of the coordinator is a group of the same type, similar action is taken.

> Right from the start today you'll be up to your eyes in *work and other responsibilities.*
> work and other responsibilities
> [work] & and & [other responsibilities]

This procedure is repeated in the case of two clauses.

> At work *things start to go your way and financial support is yours for the asking.*
> { things start to go your way } & and & { financial support is yours for the asking}

The program's objective here is currently to maintain the word/group/clause status of the element it is building. For example, if it is in the process of building a group, it tries not to disrupt this while inserting word gaps.

5. Conclusion

Drawing on the modified datasets and rules described in section 4, the system now renders a more precise analysis of texts. The difference between the level of precision achieved by Version 1 and that achieved by Version 2 is shown in Table 1 (punctuation has not been included in the calculations). The text quoted under section 3.2 above is now analysed as follows by Version 2 of the system ('+' = negative gap).

> {[A * close * friend][may * choose][the * wrong * time][to * let * slip] [a * secret][.]} {[Relationships][in * the * family][are][a * little * tense][,]}{[but][you][know][the * reason][why]}{[and][should * be][able][to * cope]}{[.]} {[In * a * career * matter][,][however][,] [you][may * have][to * go][back][to * square * one][.]} {[You] [will + not + be][happy][about * this][.]}

Table 1: Precision comparisons between Versions 1 & 2

System	Word Gap	Group Gap	Clause Gap	Word Type	Group Type
Version 1	80%	66%	60%	n/a	n/a
Version 2	100%	93%	100%	91%	97%

The current implementation of the boundary marker suffers from the fact that it is at the bottom of the pile in terms of developmental status since it is likely to be the first program of the Sinclair suite to come on-line. Although it is intended to be a marker of boundaries between words, it effectively has to be a parser in order to achieve its objective. In the design scale, the boundary marker would ideally come after a number of initial operations have been performed. These include:

(i) Lemmatisation, including morphological analysis of unknown words.

(ii) Word-sense discrimination in order to fix the exact type of each word to be parsed. The fact that each word is permitted only four entries in the word dataset necessarily results in certain nuances of word-class possibilities having to be disregarded.

(iii) Identification of possible idioms or 'idiosyncratic' collocations, some of which the program currently tries to pick up in its word dataset.

The design parameter of the 25 x 25 matrix has shown itself to be a very workable system: the matrix has proven to be an effective method of assigning word boundaries, which, given its limited size, works tolerably well, even though its assignment of gaps is at times a trifle blunt. One amendment that needs to be incorporated is in the dimensionality of the matrix. The matrix is currently 2-dimensional, so that between two word classes, only one boundary is possible. If the matrix were multi-dimensional, two contiguous word-classes could then be licensed to more possible boundaries. For example, the boundary between two nouns is currently only licensed to a word boundary. Two contiguous nouns are, however, an area where the boundaries may vary between word gap, group gap, and clause gap. The problem has not arisen in the texts that have been analysed by the system so far, but the following made up examples illustrate the point:

buy [some * baby * food]
give [the * baby] [food]
{ every time I wait for the bus } { smoke from other vehicles chokes me }

If lemmatisation and word-sense discrimination could be fixed by outside bodies, a feasible design model for the current system would involve strengthening the boundary marker's main 'brain', the category matrix, with a view to having it approach its former Version 1 design status. This model had 82 category possibilities. It incorporated a number of redundant features, and the way the system was implemented resulted in few of the categories being accessed. If the boundary marker were not required to struggle with word sense and different forms of the lemma, a multi-dimensional matrix in the order of 70 x 70 would then be a workable model and would capture the most important word-class information with little or no redundancy.

A final question to be considered is that of increasing the number of situation-specific rules that the program needs in order to identify a gap. One feature which was present in the Version 1 category dataset is that of word or, more specifically, verb valency. Knowing how a verb subcategorises, for instance, would help in the process of gap disambiguation. The program would presently parse a string such as *give the dog food* as follows:

[give] [the * dog * food]

With a multi-dimensional matrix, two analyses would be proposed:

(i) [give] [the * dog * food ; (ii) [give] [the * dog] [food

If the program could access the V+O+O information in the Version 1 category dataset, it might well be able to select (ii), the correct parse, where a group gap is required. However, once the door is opened for incorporation of rules as and when necessary, the system begins to move away from its projection as a word-boundary marker to something more akin to a pretend fully-fledged parser.

The experience of developing and modifying the prototype boundary marker has been instructive. The hypothesis that parsing can be achieved without recourse to explicit grammatical rules has proved untenable in its strongest form. On the other hand, the notion of parsing by assigning word boundaries has proved feasible and has produced impressive results. More important, the detail outlined in Sinclair (1992) promises, once the various partial parsers of the Sinclair suite have been developed, an NLP system of considerable potential.

Author's Address:
The Chinese University of Hong Kong, Faculty of Education, Shatin, NT, HONG KONG.

Notes

1. SHRDLU is an understanding system written by Winograd in 1972. It is sometimes referred to as BLOCKSWORLD. It is a system which simulates a human being at a computer terminal, interacting with the computer and manipulating blocks on a table top.
2. See Akkerman et al, 1988, for a thorough account of a similar rationalisation of the entries in the Longman Dictionary of Contemporary English.
3. The Bank of English is the archive of texts held at Cobuild. It currently consists of over 200 million words of running text.
4. Since only one received reading is ever offered, there is no need to distinguish between 'recall' and 'precision' in this case.
5. This of course has to do with the type of text being parsed. Horoscopes are unlikely to make much use of categories such as 'ergative verb' or 'adjective after noun', but other types of text will do.

References

Akkerman, E., H. Voogt-van Zutphen & W. Meijis. 1988. *A Computerized Lexicon for Word-Level Tagging: ASCOT Report No. 2.* Amsterdam: Rodopi.

Charniak, E. 1983. "Parsing, How To." *Automatic Natural Language Parsing* ed. by K. Sparck Jones & Y. Wilks, 156-63. Chichester: Ellis Horwood.

Dyer, M. G. 1983. *In-depth Understanding: a Computer Model of Integrated Processing for Narrative Comprehension.* Cambridge, Mass.: MIT Press.

Gazdar G. & G. Pullum. 1985. *Generalized Phrase Structure Grammar.* Cambridge, Mass.: Harvard University Press.

Garside, R., G. Leech & G. Sampson, eds. 1987. *The Computational Analysis of English*. London: Longman.

Hendrix, G. G. 1977. "LIFER: a Natural Language Interface Facility." *SIGART Newsletter* 61. 25-26.

Karlsson, F. 1990. "Constraint Grammar as a Framework for Parsing Running Text." *Papers Presented to the XIIIth International Conference on Computational Linguistics, Volume 3* ed. by H. Carlgren, 168-73.

Kay, M. 1982, 1986. "Parsing in Functional Unification Grammar." *Readings in Natural Language Processing* ed. by B. J. Grosz, K. S. Jones & B. L. Webber, 125-38. Berlin: Springer-Verlag.

Marcus, M. P. 1980. *Theory of Syntactic Recognition of Natural Language*. Cambridge, Mass.: MIT Press.

_______________ & B. Santorini. 1992. "Building Very Large Natural Language Corpora: the Penn Treebank." Paper Presented at the Pisa Workshop on Textual Corpora, February 1992.

Petrick, S. R. 1973. "Transformational Analysis." *Natural Language Processing* ed. by R. Rushkin, 27-41. New York: Algorithmics Press.

Riesbeck, C. K. 1978. "An Expectation-Driven Production System for Natural Language Understanding." *Pattern-Directed Inference Systems* ed. by D. A. Waterman & R. Hayes, 399-414. New York: Academic Press.

Sager, N. 1981. *Natural Language Information Processing*. Reading, Mass.: Addison-Wesley.

Sinclair, J. 1972. *A Course in Spoken English: Grammar*. London: Oxford University Press.

_________. 1991. *Corpus, Concordance, Collocation. (= Describing English Language.)* Oxford: Oxford University Press.

_________. 1992. "The Automatic Analysis of Corpora." *Directions in Corpus Linguistics. Proceedings of Nobel Symposium 82, Stockholm 4-8 August 1991* ed. by J. Svartvik, 379-97. Berlin and New York: Mouton de Gruyter.

__________ et al, eds. 1987. *Collins COBUILD English Language Dictionary*. London and Glasgow: Collins.

____________________. 1989. *Collins Cobuild Dictionary of Phrasal Verbs*. London and Glasgow: Collins.

Small, S. 1983. "Parsing as Cooperative Distributed Inference: Understanding through Memory Interaction." *Parsing Natural Language* ed. by M. King, 247-76. London: Academic Press.

Winograd, T. 1972. *Understanding Natural Language*. Reading, Mass.: Academic Press.

Woods, W. A. 1970, 1986. "Transition Network Grammars for Natural Language Analysis." *Readings in Natural Language Processing* ed. by B. J. Grosz, K. S. Jones & B. L. Webber, 71-87. Berlin: Springer-Verlag.

From Firth Principles
Computational Tools for the Study of Collocation

Jeremy Clear
Cobuild, Birmingham

Abstract
Taking inspiration from the definition of collocation proposed by Firth (1957), lexicographers and linguists have established the value of corpus-based studies of word co-occurrence. Focusing on lexicography, this paper stresses the importance of stereotyping as a linguistic phenomenon and introduces a computational approach to collocational analysis. The author analyses collocations obtained from a corpus of contemporary British English, using a software tool written as an aid to lexicographers who are working on EFL dictionaries. The statistical methods used in the program are presented and discussed and collocations are characterised in terms of their frequency, idiomaticity and positional variation.

0. Collocations and lexicography

The study of collocation as a linguistic phenomenon has not found a central place in theoretical linguistics, perhaps because its proper province is the rather ill-defined area of linguistic patterning that is neither clearly syntactic nor clearly semantic. Moreover, the Firthian framework for studying collocation is dependent upon a commitment to a model of linguistics which does not concern itself with the Chomskian competence/performance dichotomy and the consequent exclusive attention to competence. In the applied area of lexicography, however, collocation is widely acknowledged to be of central concern.

During the course of the research project which resulted in the production of the *Collins Cobuild English Language Dictionary* (Sinclair 1987a), the standard approaches to lexicographic definition became less and less convincing as models for the team of lexicographers. In particular, the availability of very many corpus instances of words in use showed up the great influence which phraseology exerts over word meaning (I use 'phraseology' loosely to mean the ways in which words occur together in actual use). The dominant dictionary model attaches meaning to words in isolation and, in the printed form available to the mass market, relies on the alphabetical ordering of single orthographic units for access to the information. One of the most difficult challenges facing a lexicographer is to deal satisfactorily with the meanings created by word combination within the straightjacket constraints of this model. Word combinations have a tendency to slip out of their appointed range of free commutation and form particular attachments: I call this process 'stereotyping'. Linguists may account for the stereotyping by labelling the combinations as idioms, set phrases, compounds, multi-word lexemes, or whatever – the mere fact of their being considered to form a distinct combination is sufficient to term them as stereotyped. Stereotyping does not entail idiomaticity (idioms have a meaning which is different from the sum of the parts, but stereotyped combinations need not) and it is syntactically neutral (words may be labelled as stereotyped irrespective of the grammatical environment).

The abstraction and the terminology can easily be reinforced by example. Consider the word *keep*, which is quite frequent in any corpus of modern English, and which defies coercion into a tidy list of discrete, enumerable senses. Each of the selected examples listed below shows a stereotyped pattern with *keep* which seems to require special treatment in a dictionary. These combinations, among many others, are each accounted for separately in the Cobuild dictionary.

The grant includes £19 for your keep during the vacations
They keep their discoveries to themselves.
...she had to hurry to keep up with me.
I ought never to have kept you up so late.

Keep to the path.	Keep going!
...keep a look-out for him.	...keep watch...
The sabbath should be kept...	I did not keep any notes.
It would be nice to keep bees.	Fish doesn't keep very well.
How is your mother keeping?	She kept on about the stupid car.
This keeps prices down.	Luckily the rain kept off.
Private property. Keep out.	We must keep to the deadlines.

It can be quickly appreciated that no single formulaic definition of *keep* as a single semantic unit will be adequate to account for the range of meanings created by the combinations above. Moreover, it should also be apparent that these combinations, perhaps through mere frequency of use, have become idiosyncratic in some way: they have particular meanings, or they do not participate in the full syntactic paradigm. Since no available grammar of the language sheds any light on the meaning or the grammatical limitations of such combinations, it falls to the dictionary maker to attempt to deal with them.

The lexicographer may choose one of three strategies in accounting for a word like *keep*. First, the lexicographer can define a generic word sense which might serve, at a stretch, to account for all the combinations illustrated above. Second, s/he can produce separate definitions for a (possibly rather large) set of word senses, each of which accounts for the sense of *keep* in each of its stereotyped combinations. Third, s/he can specify the combinations explicitly (identifying the other words which are considered privative to the combination) and produce a definition for each combination. The first method, a single generic definition, runs the risk of being so abstract that it is utterly mystifying to the dictionary user and even then may fail to cover the full range of use of *keep*. The second method suggests that all of the meanings created by the stereotyped combinations are inherent in the single word *keep* (clearly bowing to the tyranny of the 'words-as-meaning-containers' view imposed by traditional dictionary structure). The third method seems to be more true to the evidence, but sets off down a slippery slope which leads the lexicographer to attempt to write definitions for chunks of text, which are innumerable if not infinite, as if they were lexical items. Whichever of these approaches is adopted in practice, it will be necessary to consider the full range of stereotyped patterns and to classify them in some way.

In recent years, since the Cobuild project demonstrated the benefits of computational analysis of a substantial mass of real text data, the use of computerised text corpora as a resource in the compilation of dictionaries has become more widespread. One of the many consequences of adopting a corpus-based methodology is that the study of the phenomenon of stereotyping can be vastly more rigorous and revealing than it could possibly be without the basic data as a foundation. I have defined collocation as the mere recurrent co-occurrence in text of word-forms, and stereotyping as the tendency of collocations to develop a life of their own as identifiable pieces of a native speaker's lexical hoard. The data-driven corpus approach to collocation proceeds from the study of word-forms, their frequency, the frequency of their co-occurrence with other forms and the statistical measurement of the significance of such co-occurrence.

1. The importance of computational methods

Over the last decade I have been challenged on numerous occasions to justify the use of computational, and consequently algorithmic and statistical, methods in the area of lexicography. First, the use of authentic corpus material as the basis for lexicography I regard as a vital first step in making the practice of dictionary-making more rigorous, comprehensive and useful. Moreover, since a language such as English is so vast and complex it is apparent to me that a very large amount of corpus data will eventually be required in order to furnish evidence of the full range and distribution of linguistic forms and meanings. The quantity of data which can and should be gathered necessitates the use of mechanical aids. This is the weak argument for using computers. The stronger argument is that the mechanical aid of the computer does not merely allow the lexicographer or linguist to carry out analyses more quickly and easily, but opens up an approach to linguistics which is *qualitatively* different and will lead to developments in linguistics which simply could not happen without electronic processing aids. Indeed, I believe attitudes are shifting already in areas of natural language processing and computational linguistics where corpus-based approaches have yielded unpredicted and useful developments.

Collocation, being a part of the study of lexis, badly needs computational assistance since the number of lexical items is so large (relative to, say, the number of grammatical classes traditionally recognised by linguists) and the statistics of their distribution is so skewed. By far the majority of lexical items have a relative frequency in current English of less than 20 per million. The chance probability of such items occurring adjacent to each other diminishes to less than 1 in 2,500,000,000! Reliable evidence of patterning between such items can clearly be obtained only from very substantial text corpora and with the aid of computer technology.

2. Two different tools for two different needs

Lexicographers working on a particular word or set of words find it useful to have some computational tool for sifting through the mass of word combinations that might be observed in a large corpus. If they already know what word is the focus of interest, they will be interested in the collocates of that particular word-form (or set of inflected forms). I will describe a software tool in the next section which, when given a word-form or small set of forms, will retrieve collocates ordered and presented according to some statistical measure of significance.

One of the fundamental tasks for the lexicographer, in particular for the compiler of dictionaries for the native speaker, is to record and describe new words coming into a language. In modern English, the number of completely new single word-forms entering the language will be few relative to the number of combinations (compounds and phrases) that become stereotyped and recognised as new forms. In this case there is a need for a computational tool which will attempt to identify new stereotyped word combinations from a large corpus and bring these hitherto unrecognised forms to the attention of the lexicographer. A tool to achieve this task could be built by simple extension of the algorithm for extracting significant collocates of a single given word-form: one might simply iterate over all word-forms found in the corpus and discard any combinations which failed to score an arbitrary threshold value for significance. I have carried out this operation over a corpus of 25m words of varied modern English text[1] in order to produce a list of word pairs. Many new technical, slang and jargon terms are phrases or compounds, and the preparation of this list was intended to help lexicographers identify terms for inclusion in their dictionary files. The technique is however too crude to yield really useful results, since the list of candidate pairs is either too long to be reasonably scanned by the human researcher (if the statistical significance threshold is set low) or else the list is manageable but omits many pairs which a lexicographer would consider to be interesting (if the significance threshold is set higher). Information scientists use two measures of the performance of information retrieval systems: precision and recall. Precision is the measure of success of the system in retrieving *only* the items of interest to the user; precision is highest when all the items retrieved are deemed to be interesting and lowest when none of them is. Recall is a measure of the system's ability to find all the interesting items out of the information base; recall is high when the system retrieves all the interesting items and lowest when it fails to find them. We can regard the identification of 'interesting' collocations from a very large corpus as an information retrieval task. The corpus is an information base and within it are some interesting word pairs (interesting, that is, to a lexicographer). We want a computational tool which achieves high recall (i.e. picks out all the pairs that the lexicographer would want to consider) and high precision (i.e. does not retrieve a large number of pairs which are *not* of interest). Typically in such information retrieval tasks, the identification of new compounds and phrases by calculating the significance of adjacent word pairings currently trades off precision against recall, without any clearly satisfactory compromise value emerging.

3. The collocate program

I describe here the current version of a program for the identification of significant collocations, and discuss its operation and its value for lexicography.[2]

The user specifies a word of interest and under default options the program performs the following operations:

i) for every instance of the node word in the corpus, the two adjacent words to the left and two to the right are retrieved (the collocate list)
ii) the frequency of co-occurrence with the node word is calculated for each collocate
iii) the overall corpus frequency for each collocate is retrieved from a corpus index file
iv) the list of collocates, co-occurrence frequency and overall corpus frequency is passed to a routine which calculates a significance value for each collocate.
v) the collocate list is sorted in descending order of significance and output.

The important methodological aspects of this program concern the retrieval of words within a range ('span'), the frequency of co-occurrence, the treatment of inflected forms of a single lemma and the calculation of the significance value. I will comment briefly on the first three of these and deal in more depth with the last.

First, the retrieval of words within a span. Published research in this area does not reveal any clear consensus on the terms and procedural details of collocational analysis. Some researchers (for example Kjellmer 1984, Benson 1985) have considered collocation to be syntactically defined. The approach I adopt assumes that the collocating words are not necessarily contiguous, and that an initial analysis need not take account of the left-to-right ordering of the pair. Therefore, the default span for the program is taken to be a five-word window (two words left and right of the keyword). Intuitively, one would expect that a given node word would associate more strongly with immediately adjacent words, and that the associative link would be weak or non-existent the further removed are the collocating words. We expect to discover that *order* collocates significantly with *mail*, *alphabetical*, *restore* or *form*: we do not expect that it will collocate significantly with words that might be in the preceding sentence, or thirty words later in the next paragraph. This intuitive assumption is reflected in the choice of a narrow span of 2:2,[3] though it has been claimed that significant statistical association can indeed be detected between a node and collocate which are separated by as many as fifty intervening words.[4]

Second, the frequency of co-occurrence of a word and a collocate is a critical parameter in this approach to collocational analysis. Some linguists may consider every co-occurrence to constitute a *de facto* collocation. Sinclair's work in this field, and my own, defines collocation as a recurrent co-occurrence of words. The terminology is not well established yet, and it is by no means clear whether the observation of a single co-occurrence in a corpus should be ignored or whether the single instance should be taken to be one of many more which might have occurred (since the corpus is merely a minuscule sample of the totality of language in use). Unfortunately, we cannot determine from the corpus alone, though our linguistic intuition may lead us to make judgements, whether a single co-occurrence is indeed an isolated chance phenomenon or one instance from many which might have appeared in a different or larger corpus. Since the approach we take here is based on a rigorous, data-driven and bottom-up analysis, we apply a threshold value to the frequency of co-occurrence, and pairs which have been observed fewer than three times are discarded. One consequence of this step is that the collocation program is not particularly useful as a tool for identifying new compounds and phrases for inclusion in the dictionary, since it is necessary for the new item to appear three times or more even to be considered as a candidate collocation.

The third methodological point concerns the program's treatment of word-forms. The user may supply one or more word-forms to the program. If more than one form is supplied, the program generates the set of occurrences in the corpus of all specified words. If, for example, we are interested in seeing the collocates of the word *keep*, we could run the program specifying *keep*, *kept*, *keeps* and *keeping*. All instances of all these forms would be taken together and collocates retrieved from the span on either side. The lemmatisation of word-forms is not carried out by default and, in particular, it is not carried out at all on the collocates. So the list of significant collocates for the node *interest* might include *lose*, *lost* and *losing*, each treated separately with its own frequency of co-occurrence and significance value. In practice, this approach works well. Some collocations are restricted to particular inflected forms (for example *vested interest*, *crying shame*, *bodes ill*), and the lemmatisation process would hide the collocation. Moreover, it has been found through years of lexicographic practice that collocations are very rarely lost because of the lack of lemmatisation: one of the inflected forms will appear as a significant collocate, and the potential for the other forms in the paradigm to collocate will be apparent to the human analyst. If any doubt remains, the specific forms can be located in the collocate list.

4. Two measures of significance

The collocate program currently offers two statistical measures of significance: Mutual Information and a t-score. Both of these are described and presented in Church & Hanks (1990). Here, I want to describe their operation and draw conclusions about their relative effectiveness in selecting interesting collocations for the lexicographer.

Mutual Information is a fundamental statistic in information theory. It measures the amount of information that the observation of one word yields about the likely occurrence of the other. That is a rather formal definition, and an example might help to explain the principle. Suppose, in a corpus of 10 million words, we observe the word *kin* 10 times. We can estimate the probability of occurrence of the word *kin* to be 0.000001 (1 in a million). Now suppose that in the same corpus we observe that the word *kith* occurs 5 times, and that in all of these five instances *kin* follows *kith*. If we have seen *kith* we could estimate the probability of seeing the word *kin* to be 0.5 (5 times out of 10). So the appearance of the word *kith* gives us a great deal of information about the likelihood of seeing the word *kin* nearby: the probability is increased from 0.000001 (our best estimate if we have nothing else to go on) to 0.5! Similarly, if we see the word *kin* we are pretty much sure that we will see *kith* nearby. In this synthetic example, the Mutual Information between these two words is high.

The Mutual Information value (henceforth MI) is non-directional: in the example above, *kith* gives us more information about the likelihood of seeing *kin* than vice versa,[5] but the MI value averages the association in both directions. There is only one MI value for the unordered pair. MI calculates the probability of the two words co-occurring if they were independent (that is simply the product of their individual relative frequencies in the corpus) and compares this value with the observed frequency of their co-occurrence. If there is little difference between these values then we cannot say that the co-occurrence of the two words is significant. It should be noted that, for the purposes of studying collocation, we define 'co-occurrence' as co-occurrence within the span (not simply immediate adjacency). As the value of the span is increased, the independent probability of co-occurrence must increase. The span is accounted for as a parameter in the MI calculation.

Table 1 shows a list of collocates for the word *taste* (with the span set to 6:6) ordered by decreasing MI value. The corpus I have used to produce the demonstration output in this paper is made up of 30m words from the Oxford Pilot Corpus (see Note 1). The top 20 or so highest scoring collocates are listed. The output is in five columns:

1. the collocating word-form,
2. the overall frequency of this collocate in the corpus,
3. the *expected* frequency of co-occurrence with the node, if the occurrence of words in the corpus were independent of any other word,
4. the actual, observed frequency of co-occurrence of collocate and node,
5. the Mutual Information value.

Table 1. Collocates for *taste* ordered by decreasing MI value

taste	1527	0.933	1545	10.69
gilli	7	0.004	3	9.46
worm-cake	7	0.004	3	9.46
arbiters	20	0.012	5	8.68
hom	13	0.008	3	8.56
aroma	97	0.059	12	7.66
buds	109	0.067	10	7.23
decency	127	0.078	11	7.15
arbiter	59	0.036	5	7.12
seasoning	165	0.101	13	7.01
salty	64	0.039	5	7.00
juniper	39	0.024	3	6.98
citrus	40	0.024	3	6.94
lapses	70	0.043	5	6.87
sans	45	0.027	3	6.77
texture	382	0.233	22	6.56
springy	52	0.032	3	6.56
savour	72	0.044	4	6.51
pepper	465	0.284	25	6.46
salt	894	0.546	44	6.33

A cursory scan of the collocates shows that the program is identifying intuitively satisfactory items from the semantic area of food, smell and the physical action associated with the taste buds. There are few exceptions to this general pattern. The collocates with the highest MI scores are *gilli* and *worm-cake*. These appear with such prominence because they are themselves rare occurrences (all seven occurrences of these words are from one particular text source), and because three of their seven occurrences are with *taste*. Then follows *arbiters*, which highlights the cliché *arbiter(s) of taste* and a different sense of the node word. *Decency* is another collocate (*taste and decency*) which accompanies *taste* when it means discretion or refined aesthetic values. *Hom* is there because of the celebrated Chinese cook, Ken Hom.

In general, the collocates which are identified by MI are a rather motley and peculiar set. They all have a particularly marked pattern of association with *taste*: indeed this is exactly what MI measures. MI is a measure of *the strength of association between two words*. The word *arbiters* only occurs 20 times in the 30m word corpus, and in five of those instances it occurs with *taste*. We would expect these two words to co-occur only 0.012 times, given their relative frequencies and the size of the corpus.

The pair *taste buds* will almost certainly qualify for separate treatment within a dictionary. The adjectives *salty* may be useful to the lexicographer who needs to select an illustrative example of the word *taste* in use.

The problem with the MI calculation is that the measurement of the strength of association has to be made on the basis of the evidence of the corpus, but we cannot be sure whether a collocation which has been observed only a few times in the corpus is really reliable as a guide to co-occurrence patterns. So with the collocate *arbiters*, even if the pair had occurred once only in the corpus this event would be more than 60 times greater than the expected frequency of co-occurrence (0.012), yet common sense tells us that a single co-occurrence may be attributable entirely to chance or to a very peculiar turn of phrase employed by one particular writer or speaker. For this reason, collocates having a frequency of co-occurrence of less than 3 are discarded from the MI calculations. Even so, this threshold might be too low: note that many of the items appearing in the MI list have co-occurrence frequencies of 3 to 5, perhaps indicating that the items that have high MI values are idiosyncratic instances peculiar to this corpus.

Table 2. Collocates for *taste* ordered by t-score

taste	1527	0.926	1545	39.29
for	251831	152.584	292	8.18
good	22040	13.354	80	7.48
bad	4432	2.686	57	7.23
and	718005	435.037	605	6.93
salt	894	0.542	44	6.59
season	4215	2.554	33	5.35
with	188289	114.084	183	5.12
pepper	465	0.282	25	4.99
mouth	1669	1.012	24	4.75
texture	382	0.232	22	4.69
smell	887	0.538	21	4.52
first	37528	22.738	56	4.49
acquired	1185	0.718	21	4.48
popular	3626	2.197	23	4.39
sweet	1198	0.726	20	4.37
food	6202	3.758	25	4.31
my	36868	22.339	52	4.16
bitter	860	0.521	17	4.06
his	130777	79.238	121	3.83
add	2428	1.471	17	3.83
sugar	1360	0.824	15	3.73

seasoning	165	0.100	13	3.65
like	39288	23.805	48	3.55

Turning now to compare MI with a different measure of significance, Table 2 shows a list of collocates from the same corpus, ordered by t-score. The list is very different at this top end, and it will immediately be apparent that the highest scoring items have a much higher overall frequency than those found at the top of the MI list. T-score is measuring not the strength of association between two items, but *the confidence with which we can claim that there is some association.* This is rather different from MI. The topmost item of the t-score list is *for*, as in *develop a taste for scifi novels*. The collocation occurs 292 times in the corpus: it is much more frequent than *taste* + *arbiters*. The high t-score indicates that we can be very confident that an association exists between *taste* and *for*, because their co-occurrence has been observed sufficiently often to rule out the possibility that it might be a freak of chance. The strength of association (MI) between *taste* and *for* may not appear as great as between *taste* and *arbiters*, but it is more reliable.

In general, the collocates highlighted by the t-score statistic will be frequently recurring items (often grammatical words such as prepositions, personal pronouns, determiners and particles) along with those fully lexical words which are clearly associated with the node word. Both the MI and t-score lists for *taste* include the collocate *acquired*, though this item is well down the MI listing, so this is very likely to be a compound or part of a cliché, idiom or other stereotyped phrase. The MI score, on the other hand, focuses attention on the more idiosyncratic collocates of a node: technical terms, arcane or idiosyncratic phraseology and fixed compounds tend to be clearly identified by the MI statistic. The results from these two statistical calculations support the notion of 'upward' and 'downward' collocation, introduced by Sinclair (1987b, 1991). Sinclair notes that the collocates which are more frequent overall than the node word (upward collocation) "tend to be elements of grammatical frames, or superordinates", whereas those which are less frequent (downward collocation) "give us a semantic analysis of a word". While the evidence of my own research confirms Sinclair's conclusions, I do not fully concur with his claim that downward collocation provides a semantic analysis. Consider again Table 2. The words *salt*, *pepper*, *texture*, *smell* and *acquired* in this list are less frequent overall than *taste,* and the first four are clearly related semantically to the node word. Many of the pairs identified by the t-score are upward collocations, and several of these are indicators of the different senses of *taste* which one would expect to find discriminated in a dictionary. *Good* and *bad*, when they occur attributively with the nominal use of *taste*, usually signal the non-physical sense of *taste* (either aesthetic sensibility or deference to norms of social acceptability). *First taste* points to the meaning "first experience or encounter" (for example *It was his first taste of army discipline*). The collocate *acquired* is a fine example of the sort of stereotyping which is pervasive in modern English but difficult for the non-native speaker to master and for conventional dictionaries to deal with. The collocation is rather idiomatic, but not wholly opaque. It is my view that the t-score statistic, by identifying frequent and very reliable collocations, offers the lexicographer a semantic profile of the node word and a set of particular fixed phrases, grammatical frames and typical stereotyped combinations which merit investigation in compiling a dictionary entry.

The MI list is best consulted for information about pairs which, though not likely to be typical of the usage of the node word, will be very strongly associated and will tend to form idioms, proverbs and technical phrases. *Taste buds* is such a pair, which will certainly merit separate treatment in a dictionary. However, many of the collocates listed by the MI score will be untypical and peculiar to the corpus or to particular text samples within the corpus.

5. Evaluating the output

I have illustrated two statistical measures of significant collocation. As with all computational corpus analysis, there is methodological difficulty to be resolved concerning the validity of particular computational techniques and that is: how can we know whether the machine analysis is sound? One answer is to compare the output from automatic analysis with the output from manual analysis over the same corpus data. Formal testing and comparison is very time-consuming for the human analysts and cannot be undertaken lightly. Moreover, it is questionable whether this test of the success of automatic analysis is valid, since it is precisely the quantity of data furnished by large text corpora, and its consequent intractability for manual analysis, that offers new insights into linguistic patterning. The approach to collocation adopted here is founded on the quantitative analysis of the occurrence of word-forms and there is no reason to suppose that a human analyst, asked to peruse many hundreds or thousands of corpus instances of a word, will be able to collate and assimilate the quantity of evidence. On the other hand, an automatic procedure which yields results that have no apparent relevance or interpretive value to the human user is not helpful. To what extent do we put our faith in the results obtained by collocational analysis of this kind?

In the case of our example with *taste*, we can compare the collocation lists with dictionary entries compiled independently of this data. Since ELT dictionaries aimed at the foreign learner of English will typically devote more space to the description of idioms and other stereotyped combinations, we can take the entries for *taste* from the *Oxford Advanced Learners' Dictionary* (OALD), *Collins Cobuild English Language Dictionary* (CCELD) and the *Longman Dictionary of Contemporary English* (LDOCE) and collate the senses and phrases identified in those with the collocations identified by the collocate program. Table 3 lists the stereotyped features noted by the lexicographers of the dictionaries, with an indication of where these features are described (numbers are sense numbers in the dictionaries). The code in column 5 indicates which of the two statistical measures captures the relevant collocate within the top 24 items on the list it generates for *taste* (T = t-score, MI = Mutual Information). Collocates are in bold face in column 1.

Table 3: Collocations identified in dictionaries and by statistical measures

Feature	OALD	CCELD	LDOCE	Coll
taste **of** something	3	4	(Vb) 4	T
(in) **good/bad** taste	6	7.2	(N) 4	T
a taste **for**	4	5	(N) 5	T
taste **buds**	run-on entry	main entry	main entry	MI
an **acquired** taste	under *acquire*	under *acquire*	main entry	T, MI
leave **bad/nasty** taste in **mouth**	6	not listed	(N) 2	T, MI

6. Positional variation

The sample outputs shown in Tables 1 and 2 for *taste* give no information about the position of each collocate relative to the node, and all appearances of a collocate anywhere within six words either side of the node are counted as instances of the same collocation. The collocate program allows the user to specify the span, so that it is possible to broaden or narrow the focus of the collocational analysis. Tables 4 and 5 show the t-score collocate listings for the word-form *order*, with the span set to 0:6 and 6:0 respectively, from which it can be seen which collocates typically occur either to the left or to the right of the node. The software has a significant shortcoming which is relevant particularly to this node word. A stopword list of five items (*the*, *of*, *to*, *in*, *a*) is in force on the index database such that these items do not appear in the collocation lists. A revised version of the program should certainly include these items. A manual analysis of the concordances for *order* shows that *in* is by far most significant on the left and *to* on the right, from which I determine that the fixed phrase *in order to* is the massively dominant collocational pattern for this word.

Table 4. Right-hand collocates of *order*

order	7781	12.009	7845	88.44
avoid	2333	3.601	81	8.63
merit	467	0.721	71	8.37
make	21896	33.793	114	7.54
get	21334	32.926	103	6.94
keep	7177	11.077	66	6.80
maintain	1540	2.377	42	6.15
achieve	1565	2.416	40	5.98
form	8108	12.514	57	5.93
act	6157	9.503	52	5.93
their	78660	121.397	201	5.64
entry	1307	2.018	35	5.62
ensure	2296	3.544	37	5.54
survive	1141	1.762	32	5.39
create	2384	3.680	34	5.25
prevent	1954	3.016	32	5.17
for	251831	388.652	498	4.92
protect	1536	2.371	28	4.89
provide	4971	7.673	37	4.87
obtain	765	1.181	25	4.82
reduce	1914	2.955	27	4.68
please	2930	4.523	29	4.60
rule	2558	3.949	28	4.60
books	3811	5.882	31	4.56
establish	1265	1.953	24	4.56
them	43705	67.451	116	4.54
things	11821	18.244	50	4.54
offences	663	1.024	22	4.53
give	11038	17.036	48	4.52
bring	4375	6.753	32	4.52
save	2239	3.456	25	4.37
magnitude	195	0.302	19	4.35
book	7538	11.634	38	4.33
day	18088	27.916	60	4.19
understand	4024	6.211	28	4.18
st	4231	6.530	28	4.12
secure	1270	1.961	20	4.09

able	7922	12.227	36	4.02
chaos	520	0.803	17	3.99
preserve	599	0.925	17	3.96
accommodate	359	0.555	15	3.80
system	10204	15.749	39	3.78
from	125559	193.776	253	3.75
control	7199	11.111	32	3.75
bath	1252	1.933	17	3.72
improve	1575	2.431	17	3.60
explain	1990	3.072	18	3.59
placed	2046	3.158	18	3.57
or	92190	142.278	191	3.56
arranged	1044	1.612	15	3.53
complete	3450	5.325	21	3.49
coupon	125	0.194	12	3.48
which	105541	162.882	213	3.46

Among the right-hand collocates are a number of base-form verbs which seem to occur typically with the phrase *in order to*.

make	get	achieve	create	bring
provide	obtain	establish	give	
avoid	keep	maintain	ensure	secure
survive	prevent	protect	save	preserve

I have divided the verbs in the collocate list into two groups since they fall into two related sets: those sharing some meaning element of 'creation' and those having some meaning element of 'preservation'. These collocations form an interestingly homogeneous set, all semantically related to the concept of a threat or promise of a change of state. All are obligatory transitives and many are fairly formal vocabulary items, stylistically compatible with the tone of officialese conveyed by the phrase *in order to*.

A number of fixed phrases are signalled by other collocates to the right. These include: *order of magnitude, order of the day, Order of Merit* and *Order of the Bath. Order of magnitude* and *order of the day* are phrases with a technical formal meaning and a more frequent informal meaning, both of which should probably be accounted for in a dictionary. The other two fixed phrases are from a set of special public honours. Related to this second set is the right-hand collocate *St* which highlights the frame *the order of St* [Name], for example Benedict, Francis, John, and numerous other saints. Taken together, these clearly delineate a sense of *order* which relates to an historical chivalric system.

Two compound items are indicated as right-hand collocates: *order book(s)* and *order form*.

Now turning our attention to the list of left hand collocates, we find that stereotyped combinations can again be divided into sets on the basis of their semantic and grammatical properties. Most important for lexicographers are the idiomatic stereotypes whose meaning cannot be inferred from knowledge of the meanings of the individual items. *Pecking order* and *tall order* will certainly deserve a place in a dictionary intended for non-native speakers of English. Near the end of the listing in Table 5 is *house*, evidence of the idiom *put/set one's house in order*. Less idiomatic and more specialised compound units are: *mail order, compulsory purchase order, Orange order* and *postal order*. These will probably require specific treatment in a dictionary.

Table 5. Left-hand collocates of *order*

pecking	64	0.100	43	6.58
world	22242	34.327	97	6.40
restore	580	0.896	37	5.98
working	7407	11.432	48	5.32
orange	883	1.364	30	5.28
need	12478	19.258	57	5.04
customer	913	1.410	26	4.87
necessary	3909	6.034	34	4.85
tall	1177	1.817	25	4.69
natural	3851	5.944	32	4.66
under	16456	25.397	60	4.51
power	9669	14.923	44	4.43
issued	1314	2.029	23	4.43
rogers	166	0.257	19	4.36
purchase	1208	1.865	22	4.35
judge	2130	3.288	24	4.29
established	2612	4.032	25	4.25
put	14291	22.056	52	4.20
commission	4085	6.305	28	4.16
high	11566	17.851	45	4.10
must	17506	27.018	57	4.02
political	10552	16.286	42	4.02
forces	3501	5.404	25	3.98
use	12723	19.636	46	3.94
postal	176	0.272	15	3.87
probation	193	0.299	15	3.86
existing	2221	3.428	20	3.77
keep	7177	11.077	32	3.76
maintain	1540	2.377	18	3.75
refused	2327	3.592	20	3.73
placed	2046	3.158	19	3.70
chronological	91	0.141	13	3.64
compulsory	436	0.674	14	3.63
be	168552	260.127	325	3.62
good	22040	34.015	62	3.60
available	6197	9.565	28	3.55
own	21345	32.943	60	3.54
everything	5153	7.953	25	3.47
cheque	568	0.877	13	3.44

batting	228	0.353	12	3.44
your	31168	48.102	78	3.43
needed	4645	7.169	23	3.37
house	14291	22.056	44	3.36

Since *order* is significantly more frequent as a noun than as a verb, the left-hand collocates include a number of verbs (which might take *order* as an object or complement) and adjectives (which might modify *order*). The adjectives include *social* and *natural*, a neat pair of stereotypes deriving from the sense of *order* as a state in which all elements and components are arranged hierarchically. Within the social order can be distinguished the *old*, *established*, *existing order* in contrast to the *new order*, and related to these is the concept of a *world order*. *Public order* also derives from this sense of harmonious integration, but has been appropriated by the legal profession and belongs in the legal subject domain with compounds such as *court order*, *probation order*, *compulsory purchase order*, and typical collocates such as *issued an order* and *under an order*. The fixed phrase *law and order* does not however have a specialist legal sense. *Chronological* is picked out in the collocation listing and serves to represent a range of possibilities for this sense (alphabetical, size, ascending/descending, and so on).

Among the verbs to the left there is a degree of semantic clustering. Society's desire for peace and susceptibility to discord is evidenced in phrases such as *restore/keep/maintain order*. *Put* and *placed* are delexicalisations of the verb *order* meaning to arrange in a systematic way, and the potential alternation between 'to *order* something' and 'to *place/put* something *in order*' is brought into focus.

A different summary analysis of positional variation can be obtained from the collocate program, which can be instructed to produce for each collocate mean and variance figures for the position of the word relative to the node.[6] In Tables 6 and 7, I present listings of collocates of the node-word *order* with two additional statistics. Assume that, for the purposes of this analysis, a collocate occurring at the position immediately preceding the node is recorded as occurring at position −1. Similarly, the position two to the left of the node is recorded as −2, immediately following is +1 and so on. We can simply compute a mean position for a collocate relative to the node by adding the set of positions recorded and dividing the sum by the frequency of co-occurrence. This is the figure shown in the sixth column of Tables 6 and 7. As with any set of scores, we can also compute the variance (or standard deviation) from the mean: this is the figure shown in the seventh and last columns. The items listed in each table are the collocates with the highest t-scores, ordered by descending positional variance.

Table 6. Collocates of *order* (span set to 6:6)

order	7781	24.218	102	7.70	0.08	4.80
judge	2130	6.629	38	5.09	-1.13	4.01
forces	3501	10.897	33	3.85	-2.21	3.81
necessary	3909	12.166	34	3.74	-1.18	3.74
under	16456	51.218	93	4.33	-0.70	3.59
made	26782	83.356	123	3.57	0.22	3.49
chaos	520	1.618	22	4.35	1.45	3.46
aldington	426	1.326	15	3.53	-1.60	3.38
act	6157	19.163	61	5.36	-0.30	3.35
make	21896	68.149	169	7.76	0.47	3.24
give	11038	34.355	71	4.35	0.45	3.22
discipline	793	2.468	18	3.66	0.28	3.21
court	7163	22.294	119	8.87	-1.26	3.15
placed	2046	6.368	34	4.74	-0.35	3.13
established	2612	8.130	34	4.44	-0.24	3.10
an	99821	310.683	494	8.25	-0.39	3.09
offences	663	2.064	26	4.69	1.04	3.05
world	22242	69.226	112	4.04	-0.54	3.02
new	40708	126.700	235	7.06	-0.51	3.01
existing	2221	6.913	27	3.87	-0.81	3.00
restored	679	2.113	20	4.00	0.05	2.98
keep	7177	22.338	98	7.64	0.52	2.96
bring	4375	13.617	40	4.17	0.82	2.90
establish	1265	3.937	34	5.16	0.56	2.90
rule	2558	7.962	32	4.25	2.09	2.90
get	21334	66.400	130	5.58	1.15	2.89
preserve	599	1.864	25	4.63	0.36	2.89
protect	1536	4.781	23	3.80	1.57	2.83
please	2930	9.119	28	3.57	0.54	2.81
obtain	765	2.381	26	4.63	0.54	2.75
improve	1575	4.902	22	3.65	1.09	2.74
customer	913	2.842	31	5.06	-0.77	2.58
form	8108	25.235	65	4.93	1.23	2.57
arranged	1044	3.249	19	3.61	1.16	2.52
natural	3851	11.986	33	3.66	-0.97	2.52
old	15582	48.497	89	4.29	-0.93	2.49
maintain	1540	4.793	60	7.13	1.07	2.48
create	2384	7.420	35	4.66	1.43	2.46
entry	1307	4.068	41	5.77	0.20	2.45
issued	1314	4.090	35	5.22	-0.91	2.39
social	8412	26.182	112	8.11	-0.36	2.34
secure	1270	3.953	21	3.72	1.33	2.33
ensure	2296	7.146	45	5.64	0.98	2.31
prevent	1954	6.082	26	3.91	1.38	2.28
restore	580	1.805	34	5.52	-1.12	1.97
public	12197	37.962	141	8.68	-0.74	1.95
purchase	1208	3.760	22	3.89	-1.09	1.87
achieve	1565	4.871	32	4.80	2.16	1.85
law	6916	21.525	202	12.70	-1.86	1.75

probation	193	0.601	17	3.98	-0.47	1.59
merit	467	1.453	65	7.88	1.78	1.17
orange	883	2.748	28	4.77	-1.29	1.08
survive	1141	3.551	23	4.06	2.30	1.02
avoid	2333	7.261	71	7.56	1.99	0.95
reduce	1914	5.957	23	3.55	2.22	0.85
mail	933	2.904	194	13.72	-1.00	0.81
pecking	64	0.199	42	6.45	-1.12	0.77
magnitude	195	0.607	19	4.22	2.05	0.23

Looking at Table 7 below, we see that the bottom of the listing shows collocates whose positional variance is zero (they *always* occur at the same position relative to the node), while items at the top of the listing are significant collocates whose position is more variable. The value of the mean position gives an indication of the positional tendency of the collocation. In the case of the fixed pairs, the mean position is easily interpretable as an absolute value. In the case of the more variable combinations, the mean is less helpful, but nevertheless gives an indication as to whether the collocate seems to appear more to the left, to the right or equally either side.

Table 7. Collocates of *order* (span set to 2:2)

placed	2046	2.123	20	4.00	-0.15	1.93
made	26782	27.785	58	3.97	-0.12	1.86
issued	1314	1.363	28	5.03	-0.75	1.73
preserve	599	0.621	20	4.33	1.00	1.59
maintain	1540	1.598	44	6.39	1.23	1.46
make	21896	22.716	95	7.42	1.41	1.40
bring	4375	4.539	30	4.65	1.40	1.38
create	2384	2.473	27	4.72	1.56	1.28
give	11038	11.452	43	4.81	1.56	1.24
points	4545	4.715	21	3.55	-1.43	1.21
restore	580	0.602	34	5.73	-0.68	1.17
produce	2821	2.927	18	3.55	1.61	1.14
establish	1265	1.312	19	4.06	1.63	1.12
obtain	765	0.794	26	4.94	1.69	1.09
keep	7177	7.446	60	6.78	1.63	1.02
get	21334	22.133	78	6.33	1.74	0.99
new	40708	42.233	159	9.26	-1.26	0.98
achieve	1565	1.624	33	5.46	1.76	0.97
please	2930	3.040	26	4.50	1.69	0.97
things	11821	12.264	37	4.07	1.70	0.94
established	2612	2.710	19	3.74	-0.84	0.90
act	6157	6.388	46	5.84	0.91	0.84
provide	4971	5.157	34	4.95	1.79	0.84
st	4231	4.390	23	3.88	1.83	0.83
ensure	2296	2.382	36	5.60	1.81	0.82
court	7163	7.431	62	6.93	-1.05	0.80
an	99821	103.561	295	11.15	-0.97	0.79
customer	913	0.947	22	4.49	-1.05	0.79

reduce	1914	1.986	26	4.71	1.85	0.78
form	8108	8.412	42	5.18	1.00	0.77
purchase	1208	1.253	21	4.31	-0.76	0.77
probation	193	0.200	16	3.95	-0.81	0.75
world	22242	23.075	78	6.22	-0.95	0.75
book	7538	7.820	30	4.05	0.77	0.73
merit	467	0.484	70	8.31	1.90	0.59
books	3811	3.954	28	4.54	0.89	0.57
social	8412	8.727	97	8.96	-0.94	0.43
law	6916	7.175	146	11.49	-1.95	0.41
natural	3851	3.995	20	3.58	-1.15	0.37
old	15582	16.166	77	6.93	-1.16	0.37
offences	663	0.688	14	3.56	1.07	0.27
mail	933	0.968	175	13.16	-0.99	0.24
working	7407	7.685	40	5.11	-1.05	0.22
public	12197	12.654	134	10.48	-1.01	0.09
accommodate	359	0.372	14	3.64	2.00	0.00
avoid	2333	2.420	68	7.95	2.00	0.00
entry	1307	1.356	34	5.60	1.00	0.00
law	6916	7.175	30	4.17	-2.00	0.00
magnitude	195	0.202	17	4.07	2.00	0.00
mail	933	0.968	14	3.48	-1.00	0.00
orange	883	0.916	26	4.92	-1.00	0.00
pecking	64	0.066	36	5.99	-1.00	0.00
prevent	1954	2.027	28	4.91	2.00	0.00
protect	1536	1.594	23	4.46	2.00	0.00
save	2239	2.323	20	3.95	2.00	0.00
secure	1270	1.318	18	3.93	2.00	0.00
survive	1141	1.184	30	5.26	2.00	0.00
tall	1177	1.221	25	4.76	-1.00	0.00

I have carried out the analysis twice with the span setting altered from 6:6 to 2:2. Comparison of the two listings obtained highlights an interesting feature of the collocational behaviour of English words in text and of the statistical methods being employed. In Table 7, fourteen fixed position collocations are clearly identified, but none are present in Table 6. This seems odd at first, since we might presume that all collocations identified within a span of 2:2 will necessarily be equally well represented in a wider span of 6:6. Indeed they are represented, but their statistical peculiarity is lost amongst the increased volume of data gathered by the wider span setting. In particular, the statistical method used causes the lexicographically important collocation *tall order* to fall much further down the significance ranking, such that it does not appear in the first pageful of the listing in Table 6. Similarly, *working order* does not appear in the 6:6 listing.

The 6:6 span should extend the scope of the collocational analysis so that pairings between words are not necessarily identified within local grammatical patterns (adjective+noun, noun modifiers, adverb+adjective are the most typical collocational frames), and this effect is noticeable by comparison of Tables 6 and 7. However, it is clear from lexicographic practice, and from my own commentary on the collocational listings obtained, that long-range collocations (those which extend beyond one or two words) are not so susceptible of explanation by a model of stereotyping. The items which appear in Table 6 but not in Table 7 include ones which are intuitively appealing, such as *judge* (the person who issues a legal order), *chaos* (an antonym of *order*) and *discipline* (a near-synonym of *order*). These pairings are neither fixed (they have moderately high variance scores in the tables) nor idiomatic. Indeed, they seem hardly stereotyped at all. The items which are lost from the 2:2 listing in the wider span (*tall* and *working*) seem to be worth more than the extra items which are gained. The great majority of idiomatic stereotypes recorded in dictionaries are contained within two or three word sequences. Only a relatively few proverbs, for example *Don't count your chickens before they are hatched* and *I'll cross that bridge when I come to it,* extend beyond this range, and these frozen metaphors are not identifiable at all by the statistical methods I have employed in the collocate program.

7. Conclusion

The statistics of lexis in English have traditionally received little attention from theoretical linguists or lexicographers. A corpus-based analysis of word co-occurrence yields a wealth of data for the benefit of lexicographers, and the collocate program illustrates one approach to the study of word combination. The stereotyping of word combinations is a pervasive feature of the language, and it falls to the dictionary makers to attempt some systematic account of its scope and relation to grammar and meaning. Using one or two appropriate measures of statistical significance, computational analysis can begin to shed light on this complex area of linguistic patterning. Collocations can be characterised in terms of their frequency, their position and the variation in their composition, leading to a fuller and more satisfying account of the full meaning potential of lexical items.

Author's Address:
Cobuild, Westmere, 50 Edgbaston Park Road, Birmingham, B15 2RX. U.K.

Notes

1. This corpus of 25 million words formed part of the Oxford Pilot Corpus, compiled for lexicographic purposes by the Oxford University Press. During 1989-91, this corpus was growing to form a pilot for the British National Corpus, and at the time the bigram analysis was carried out the corpus contained around 250 different text samples, covering the period 1989-91 and including newspapers (approximately 11m words), books, magazines, transcribed meetings, radio programmes, lectures and ephemera.
2. The program was written for the OUP while I was employed as Corpus Project Manager.
3. I will use this notation hereafter to denote a span of two words left of the node and two words right.
4. Ken Church (personal communication) has carried out collocation analyses at AT&T Bell Labs, New Jersey, calculating Mutual Information as the measure of association. He maintains that very long-distance associations have shown up in his results.
5. The corpus evidence indicates that *kith* always occurs with *kin*, but the reverse is not true; there are occasions when *kin* appears with other collocates without the appearance of *kith*.
6. This metric was suggested to me by Nicoletta Calzolari of the Istituto di Linguistica Computazionale in Pisa.

References

Benson, M. 1985. "Collocations and Idioms". *Dictionaries, Lexicography and Language Learning* ed. by R. Ilson, 61-68. Oxford: Pergamon.

Church, K. W. & P. Hanks. 1990. "Word Association Norms, Mutual Information & Lexicography". *Computational Linguistics* 16,1. 22-29.

___________, W. Gale, P. W. Hanks & D. Hindle. 1991. "Using Statistics in Lexical Analysis". *Lexical Acquisition: Using On-line Resources to Build a Lexicon* ed. by U. Zernik, 115-64. Englewood Cliff, New Jersey: Lawrence Erlbaum.

Firth, J. R. 1957. "Modes of Meaning". *Papers in Linguistics* ed. by F. R. Palmer, 190-215. London: Oxford University Press.

Kjellmer, G. 1984. "Some thoughts on Collocational Distinctiveness". *Corpus Linguistics* ed. by J. Aarts & W. Meijs, 163-71. Amsterdam: Rodopi.

Longman Dictionary of Contemporary English. 1978 (P. Procter, ed.), 1987 (D. Summers, ed.). Longman: Harlow.

Oxford Advanced Learner's Dictionary. 1989 (A. Cowie, ed.). Oxford: Oxford University Press.

Sinclair, J. McH, ed. 1987a. *The Collins Cobuild English Language Dictionary.* London & Glasgow: Collins.

____________. 1987b. "Collocation: a Progress Report". *Language Topics: Essays in Honour of Michael Halliday* (Volume II) ed. by R. Steele & T. Threadgold, 319-31. Amsterdam: John Benjamins.

____________. 1991. *Corpus, Concordance, Collocation.* Oxford: Oxford University Press.

Statistical Methods and Large Corpora
A New Tool for Describing Text Types

Junsaku Nakamura
University of Tokushima, Japan

Abstract
This paper describes an attempt to classify text types in large corpora by means of a statistical method called Hayashi's Quantification Method Type III.[1] Quirk et al (1985:1180-83) proposed three semantic classes of verbs: private verbs, public verbs and suasive verbs. In the present study, occurrences of 149 of these verbs, including their inflected forms, were counted in four sub-corpora of the Bank of English[2] and fed into the statistical procedure. The resulting analysis enabled me to describe the structure of the Bank of English from the viewpoint of the use of these particular semantic classes of verbs. In so doing, I was able to make explicit the relationships among the four corpora, the relationships among verbs, and, more important, the relationship between corpora and verbs. The major factors which seem to play a significant role in determining the structure of the corpora are: (a) the private versus public dichotomy proposed by Quirk et al (ibid), and (b) somewhat inconclusively, the spoken versus written dichotomy which one might expect. The results of this study also suggest that some recategorisation of the verbs involved may be required.

0. Introduction

Various methods have been developed to analyse and classify English texts on the basis of the frequency distribution of lexical items and grammatical tags. Nakamura (1985) attempted to compare and evaluate several such methods, including Miyajima's C and Muzutani's D, D' and D'', which were originally proposed to measure the similarities between Japanese texts. Another kind of similarity measure included in the analysis involved the correlation coefficients[3] among texts. Similarity indices[4] obtained by these methods were further processed by cluster analysis and Hayashi's Quantification Method Type IV, both of which process a similarity or distance matrix and produce figures which show the relationships among texts graphically. A matrix of correlation coefficients alone was further fed into factor analysis, a statistical technique which attempts to identify underlying traits or factors hidden in the given frequency data by extracting them from a correlation coefficient matrix. But the results were not really conclusive, yielding more or less the same distribution of texts despite the differences in the methods employed.

In Nakamura (1986), a classification of the same texts was attempted again, but this time a statistical technique called Hayashi's Quantification Method Type III was used to process the vocabulary frequency table directly, without intervening steps such as similarity matrices or correlation coefficients. The results were more satisfactory than the ones discussed in Nakamura (1985), in the sense that the three-dimensional plotting of the texts revealed a distribution which was intuitively more meaningful and reasonable. Nakamura (1987) excluded ubiquitous word types such as *the* and *a* and other function words which are more or less evenly distributed across texts and therefore do not contribute very much to their classification. Also excluded were hapax legomena, which are statistically uninteresting in the sense that they have a high probability of being used by chance. The distribution of the texts thus obtained was again more satisfactory than Nakamura (1986). It yielded a sharper and clearer separation of texts, with an explanation of their distribution being given in more general terms.

As noted in Kučera and Francis (1967) and Francis and Kučera (1982), the use of linguistic categories such as lexis or grammatical tags is quite context-dependent. In other words, there is a strong correlation between types of texts and the lexis used in them, as well as between types of texts and the distribution of grammatical tags in them. This being so, and if the frequency distribution table of a category is given across samples, Hayashi's Quantification Method Type III would seem to be ideal for analysing relationships among items within a given category (e.g. pronouns) and samples (e.g. different types of texts) as well as relationships between categories and samples. Later studies carried out by Nakamura, including Nakamura 1989b, 1990, 1991a, 1991b and 1992, have all been successful in describing the structure of the Brown and the LOB corpora on the basis of the distribution of pronouns and of grammatical tags. One could see very clearly the relationships among 15 genres set up in these two corpora and also the relationships among pronouns and among grammatical tags, and the distribution of genres was neatly explained in terms of the distribution of these two categories. Nakamura (1992b) also used pronominal forms to classify a total of 728 individual texts extracted from the Bank of English. The structure of the corpus was again successfully described on the basis of pronominal forms and the major factor of classification could be attributed to personal involvement or detachment.

It is time now to go beyond pronouns to major categories such as nouns, verbs and adjectives. The main purpose of this paper is therefore to apply the same methodology to the textual data stored in the Bank of English from the viewpoint of particular semantic groups of verbs. In the following sections, I will attempt to show how the various sub-corpora of the Bank of English are related to one another on the basis of the distribution of the verbs in question, and how the verbs are related to one another in the entire corpus. It is hoped that the results may provide further validation of the current methodology in the field of corpus linguistics.

1. The data for the current study

This section describes the data which formed the basis for the current study and which consists of (a) the corpus to be analysed in terms of text types, and (b) the categories selected as input to the statistical method and used as a basis for identifying text types in the corpus.

1.1. The corpus

The corpus on which the study is based consists of four large sub-corpora of the Bank of English, namely the Book Corpus, the Times Corpus, the Spoken Corpus and the BBC 1991 Corpus. The Book Corpus consists of 318 books published in the years 1983 to 1992, mainly by HarperCollins and Thorsons, and covering various genres and topics from mystery and crime fiction to health, fitness and astrology. This corpus contains 236,856 types and 19,333,737 tokens,[5] the average size of each text being 60,000 words. The Times Corpus consists of 122 issues of *The Times*, a daily newspaper published in Britain. These issues cover the months of April through to July 1991 and consist of 188,522 types and 11,089,548 tokens. The average size of each issue is approximately 90,000 words. The Spoken Corpus includes various types of spontaneous spoken English, such as face-to-face interaction, telephone calls and radio broadcasts. It covers such events as friendly conversations, radio phone-in programmes, seminars and lectures. Topics range from current affairs to highly specialised subjects such as 'Coleridge as a critic'. This corpus consists of 438 texts, with an average size of approximately 8,300 words: a total of 49,200 types and 3,630,446 tokens. Finally, the BBC 1991 Corpus consists of programmes broadcast by the British Broadcasting Corporation during the year 1991 and covers news, current affairs, features, and sports reporting among other topics. These are mainly written to be spoken, but some are spontaneous spoken texts. This corpus consists of 87,936 types and 7,205,625 tokens.

The overall size of each corpus, in terms of number of tokens, is as follows. the Book Corpus: 19,333,737; the Times Corpus: 11,089,548; the Spoken Corpus: 3,630,446; the BBC91 Corpus: 7,205,625. This gives a total of 41,259,356 tokens and an average corpus size of 10,314,839 tokens.

1.2 The categories

The categories used as input to the statistical method in this study are the three semantic classes of verbs identified by Quirk et al (1985:1180-83).

Verbs can be classified syntactically as transitive, intransitive or middle verbs, depending on whether or not they take an object. Transitive verbs can be further classified according to whether or not they take a *that*-clause. Among the transitive verbs which take a *that*-clause, a further classification can be made according to the construction allowed in the *that*-clause which follows the verb. For example, in Quirk et al (ibid), the verbs with *that*-clause complementation are divided into factual verbs, suasive verbs, emotive verbs and hypothesis verbs. Factual verbs are followed by a *that*-clause which contains the indicative construction introducing factual or propositional information, as in *They claim that she was misled*. Suasive verbs are followed by a *that*-clause with putative *should* or the mandative subjunctive, as in *People are demanding that she should leave/that she leave*. In British English, a suasive verb can also be followed by a *that*-clause with an indicative verb: *People are demanding that she leaves*. Emotive verbs are followed by a *that*-clause which contains the indicative or putative *should* construction, but are not followed by the mandative subjunctive construction: *I regret that she worries about it/that she should worry about it* but not **I regret that she worry about it*. Finally, hypothesis verbs are followed by a *that*-clause containing the hypothetical past or the *were*-subjunctive: *Suppose that one of us died*; *I wish that she were here*.

Quirk et al further subdivide factual verbs into public and private verbs on the basis of the semantic characteristics of the verb itself. The public type consists of speech act verbs which introduce indirect statements. The private type expresses intellectual states such as belief or intellectual acts such as discovery, which are not observable. The present study uses as input to the statistical method these two types of public and private verbs plus suasive verbs, which express the speaker's intention to cause certain events to happen in the future. The reason for choosing these three types is that, according to Quirk et al, they each comprise a fair number of exponents: 49, 64 and 36 respectively. This makes the results more reliable, by contrast with the very limited number of emotive and hypothesis verbs, namely four and two, which are categorised as minor types by Quirk et al. Moreover, the three semantic classes of verbs selected in this study were included in Biber (1988) who also tried to determine the typology of genres by means of factor analysis based on the frequency distribution of 67 linguistic features across 23 genres. These verbs seemed to play an important role in determining text typology in Biber's study.

Table 1 lists all the verbs included in the present study and gives their raw frequency in each of the four corpora described in 1.1 above. The verbs are ordered alphabetically and each verb is followed by the classification given to it by Quirk et al, as a private (PV), public (PB) or suasive (SU) verb.

Table 1. Frequency of verbs across corpora

NO.	VERB	TYPE	BOOK	TIMES	SP'KN	BBC91	TOTAL
1	accept	PV	3837	2276	495	2062	8670
2	acknowledge	PB	798	532	37	405	1772
3	add	PB	8484	3493	440	1617	14034
4	admit	PB	1572	1686	228	905	4391
5	affirm	PB	145	45	3	22	215
6	agree	PB&SU	2266	2848	1320	3795	10229
7	allege	PB	180	848	26	1329	2383
8	allow	SU	5408	3854	690	3298	13250
9	announce	PB	651	2547	77	3588	6863
10	anticipate	PV	345	280	34	103	762
11	argue	PB	865	1329	344	973	3511
12	arrange	SU	1413	491	98	348	2350
13	ascertain	PV	163	41	0	15	219
14	ask	SU	11754	4388	1950	2903	20995
15	assert	PB	367	181	20	101	669
16	assume	PV	1875	743	398	361	3377
17	beg	SU	322	158	61	39	580
18	believe	PV	6570	5266	1170	4111	17117
19	bet	PB	329	462	272	28	1091
20	boast	PB	162	273	11	28	474
21	calculate	PV	877	291	146	131	1445
22	certify	PB	49	44	1	5	99
23	check	PV	2616	476	349	276	3717

24	claim	PB	1715	2876	152	1707	6450
25	command	SU	386	373	23	135	917
26	comment	PB	519	486	313	466	1784
27	complain	PB	731	823	209	397	2160
28	concede	PB&SU	115	514	6	166	801
29	conclude	PV	689	667	57	459	1872
30	confess	PB	260	191	46	53	550
31	confide	PB	166	49	5	5	225
32	confirm	PB	1200	1341	68	1421	4030
33	conjecture	PV	32	3	0	1	36
34	consider	PV	5293	2855	383	2074	10605
35	contend	PB	104	149	6	27	286
36	convey	PB	381	217	45	93	736
37	decide	PV&SU	3746	2913	735	2120	9514
38	declare	PB	666	1097	52	1621	3436
39	decree	SU	61	43	1	39	144
40	deduce	PV	99	25	10	11	145
41	deem	PV	135	205	12	43	395
42	demand	SU	1452	1198	55	1354	4059
43	demonstrate	PV	1249	667	88	384	2388
44	deny	PB	801	1324	114	1572	3811
45	desire	SU	771	85	12	57	925
46	determine	PV&SU	2052	1125	134	764	4075
47	discern	PV	195	42	8	5	250
48	disclose	PB	122	552	6	266	946
49	discover	PV	2833	1187	235	567	4822
50	doubt	PV	573	342	84	157	1156
51	dream	PV	680	229	56	45	1010
52	enjoin	SU	15	7	0	0	22
53	ensure	PV&SU	1460	1315	44	800	3619
54	entreat	SU	17	8	0	0	25
55	establish	PV	2376	1588	148	1476	5588
56	estimate	PV	594	981	49	590	2214
57	exclaim	PB	221	30	0	1	252
58	expect	PV	4415	5586	701	4426	15128
59	explain	PB	3689	1371	431	1263	6754
60	fancy	PV	256	150	134	17	557
61	fear	PV	838	976	42	800	2656
62	feel	PV	20242	4302	2753	1978	29275
63	find	PV	26642	8512	3323	4729	43206
64	forecast	PB	174	470	6	79	729
65	foresee	PV	123	124	4	56	307
66	foretell	PB	73	6	1	6	86
67	forget	PV	2646	783	552	251	4232
68	gather	PV	1440	621	94	886	3041

69	grant	SU	681	690	49	602	2022
70	guarantee	PB	341	603	87	393	1424
71	guess	PV	1194	183	184	31	1592
72	hear	PV	7544	2563	2042	1631	13780
73	hint	PB	147	157	4	157	465
74	hold	PV	8271	5725	875	6901	21772
75	hope	PV	2495	2642	967	2470	8574
76	imagine	PV	2278	517	465	174	3434
77	imply	PV	752	365	87	167	1371
78	indicate	PV	3476	980	100	972	5528
79	infer	PV	118	29	14	18	179
80	insist	PB&SU	947	1336	51	1132	3466
81	instruct	SU	378	151	13	81	623
82	insure	PV	43	92	60	56	251
83	intend	SU	1356	1306	149	1029	3840
84	judge	PV	732	574	78	170	1554
85	know	PV	36101	7974	28996	4868	77939
86	learn	PV	6178	1698	614	539	9029
87	maintain	PB	2367	1466	113	1206	5152
88	mean	PV	11537	3135	17500	1959	34131
89	mention	PB	2443	612	606	300	3961
90	move	SU	7404	3467	1310	2599	14780
91	note	PV	3333	534	76	513	4456
92	notice	PV	2548	380	353	113	3394
93	object	PB	190	216	86	154	646
94	observe	PV	1579	432	71	421	2503
95	ordain	SU	81	37	8	5	131
96	order	SU	1036	857	82	1107	3082
97	perceive	PV	857	233	72	108	1270
98	pledge	SU	42	254	19	329	644
99	pray	SU	458	120	29	67	674
100	predict	PB	501	696	75	503	1775
101	prefer	SU	1669	925	141	264	2999
102	presume	PV	191	59	130	35	415
103	presuppose	PV	31	7	4	3	45
104	pretend	PV	509	193	83	55	840
105	proclaim	PB	360	170	9	141	680
106	promise	PB	1603	1180	98	856	3737
107	pronounce	PB&SU	391	170	105	39	705
108	prophesy	PB	71	21	4	1	97
109	propose	SU	587	1443	78	1385	3493
110	protest	PB	266	230	33	537	1066
111	prove	PV	2650	2301	261	882	6094
112	realise	PV	3471	1134	602	412	5619
113	reason	PV	357	103	33	16	509

114	recall	PV	1089	690	76	175	2030
115	reckon	PV	588	283	195	71	1137
116	recognise	PV	2718	1092	244	1249	5303
117	recommend	SU	2470	664	94	330	3558
118	reflect	PV	1725	1102	110	623	3560
119	remark	PB	380	175	7	39	601
120	remember	PV	6221	1238	1577	242	9278
121	repeat	PB	2394	671	230	792	4087
122	reply	PB	975	502	46	165	1688
123	report	PB	2648	2172	165	5946	10931
124	request	SU	279	200	15	176	670
125	require	SU	3897	1911	202	773	6783
126	resolve	SU	610	508	63	960	2141
127	retort	PB	72	47	0	13	132
128	reveal	PV	1748	1285	65	679	3777
129	rule	SU	1197	1374	101	2068	4740
130	say	PB	47666	45879	19796	58143	171484
131	see	PV	41642	10312	10313	6768	69035
132	sense	PV	513	151	26	46	736
133	show	PV	8234	5425	1075	2233	16967
134	signify	PV	746	37	2	28	813
135	state	PB	1233	679	91	253	2256
136	stipulate	SU	35	54	5	39	133
137	submit	PB	227	404	26	236	893
138	suggest	PB&SU	5426	3023	501	1837	10787
139	suppose	PV	3249	759	1946	368	6322
140	suspect	PV	1030	587	180	674	2471
141	swear	PB	488	134	56	134	812
142	testify	PB	130	89	2	82	303
143	think	PV	24154	7056	30795	5596	67601
144	understand	PV	7216	1875	1329	1251	11671
145	urge	SU	451	858	25	1157	2491
146	vote	SU	140	1203	313	1574	3230
147	vow	PB	89	90	2	41	222
148	warn	PB	1481	1745	111	2527	5864
149	write	PB	8180	4858	3232	1195	17465
TOTAL			**444200**	**228052**	**149199**	**193694**	**1015145**

2. Methodology and statistical technique

2.1 Basic procedures: lemmatising, disambiguating & adjusting figures

The 149 verbs listed in Table 1 above were lemmatised prior to counting their occurrences in the four corpora. The 8,180 occurrences of *write* in the Book Corpus for instance include all the occurrences of *write, writes, wrote, written* and *writing* in that corpus.[6] Taking the lemma as the sole basis for counting the occurrences of these verbs has certain implications. The most important implication is that the data which was fed into the statistical procedure included not only instances where the verbs in question are followed by a *that*-clause as an object, but also instances where an object clause is omitted, where the object of a verb is not a clause but a noun, or where the verb is used adjectivally in the participle form, either present or past. In addition, cases where the verb is followed by a particle or a preposition, for example *conjure up* or *confide in*, were also included. The latter examples involve intransitive uses of the verbs in question. These uses were still included because the verbs clearly maintained the semantic characteristics discussed in 1.2 above. Homonyms of the base and the past forms which are not verbs, for example *mean* as noun or adjective and *thought* as noun, were carefully excluded from the counting.[7] Some nouns ending in *-ing*, such as *remembering*, were however counted as verbs. It is generally very difficult to distinguish between nominal and verbal uses in the case of *-ing* forms.

In addition to lemmatisation and some basic disambiguation of grammatical classes, the raw frequency figures given in Table 1 were also adjusted to the average corpus size, i.e., 10,314,839 words. The adjusted frequency figures were then fed into the statistical procedure known as Hayashi's Quantification Method Type III.

2.2 *Hayashi's quantification method type III*

Generally speaking, Hayashi's Quantification Method Type III is used for quantifying qualitative categories and samples simultaneously. The distinctive feature of the method is that it can classify or quantify both categories and samples merely by looking at qualitative dichotomic response patterns, i.e., whether each sample reacts positively or negatively to several categories. Given a data matrix the size of which is the number of samples by the number of categories, the basic principle is to rearrange the rows and the columns so that the positive responses can converge around the diagonal. The categories close to one another, and the samples close to one another, are considered to be qualitatively similar; those distant from one another are said to be qualitatively different.

Hayashi's Quantification Method Type III performs the task of rearranging the data matrix, not physically but numerically, giving the categories close to one another numerical values which are close to one another; these are called category weights. It also gives the samples close to one another numerical values which are close to one another; these are called sample scores. Using the quantities *x* and *y* which represent the quantities given to samples and categories respectively, the procedure is to determine the numerical values of x_i and y_j which maximise the correlation coefficient between *x* and *y*. The algorithm extended to deal with ordinary cross tables containing frequency figures was used to process the adjusted frequency figures mentioned above, with the four corpora treated as categories and the 149 verbs as samples.

If there are '*n*' categories and more than '*n*' samples, Hayashi's Quantification Method Type III can produce *n-1* sets, or *axes* to use a statistical term, of category weights and sample scores. Consequently, in the present case there are three ways of assigning numerical values to corpora and verbs, each of which yields a correlation coefficient lower than that of the set produced in the previous stage. The set of values producing the highest correlation coefficient is called the first axis or Axis 1, the second set is called Axis 2, and so on. Only the first two sets were used here since the cumulative proportion accounted for by these two axes amounted to 96.4%, leaving only 3.6% of the information contained in the adjusted frequency figures unaccounted for. Table 2 gives the correlation coefficient, the proportions accounted for and the cumulative proportion accounted for by each of the three axes.

Table 2. Correlation coefficient, proportion accounted for, and cumulative proportion accounted for for each axis.

AXIS	COR. CO.	PROP. A/C FOR	CUM. PROP. A/C FOR
1	0.5376905D+00	0.7520587D+02	0.7520587D+02
2	0.2853853D+00	0.2118606D+02	0.9639193D+02
3	0.1177725D+00	0.3608068D+01	0.1000000D+03

The two sets of category weights and sample scores calculated for producing the above correlation coefficients were then normalised, with the means equal to 0 and the variances equal to 1.0. Normalised category weights thus calculated are given in Table 3. Table 4 shows, by way of exemplification, sample scores given to the first ten of the 149 verbs listed alphabetically in Table 1.[8]

Table 3. Category weights given to each corpus

CORPUS	AXIS 1	AXIS 2
BOOK	0.1333368D+00	0.1767771D+01
TIMES	0.8495521D+00	0.2782209D+00
SPOKEN	-.1232992D+01	-.4307540D+00
BBC91	0.1121152D+01	-.1065223D+01

Table 4. Sample scores given to each verb

NO.	VERB	AXIS 1	AXIS 2
1	accept	0.7958262D+00	0.1881962D+00
2	acknowledge	0.1155567D+01	0.4964046D+00
3	add	0.7244712D+00	0.1823449D+01
4	admit	0.8968765D+00	0.2093224D+00
5	affirm	0.8254461D+00	0.2446622D+01
6	agree	0.5530824D+00	-.1216705D+01
7	allege	0.1771166D+01	-.2044218D+01
8	allow	0.8917558D+00	0.5977609D-01
9	announce	0.1739141D+01	-.1863443D+01
10	anticipate	0.7887033D+00	0.1013577D+01

3. Discussion of results

The numerical values shown in the above tables indicate the relative positions of each corpus or each verb along two axes, but they do not, in their raw state, enable us to grasp the overall picture. Mere numerical values are always difficult to process. The values in Table 3 and in the extended version of Table 4 are therefore plotted in Figure 1 and Figure 2 below to show the relative position of each item in a two-dimensional space. The plotting of the values was achieved by arbitrarily assigning the numerical values which produced the highest correlation coefficient to the horizontal axis of the figure, hence this axis is called Axis 1. The vertical axis of the figure, which represents the second set of numerical values, is called Axis 2 for the same reason. These figures also contain another piece of information obtained from the adjusted frequency table, namely the relative size of the item in question indicated by the size of the circle. For example, Figure 1 shows that the Spoken Corpus uses the largest number of the verbs in question, in contrast to the Times Corpus which has the smallest number. Now that the numerical values are all transformed into points in a two-dimensional space, it becomes possible to grasp visually the relationships of items to one another.

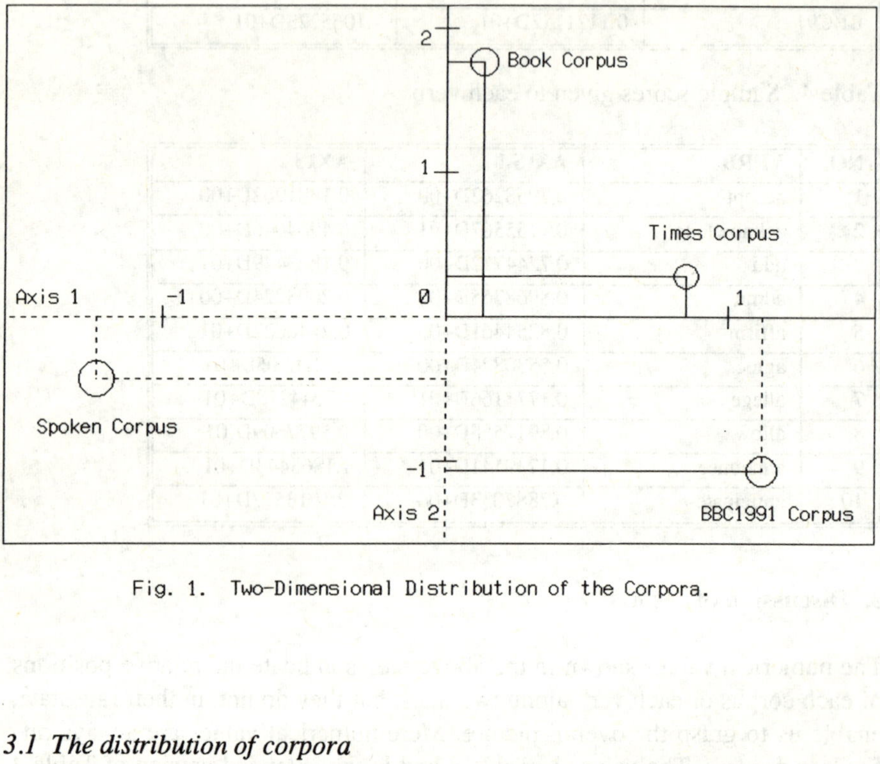

Fig. 1. Two-Dimensional Distribution of the Corpora.

3.1 *The distribution of corpora*

A glance at Figure 1 shows that the four corpora contrast sharply with one another. Along Axis 1, which explains more than three quarters of the information contained in the original frequency table, the Spoken Corpus assumes an extreme position in the negative range, with the other three corpora located in the positive range.[9] The BBC Corpus is located at the opposite extreme of this axis, with the Times Corpus close to it and the Book Corpus close to the origin of the axis. Along Axis 2, the Book Corpus is quite conspicuous, assuming a very high positive value in contrast to the BBC Corpus with a high negative value. The Times and the Spoken Corpora are located in between, with the Times in the positive range and the Spoken in the negative range.

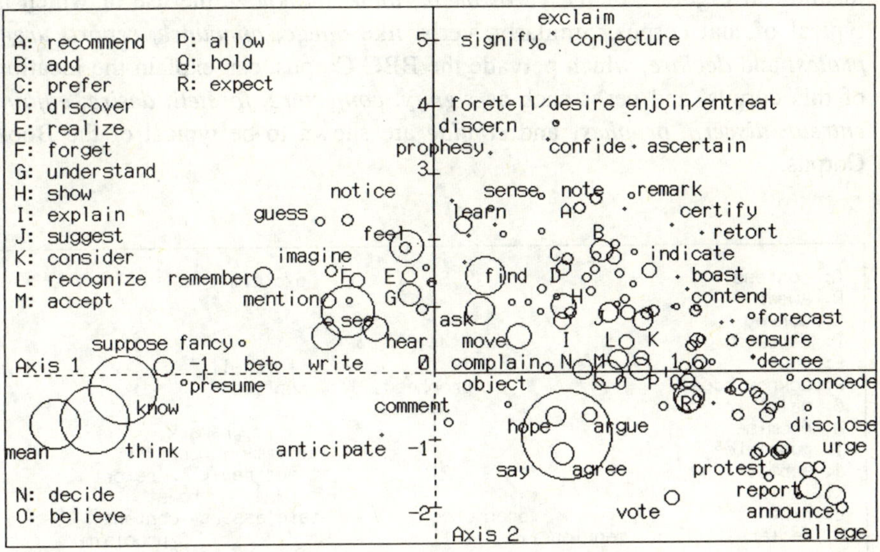

Fig. 2. Two-Dimensional Distribution of the Verbs.

The kind of distribution we see in Figure 1 is in line with our intuition. The four kinds of texts are generally assumed to be significantly different from each other, which is precisely why these four groups have been set up in the Bank of English. The Spoken texts in particular are expected to be very different from other kinds of texts and the fact that the newspaper texts are situated between books and BBC broadcasts seems to be intuitively reasonable. But this intuitively satisfying distribution can be explained more objectively in terms of the distribution of our 149 verbs, as shown in Figure 2. As this figure contains the relative locations of as many as 149 verbs, there is inevitably a great deal of overlap which obscures the individual verbs. Figures 3, 4, 5 and 6 below give a clearer view of the distribution of the three types of verbs: public, private and suasive.

Patterns of distribution relating to corpora, as observed in Figure 1, are clearly reflected in the distribution of verbs as shown in Figure 2, and vice versa, since both category weights and sample scores are given to maximise the correlation coefficient between the two. In other words, the distribution of corpora in Figure 1 may be explained with reference to Figure 2. For example, the location of the Spoken Corpus in Figure 1 coincides more or less with the location in Figure 2 of the verbs *mean, think* and *know*, the use of which is typical of that corpus. Similarly, verbs like *allege, announce, report, urge, protest* and *declare*, which pervade the BBC Corpus, can explain the location of this corpus, and verbs such as *signify, conjecture, foretell, desire, enjoin, entreat, discern, prophesy* and *confide* are shown to be typical of the Book Corpus.

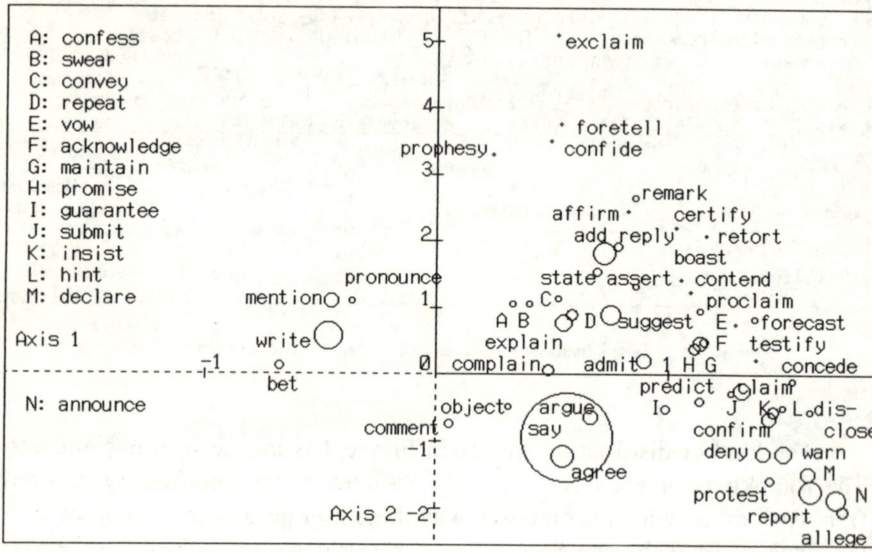

Fig. 3. Distribution of Public Verbs.

The patterns of distribution become clearer if we extract one type of verb at a time from Figure 2. For example, Figure 3 shows only the locations of Quirk et al's category of public verbs. Here, most of the public verbs are located in the positive range along Axis 1, with only four exceptions: *pronounce, mention, write* and *bet.* In fact, no verb is located beyond -1, the range within which one would expect to find verbs which are used pervasively in the Spoken Corpus. Even *say*, which represents the largest sample within this group of verbs and is naturally expected to occur with very high frequency in the Spoken Corpus, is situated towards the BBC Corpus. It is in fact the BBC Corpus which shows the highest frequency of use of this particular verb.

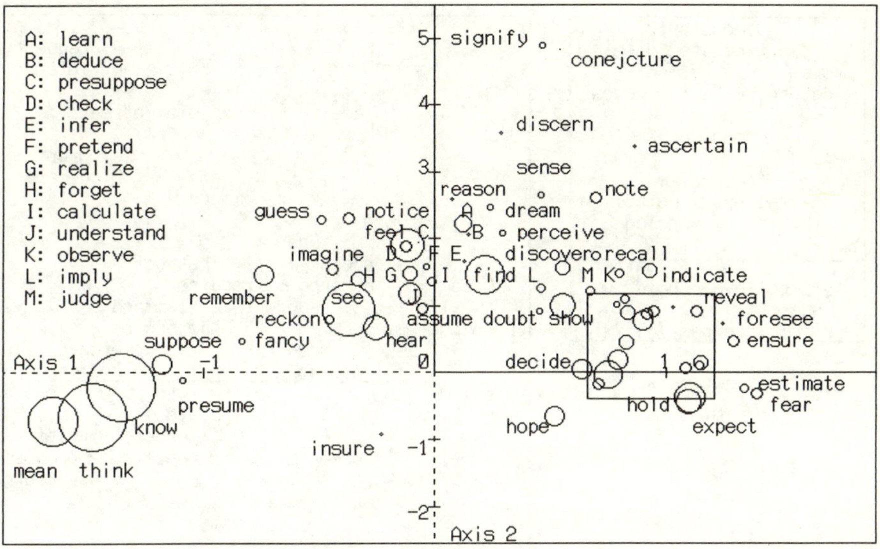

Fig. 4. Distribution of Private Verbs.

In contrast to public verbs, private verbs are scattered right across Figure 4, except where we would expect to find strongly BBC-oriented verbs. A small number of verbs such as *fear, estimate, expect, hold* and *hope* are located rather close to the BBC Corpus. However, major verbs in this category, for example *think, know, mean, suppose, remember, see, hear, feel* and *understand* are all located in the negative range of Axis 1. What is worth noting in this context is that some of the verbs classified as private types by Quirk et al, for example *show, indicate, demonstrate, reveal* and *prove* may need to be reassessed, since both their semantic content and the sheer fact of their distribution in the corpora would seem to indicate that they are public rather than private verbs.

As for suasive verbs, Figure 6 shows that they are all located in the positive range of Axis 1, with the exception of two tiny samples of *pronounce* and *beg*, both of which are located in the negative range, with *beg* very close to Axis 2. This is quite natural since many of these verbs are public in the sense that they involve someone doing something in the future. However, some verbs which are purely private, such as *intend*, are also located in the positive range. This sort of tendency is also evident in the case of private verbs.

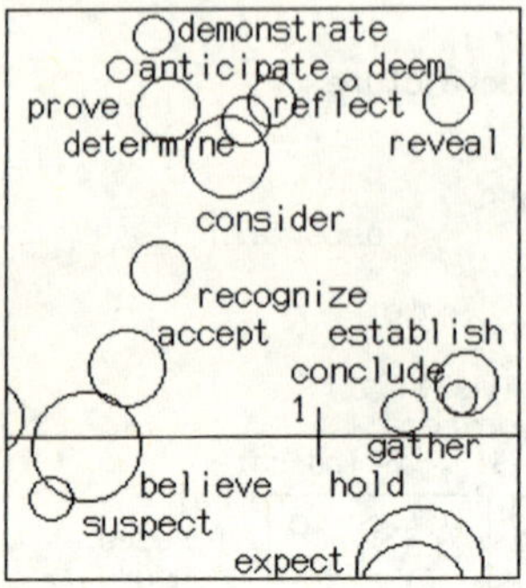

Fig. 5. Extract from Fig. 4.

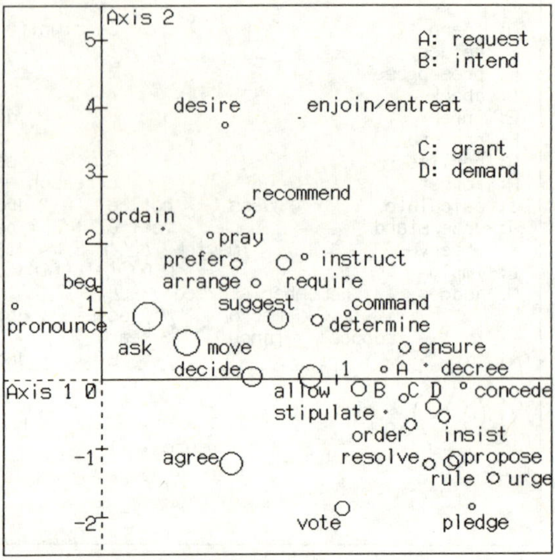

Fig. 6. Distribution of Suasive Verbs.

3.2 The interpretation of the axes

The axes which organise the above figures can normally be given specific interpretations in context. The interpretation of Axis 1 in the present case seems quite straightforward. The distribution of verbs indicates that the distinction between public and private domains is clearly shown along this axis, the only reservation being that some private verbs are used more frequently in public domains such as newspaper texts and broadcasting. So we may safely say that the BBC Corpus is the most publicly-oriented of the four corpora, followed by the Times. By contrast, the Spoken Corpus is the most privately-oriented and the Book Corpus occupies a middle position between the public and private domains.

The interpretation of Axis 2 is not quite so straightforward, but there is some evidence to suggest that the written vs. spoken dichotomy is in operation here. Verbs which are typically used in the Spoken Corpus, for instance *think*, *know, mean* and *suppose*, appear in the negative range of Axis 2 or close to its origin. The same is true of verbs such as *say, hope, agree, argue, expect* and *believe*. Those verbs which are located close to the origin of the axis are the ones which are used more or less evenly across the corpora, which simply means that spoken texts tend to use their share of such high frequency verbs. The Book Corpus, by contrast, assumes an extreme positive position along Axis 2. But the problem here is how to account for the BBC Corpus, which assumes an extreme negative position beyond the Spoken Corpus. Some of the BBC texts, such as interviews on current affairs, are spontaneously spoken, which explains the BBC's location close to the Spoken corpus. But many of the BBC texts, including news broadcasts, are carefully scripted to be spoken. Verbs which are typical of the BBC Corpus, such as *allege, announce, report, declare, protest, pledge, urge, rule* and *propose*, are not used that frequently in the Spoken Corpus, but they are also rare in the Book Corpus. So we may tentatively propose that the BBC Corpus has its own unique style, partly close to spoken texts but different from them because of its publicly-oriented written-to-be-spoken nature. At any rate, along Axis 2, written and written-to-be-spoken texts are located at the extremities, both positive and negative. Purely spoken texts are located in the negative range, close to the origin, where large numbers of high frequency verbs are also located. Although written, the Times Corpus is located between the Book Corpus and the Spoken Corpus. This is due to its use of publicly-oriented verbs.

4. Conclusion

This paper has argued that our intuitive knowledge concerning the distribution of texts and corpora and the use of such categories as Quirk et al's semantic classes of verbs can be tested objectively. A methodology for achieving this has been explained and its usefulness demonstrated. Hayashi's Quantification Method Type III has been shown to provide an effective tool for identifying and describing text types in large corpora. This statistical method assumes, as its starting point for analysis, that there is mutual dependency between types of texts and the linguistic categories used in them, an assumption which reflects the facts of natural languages. Mutual dependency is presupposed because the values are obtained in such a way as to maximise the correlation coefficient between types of texts and the categories chosen for analysis. This contrasts sharply with other statistical methods discussed in Nakamura (1985), including factor analysis which is also used by Biber (1988), where dependency is discovered only as a result of the analysis performed on similarity matrices which consist of correlation coefficients. Moreover, Hayashi's Quantification Method Type III offers the added advantage of showing us the relationships among linguistic categories as well as the relationships among texts. This enables the analyst to compare the two directly and to explain the distribution of texts in terms of the distribution of linguistic categories and vice versa.

It is hoped that the strengths of this method will recommend it to scholars in the field of corpus linguistics and that it will help us cope more effectively in future with the ever-increasing amount of textual data which is gradually becoming available to linguists.

Author's Address:
College of Integrated Arts & Sciences, University of Tokushima, 1-1 Minamijosanjima-cho, Tokushima-shi 770, JAPAN.

Notes

1. This research was supported by a grant from the Japanese Ministry of Education to study abroad for the fiscal year of 1991 (Grant No. 3-Kou-210). I would also like to express my gratitude to Zoe James and Tim Lane at Cobuild for their patient assistance and support on the computing side, without which this work would not have been completed.
2. The Bank of English is the title of the current archive of texts held at Cobuild, Birmingham. This consists of several corpora totalling over 200 million words of running text.
3. This is a statistical technique which is used to measure the degree of correlation between two variables. If the coefficient is zero, the correlation is random. But the closer it is to 1 or −1, the greater the correlation. For simple explanations of this and other statistical techniques mentioned in this paper, see Nakamura (1985).
4. Similarity indices indicate numerically how close the texts are to one another or how distant they are from each other.
5. Individual occurrences of the words are referred to as tokens in corpus linguistics, whereas a class of tokens having the same formal characteristics is referred to as a type. For example, the word *the* is a type consisting of 1,114,320 tokens in the Book Corpus.
6. The spelling variants of *realise* and *recognise*, namely *realize* and *recognize*, were treated together.
7. Disambiguation of the word form *mean* in the Spoken Corpus was performed proportionately because the present system installed in Cobuild restricts the number of citations which can be seen in a KWIC format to 16,000. This is less than the actual number of citations of *mean* in the Spoken Corpus, namely 17,642. So, 16,000 citations were inspected and the rest were calculated proportionately.
8. For reasons of space, the full list of sample scores could not be shown here. This is however available from the author on request.

9. The word *negative* is used here merely to indicate that a given numerical value is less than 0 and the word *positive* to indicate that a given value is more than 0. Neither is used as an evaluative term in this context.

References

Biber, D. 1988. *Variations Across Speech and Writing.* Cambridge: Cambridge University Press.

_________. 1989. "A typology of English texts." *Linguistics* 27. 3-43.

Francis, W. N. & H. Kučera. 1982. *Frequency Analysis of English Usage: Lexicon and Grammar*. Boston: Houghton and Mifflin Co.

Kučera, H. & W. N. Francis. 1967. *Computational Analysis of Present Day American English.* Providence: Brown University Press.

Miyajima, T. 1970. "Goi no Ruijido" (Similarity of Vocabulary). *Kokugogaku (Japanese Linguistics)* 82. 42-64.

Muzutani, S. 1977a. "Keiryo-goiron kara mita Myojoha to Negishiha" (Myojo School and Negishi School from the Viewpoint of the Quantitative Analysis of Vocabulary). *Keiryo-Kokugogaku (Quantitative Japanese Linguistics)* 11. 30-7.

__________. 1997b. *Suri-Gengogaku (Mathematical Linguistics).* Tokyo: Baifu-kan.

Nakamura, J. 1985. "On the Methodologies of Quantitative Groupings of English Texts." *JACET Bulletin* 16. 133-48.

__________. 1986. "Classification of English Texts by Means of Hayashi's Quantification Method Type III." *Journal of Cultural and Social Science* (University of Tokushima) 21. 71-86.

__________. 1987. "Notes on the Use of Hayashi's Quantification Method Type III for Classifying English Texts." *Journal of Cultural and Social Science* (University of Tokushima) 22. 127-45.

__________. 1989a. "Creation of a Vocabulary Frequency Table from the Brown Corpus." *Journal of Cultural and Social Science* (University of Tokushima) 24. 171-82.

__________. 1989b. "A Quantitative Study on the Use of Personal Pronouns in the Brown Corpus." *JACET Bulletin* 20. 51-71.

__________. 1990. "A Study on the Structure of the Brown Corpus Based upon the Distribution of Grammatical Tags." *Journal of Foreign Languages and Literature* (University of Tokushima) 1. 13-35.

__________. 1991a. "A Study on the Structure of the Brown Corpus Based upon the Distribution of its Vocabulary Items." *Journal of Foreign Languages and Literature* (University of Tokushima) 2. 27-47.

__________. 1991b. "The Relationships among Genres in the LOB Corpus Based upon the Distribution of Grammatical Tags." *JACET Bulletin* 22. 55-74.

__________. 1992a. "The Comparison of the Brown and the LOB Corpora Based upon the Distribution of Grammatical Tags." *Journal of Foreign Languages and Literature* (University of Tokushima) 3. 43-58.

__________. 1992b. "Hayashi's Quantification Method Type III: A Tool for Determining Text Typology in Large Corpora." Unpublished MS, University of Tokushima, Japan.

__________. Forthcoming. "Quantitative Comparison of Modals in the Brown and the LOB Corpora." To appear in *ICAME Journal* 16.
Sinclair, J. 1991. *Corpus, Concordance, Collocation*. Oxford: Oxford University Press.
Quirk, R., S. Greenbaum, G. Leech and J. Svartvik. 1985. *A Comprehensive Grammar of the English Language*. London: Longman.
Yasuda, S. & M. Umino. 1968, 1977. *Shakai Tokeigaku (Social Statistics)*. Tokyo: Maruzen.

The Automatic Analysis of Dictionaries
Parsing Cobuild Explanations

Geoff Barnbrook
School of English, The University of Birmingham

Abstract
Dictionaries provide information about the semantics of their headwords and the lexical relations between these and other words. If dictionary definitions were susceptible to automatic analysis, this information could be made directly available to a computer and would provide an immensely powerful resource for natural language processing applications. Cobuild dictionaries are particularly suited to this purpose because they use natural English sentences to explain both the meanings and the typical linguistic environments of their headwords. A parser capable of producing an automatic analysis of Cobuild explanations would therefore be more powerful and more flexible than one which operates on a conventional dictionary. Such a parser is currently being developed by the writer. This paper describes the strategies used in developing and enhancing the Cobuild parser and outlines some of its main applications in lexicography and natural language processing.

0. Introduction

Dictionary definitions contain a great deal of valuable information. This information can normally only be accessed in a limited, linear way, by looking up headwords which are arranged alphabetically. If the different functional components of definitions could be identified in a way that makes them amenable to machine analysis, the computer could access and interpret them directly. The output from this functional analysis could then be used for a wide range of natural language processing applications.

This paper describes the development of a parser which analyses the functional components of dictionary definitions. It should be emphasised at the outset that this is not a syntactic parser; it is a tool for identifying functional elements such as 'superordinates' and 'discriminators', irrespective of their syntactic status. This work is being carried out by the writer, under the supervision of John Sinclair. It initially began as part of the Chamberlain Project, a research programme supported by IBM. Current work on the parser forms part of a Eurotra project, carried out under the auspices of the European Commission.

0.1 Cobuild explanations vs. conventional definitions

Dictionary definitions set out to describe what their headwords mean. They usually take the form of a complex, highly structured explanation, which is buried in a densely encoded system of abbreviations, giving various types of information about the headword, including its register, grammar and pronunciation. The concept of meaning itself can vary from a prescriptive notion of correctness, based on the lexicographer's intuition, to a more neutral description of observed usage. In the Collins Cobuild range of dictionaries, meaning is taken to be inseparable from usage. Usage, in turn, is ascertained by access to the corpora that have been collected at Birmingham since the project began in 1980. Moreover, in order to convey the meaning associated with observed usage as clearly as possible, Cobuild explanations are expressed in natural language rather than in the specialised, highly compressed forms encountered in other dictionaries.

Hanks (1987) points out that conventional dictionaries attempt to produce a precise and substitutable expression of the meaning of each headword, without making any allowance for the inherent imprecision of natural language. The Cobuild approach, by contrast, attempts to explain headwords in such a way as to be "both true to the facts and informative about lexical details" (ibid:119). For example, the word *establish* is explained in the *Collins Cobuild Student's Dictionary* (Sinclair 1990) as follows:

1. If you **establish** an organization or a system, you create it.
2. If you **establish** contact with a group of people, you start to have discussions with them.
3. If you **establish** that something is true, you prove that it is definitely true.

The first two explanations contain a direct object immediately after the headword, and both relate to different ways of *establishing* the type of entity or process described by that object. The third explanation is followed by *that* and gives a completely different category of meaning. A comparison with the definitions provided by other learner's dictionaries may be useful here. From the *Oxford Advanced Learner's Dictionary* (1989) we have:

1. set (sth) up on a firm or permanent basis.
2. **~ sb/oneself (in sth) (as sth)** place sb/oneself in a position, office etc, usu on a permanent basis.
3. show (sth) to be true; prove.
4. cause people to accept (a belief, custom, claim, etc).

and from the *Longman's Dictionary of Contemporary English* (1987):

1. to set up; to begin; CREATE.
2. [(**as, in**)] to cause to be firmly settled or accepted in a particular state or position; put beyond doubt.
3. to find out or make certain of (a fact, answer, etc.).
4. to make (a religion) official for a nation.

The difference between the two approaches should now be evident. The explanations in the Cobuild dictionary contain the information needed to disambiguate the word *establish* in a given context, in sentences which contain that context; in this case, the explanation encodes the complementation patterns of the verb. The Longmans and Oxford dictionaries provide some of the information but do not fix it in a clear contextual relationship with the headword. In both dictionaries, sense 3, which corresponds to Cobuild's sense 3, lacks the context needed to identify it. This context is clearly included in the explanation from the *Student's Dictionary:* "If you **establish** *that* ...". It is this unique property of the Cobuild explanations which makes them more useful as a subject for automatic analysis than conventional definitions.

0.2 The Student's Dictionary and its usefulness as a sample

The strategies which were developed to enable the lexicographers who worked on the original Cobuild Dictionary (Sinclair 1987) to select the most appropriate form of explanation for each word are described by Hanks (1987). The actual structures selected during the compilation of the various Cobuild dictionaries are the result of an interaction between general editorial policy and the lexicographers' own decisions on individual instances. Some examples of the main explanation patterns, taken from the *Student's Dictionary,* may be useful to illustrate how these strategies were eventually codified.

The most common basic pattern for transitive verbs is illustrated by the explanation of *establish* given above. Transitive and intransitive senses of the same headword are clearly distinguished by the co-text, as can be seen in the following examples:

If someone **develops** a new machine, they produce it by improving the original design.
When something **develops**, it grows or changes over a period of time into a better, more advanced, or more complete form.

Count nouns are often dealt with in a similar way to *packing case*, while *needlework* illustrates the typical pattern adopted for uncount nouns:

A **packing case** is a large wooden box in which things are stored or taken somewhere.
Needlework is sewing or embroidery that is done by hand.

Adjectives are explained using a variety of patterns, depending, for example, on whether they are typically predicative or attributive:

If you are **disinclined** to do something, you do not want to do it.
Close contact or co-operation involves seeing or communicating with someone often.

This general style of explanation, and the rationale behind it, were gradually developed in the course of compiling the *Collins Cobuild English Language Dictionary* (Sinclair 1987). The patterns established during the production of this work were refined, in some cases simplified, and perhaps applied more consistently in the dictionaries which followed. At the time of writing, the *Students' Dictionary* is the most recent, and the smallest. The list of headwords and the number of senses given for each headword are both restricted by comparison with the larger dictionaries which preceded it. This may suggest that an exploration of the Cobuild language of explanation based on the *Student's Dictionary* could be incomplete in some senses. In practice, however, the *Student's Dictionary* has proved ideal as a starting-point for the development of the parser.

The philosophy underlying the form of explanations chosen for the Cobuild range was developed before the production of the first edition, but experience with the production of subsequent versions of the dictionary has inevitably modified the detailed implementation of that philosophy in explanation structures. To some extent, this means that the forms of explanation used in the *Student's* version are likely to be more suitable as a basis for the development of a generally useful explanation parser. It is clearly pointless to develop algorithms which deal with problems that will not be found in subsequent editions of the dictionary.

1. The analysis of Cobuild explanations: structural patterns

In most dictionaries, the headword (or 'definiendum', the item to be defined) is completely separate from the text which defines it (its 'definiens') and acts as a label for it.

establish	set (sth) up on a firm or permanent basis
(definiendum)	(definiens)

The computer analysis of a set of explanations which follows this basic pattern, taken from the *Longman Dictionary of Contemporary English*, has been described by Alshawi (1989), but the equivalent task for Cobuild explanations is rather different. In the Cobuild dictionaries, the explanation contains both the definiendum and its definiens.

If you **establish** an organization or a system, you create it.

A functional analysis must begin by distinguishing these two components from each other and identifying any material in the explanation which cannot be allocated to either component.

The ultimate purpose of analysing dictionary definitions, or explanations in the case of Cobuild dictionaries, is to identify their major functional components so that they can be used to extract or generate information about the headword. To start with then, the individual components of the definiendum need to be identified. In Cobuild explanations, the definiendum typically includes other items beside the headword, such as *you* in the above example. A similar analysis must then be performed for the definiens, an analysis which would separate extraneous items such as *you* and *it* from the superordinate *create* and the typical objects: *organisation* and *system.* But although the aim is to perform this type of functional analysis eventually (because this type of analysis can inform natural language processing and similar applications), the starting point has to be a structural analysis of the explanation patterns used in the dictionary. It is only by developing a taxonomy based initially on linear word patterns, and even on such things as the position of typesetting codes (see 1.1 below), that we can develop a functional taxonomy of explanation types, and hence be able to retrieve the relevant information automatically from any given explanation.

1.1 Basic structures of Cobuild explanations

The first characteristic of the explanations that becomes apparent from the most cursory examination is the fact that most of them have the structure:

> explanation text **headword** more explanation text

in which the dictionary typesetting codes which turn bold text on and off divide the explanation into three fields: the text before the headword, the headword itself, and the text which follows the headword. This division has been extremely useful in the development of the parser: the patterns found in these two areas of text (before and after the headword) differ significantly, and the division therefore provides a rough but effective method for localising pattern analysis. Most of the explanations divide into three fields on this basis, but a few consist of only one field while some have as many as twelve.

1.1.1 Standard (three field) explanations

On the basis of this initial crude distinction of bold/non-bold type, 91% of the explanations in the *Student's Dictionary* divide into three fields. There is however sufficient variation within this pattern to make further consideration worthwhile.

In many cases, especially the explanations beginning with *a* or *an*, the definiendum simply corresponds to the first two fields of the explanation, so that in explanation 1 of *label:*

> a **label** is a piece of paper or plastic that is attached to an object to give information about it.

the definiendum is *a label*, the definiens is *a piece of paper or plastic that is attached to an object to give information about it* and the word *is* at the start of the third field simply joins the two halves together. This pattern accounts for 17% of the explanations. Some of the other patterns identified in the analysis behave similarly, but in most cases there are elements of the definiendum in the third segment or field of the explanation. For example, explanation 2 of *arrival* is:

> if you talk about the **arrival** of something new, you are referring to the fact that it has begun to exist or happen.

This has a definiendum consisting of *you talk about the* ***arrival*** *of something new*, with the definiens *you are referring to the fact that it has begun to exist or happen.* The point in the third field at which the definiens begins is marked by the word *you*, which echoes *you* in the first field. This high degree of symmetry, forced on the explanations by the philosophy adopted at the outset of the Cobuild project, provides a simple mechanical basis for the initial division of the explanation into definiendum and definiens.

1.1.2 Non-standard explanation types

Explanations which consist of only one field have no headword indicators and are simply a special form of usage note, as in

> Also used in questions and statements about size

under sense 1 of *big*. All other explanations have a similar structure to the standard three-field pattern, but either use it in a different sequence or contain other elements which require a more complex arrangement of the explanation text. There are those with only two sections, such as *listener*:

> people who listen to the radio are often referred to as **listeners.**

This pattern uses a reversed form of explanation in which the definiens precedes the definiendum. It seems to be a variation of the form "you use **headword** to refer to..." found, for example, in sense 1 of *mess*:

> you use **mess** to refer to something that is very untidy and dirty or disorganized.

The parsing algorithms for these explanations will simply involve rearranging the techniques used for the more common three-field versions of the same explanation types.

Explanations with more than three fields are generally simple extensions of the basic structure in which the headword is split for some reason. For example, the explanation of *acquaintance* in the phrase *make acquaintance*:

> when you **make** someone's **acquaintance**, you meet them for the first time.

is split by *someone's*, which does not form part of the phrasal headword. This produces five fields (split by the bold markers) and highlights the need to separate *someone's* from the surrounding phrase. Explanations which produce the maximum number of fields (twelve) are similar but more involved:

> if something **bores** you **to tears**, **bores** you **to death**, or **bores** you **stiff**, it bores you very much indeed

Many of the four-field explanations consist of cross-references, as in:

> **met** is the past tense and past participle of **meet.**

1.2 Initial words as a starting-point for pattern recognition

I wanted to explore the nature of the explanation structures in as objective a way as possible, allowing the computer itself to identify patterns as far as possible. I therefore began by using a standard utility to list the first words of the explanations in order of frequency of occurrence in the dictionary. The ten most frequent initial words are:

10189	if	1461	when
7749	a/an	1346	the
5139	headword	670	to
1810	someone/thing	456	your
1702	you	262	in

Total: 30,784 (97.98% of the total 31,418 explanations)

Words which are exact or near equivalents are amalgamated, and headwords which introduce their own explanations are treated as a single category. These top ten words (including the category 'headword') account for nearly 98% of all the explanations in the *Student's Dictionary*, which shows that there are relatively few ways of introducing an explanation.

The initial word frequency list provides a starting-point for the construction of a taxonomy of explanations based on simple linear patterns. Although each initial word can introduce more than one type of explanation, the number of patterns associated with each word seems to be fairly limited. The refinement of this initial classification into a structural taxonomy which can provide a basis for the development of the parser involved an attempt to identify the recurring structural patterns associated with a functional distribution of elements in each explanation type.

1.3 From initial word to recurring structural patterns

The list of initial words formed a crude framework for the next stage of exploration, the identification and analysis of structural patterns to determine the best parsing strategies. Some of the patterns were immediately obvious on a cursory examination of the data. For example, the explanation of *abattoir*

> an **abattoir** is a place where animals are killed for meat.

follows a very common pattern which can be generalised as:

> a/an **noun headword** is a/an ...

The simplicity of this pattern and its frequency in the dictionary (over 13% of all explanations) made it one of the first candidates for separation into its own parsing category.

Being able to identify patterns in this way was both efficient and rewarding, but a major problem became apparent once the first few obvious patterns had been identified. If the analysis were to be rigorously controlled, all patterns of explanation needed to be fully accounted for, but many patterns were too infrequent to be identified by such a superficial examination.

To ensure that this problem was properly dealt with, a routine was developed to isolate explanations which followed a perceived pattern, and to keep track of the rest. So, at any given time there was a file of all explanations within a particular subcategory of the dictionary whose patterns had not yet been fully identified. For example, in exploring the pattern:

> if you are **headword** you are...

a routine was developed to:

(i) extract explanations which begin with *if you are*, followed by any number of characters in the headword, followed by *you are*, from the main file and put them into a separate file, and

(ii)) extract the explanations which do not match this pattern from the original file into a new file of unanalysed explanations.

Constant line count reconciliations were carried out to ensure that no explanations had been lost because of incorrect command entry, poor pattern specification or other errors.

The identification of less frequent patterns proved harder to solve. However, once the most obvious patterns were eliminated, the more obscure ones began to reveal themselves. In cases of doubt, it was easy enough to go back to word frequency analysis, now applied to second or subsequent words in the explanations, for further clues.

2. Refining the methodology for identifying structural patterns

The patterns identified through the techniques discussed in the previous section were often adequate by themselves as a basis for developing parsing strategies. For example, the pattern

> if you are **headword** you are...

identifies a small group of explanations which match it exactly and which could therefore be subjected to a single parsing strategy.

The main failing of this simple pattern identification method as a complete basis for the development of parsing strategies is that it cannot distinguish between differences in patterns which are incidental to the project's objective and those which fundamentally affect the approach to be adopted.

For example, the pattern shown above accounts for only 130 explanations, but there is a similar pattern which only differs in the interpolation of material between the headword and the second *you are*, as in:

> if you are **adamant** about something, you are determined not to change your mind.

This pattern accounts for a further 482 explanations and requires only a slight alteration to the parsing strategy. All that is needed is to adapt the parsing strategy for the pattern "if you are **headword** you are ..." to allow an optional slot for non-matching cotext after the headword (see section 3.1).

The investigation of differences between patterns is essentially a manual process, but it is made much easier by the systematic arrangement of explanation types, which is required for the pattern-matching exercise.

2.1 The role of grammatical information

Many of the explanation types can be successfully parsed using strategies developed purely from the process of identifying patterns described above. In some, however, the correct approach to parsing can only be determined if the grammatical function of the headword is taken into account. For example, the general pattern "a/an **headword** text" can be used for both noun and adjective headwords and needs to be analysed in a different way in each case. This is not a major problem, however, since the dictionary contains full grammatical details for each headword. In the case of an explanation such as that of sense 3 of *absolute*:

> an **absolute** ruler has complete power and authority over his or her country.

the knowledge that the headword is an adjective makes it possible to distinguish the pattern of the text which follows the headword from its equivalent in an explanation such as that of *village*:

> a **village** consists of a group of houses, together with other buildings such as a church and school, in a country area.

Even if pattern-matching can provide the necessary analysis unaided, it is always useful to resort to grammatical information of this type as a further check.

3. From structural to functional analysis

I have already explained that the main objective of developing an automatic parser is to provide a functional rather than a structural analysis of Cobuild explanations. Having dealt with the way in which the main structural patterns were identified, I would now like to move on to the functional elements revealed by these patterns.

3.1 Functional elements

The strategies developed for writing Cobuild explanations produce a high degree of semantic equivalence between the definiendum and the definiens. However, the explanation text includes elements which form part of the 'framework' for explaining the headword, but which do not properly belong to either the definiendum or the definiens. For example, sense 1 of *label* is:

> a **label** is a piece of paper or plastic that is attached to an object to give information about it.

Here, the word *is* joins the definiendum *a label* to its definiens *a piece of paper or plastic that is attached to an object to give information about it* without forming part of either. Similarly, the explanation of *lunar* is:

> **lunar** means relating to the moon

In this case, the word *means* performs an equivalent function. These words can be seen as 'hinges' between the two halves of the explanation, with the headword preceding the hinge and the definiens following it being semantically equivalent in a straightforward way. Some explanations are more complex, for example sense 1 of *dreadful* :

> if you say that something is **dreadful**, you mean that it is very unpleasant or very poor in quality

Here, semantic equivalence can be considered on two levels. The headword *dreadful* is equivalent to *very unpleasant or very poor in quality*, but the whole stretch *you say that something is dreadful* is also equivalent to *you mean that it is very unpleasant or very poor in quality*. The expressions *you say* and *you mean* occupy similar positions and mirror each other, and so do the expressions *that something is* and *it is*. These components in themselves signal more semantic information than the simple *is* and *means* of the earlier explanations (see Hanks 1987). More important, however, they also provide an identifiable framing mechanism which enables us to establish the boundaries of the other components of the explanations, namely the headword and that part of the definiens which is semantically equivalent to it.

In order to proceed with the functional analysis of Cobuild explanations, the parser has to be able not only to identify the boundaries of the three major components of each explanation (the framework, the definiendum and the definiens), but also to identify the precise contribution of every element in each of the three functional components. Take for example the explanation of *astonished*, which follows the pattern "if you are **headword**, you are ...":

If you are **astonished**, you are very surprised about something.

This can be parsed as follows:

Hinge	: if
Cotext 1	: you are
Headword	: astonished
Matching cotext 1	: you are
Explanation	: very surprised about something.

The extended form of this pattern, discussed in section 2, includes a further category of cotext, as can be seen in the analysis of *adamant*:

Hinge	: if
Cotext 1	: you are
Headword	: adamant
Cotext 2	: about something
Matching cotext 1	: you are
Explanation	: determined not to change your mind

This analysis, which is more general than the one offered for *astonished* above, can be extended to cover the basic form of the first pattern. The output from a parse of *astonished* would then be:

Hinge	: if
Cotext 1	: you are
Headword	: astonished
Cotext 2	:
Matching cotext 1	: you are
Explanation	: very surprised about something.

The slot for cotext 2 would simply be left blank to show that this potential component is not present in this case. This principle of developing parsing strategies which can deal with all the potential functional components of a generalised explanation pattern and record their non-realisation in specific examples can make the dictionary an extremely powerful source of linguistic information.

Sinclair (1991) gives a more detailed account of the functional analysis of common explanation patterns.

3.2 Functional components v. literal elements

The starting point for the analysis was originally the initial word plus the most significant characteristics of the remainder of the text, often roughly in order of occurrence. However, some adjustment of literal text patterns had already crept in at a very early stage. For example, when the pattern-matching routine treats *a* and *an*, or *is* and *are*, as equivalent, it is concerned more with the type of text contained in the explanation than with the exact form of words. Similarly, the choice of those elements which are considered in pattern-matching and those which are ignored (for instance *about something* in the explanation of *adamant*), shows that there are several different forms of explanation text which can be seen as basically belonging to the same type of pattern.

This process of collating forms which appear to be different on the surface and grouping them together into functionally equivalent patterns extends to some explanations with different initial words, on the basis that they can be parsed using the same strategy. For example, *if* and *when* are both used to introduce explanations. They are not treated as equivalent in the metalanguage of explanation: *when* implies that a process or event is normal or even inevitable (*when the sun rises*), whereas *if* implies that an action or event is either less likely or less desirable. Apart from that, the two items basically introduce the same type of pattern:

> if you **illustrate** a point, you make it clear by using examples or stories.
> when you **laugh**, you make the sound which shows that you are happy or amused.

A single parsing strategy which treats *if* and *when* as variants of the same functional component can therefore be used for both explanation types.

4. Overview and evaluation of parsing strategies

Using the methodology described in sections 2 and 3, a number of basic parsing strategies have now been developed for Cobuild explanations. For reasons of space, they cannot all be discussed in this paper. However, I will attempt in this section to give a rough idea of the coverage achieved by some of the main strategies which are currently available and to suggest ways in which this coverage may be extended by refining the strategies in question.

4.1 Coverage

A single parsing strategy is capable of dealing with most variants of the explanation types "a/an/the **headword** is ..." and "**headword** is ...", which amount to nearly 11,000 dictionary entries (or 35%). Other strategies have also been developed for the following patterns:
(i) "if/when cotext1 **verb headword** ..." (covering another 5,700 entries, or 18%);
(ii) "someone/something who/that is **adjective headword** ..." (3,100 entries, or 10%);
(iii) "if subject is/are **adjective headword** ..." (2,000 entries, or 6%).
These four basic types, then, cover over 21,800 individual explanations, or nearly three quarters of the dictionary.

The remaining explanations include many forms which look dissimilar from each other but which have similar parsing requirements. Consider, for instance, the following explanations:

> if you say that a proposal or suggestion is **out**, you mean that it is unacceptable.
> if you describe a situation or event as **monstrous**, you mean that it is very shocking or unfair.
> if you refer to something bad as **glaring**, you mean that it is very obvious.

With minor variations, these can all be dealt with in the same way, covering another 400 explanations. As successive strategies deal with fewer explanation patterns, however, the efficiency of the process will clearly be reduced. It nevertheless seems likely that relatively little effort will be needed to cover most patterns.

4.2 Refinements

Two main refinements are proposed to enable more effective parsing to be achieved in the future.

4.2.1 The refinement of parsing strategies

Existing parsing strategies make assumptions about the nature of all the explanations which follow a particular type of pattern, on the basis of a sample of explanations of that type. These assumptions need to be tested to ensure that the basis of each parsing strategy is valid. For example, careful examination reveals that the explanation type:

a/an **noun headword** is ...

normally contains two elements in its third field: a superordinate, describing the general type of entity represented by the headword, and one or more discriminators which distinguish it from its co-hyponyms. Discriminators which follow the superordinate are introduced by one of a range of words, usually those which can introduce relative clauses. An initial list of these potential 'separators' was compiled on the basis of a mixture of observation and intuition. This list was then refined by searching all the explanations which followed the above pattern to determine which, if any, did not contain one of the words in the list. The new words brought to light by this process were used to augment the initial list. Explanations which did not prove amenable to parsing by this basic strategy were isolated. Either alternative strategies will be developed to deal with these explanations, or they will be rewritten to fit into one of the common patterns which are already amenable to parsing.

4.2.2 The refinement of explanations

Some forms of explanation are difficult to analyse automatically. Given that the purpose of this research is to make the dictionary accessible to the computer, it seems reasonable to consider the possibility of rewriting a small number of problematic, one-off explanations to make them more usable by the machine. For example, some explanations do not use a separator between the discriminator and the superordinate. The discriminator in the explanation of *artefact*:

an **artefact** is an ornament, tool, or other object made by a human being

begins with the past participle *made*, rather than with one of the separators used in explanations of this type, for example *which is*. Modifying the parsing strategy to detect past participles as separators could be both difficult and unreliable. Finding the explanations which pose this problem and rewriting them, on the other hand, may prove easier and more efficient in the long run.

4.3 Overlapping patterns and strategies

The analysis described so far uses a linear pattern-matching technique to identify and separate explanation types. This means that the first field of data in the explanation, the part before the headword, carries the most weight in determining the parsing strategy. This method is suitable for the analysis of the first components of the explanation, but it tends to separate explanations which have similar requirements in their later, third field stages. This issue needs further investigation to determine whether different parsing strategies can be adopted for the definiens (which typically, though not invariably, occurs in the third field) and the definiendum of the same explanation. This would involve generating a further set of patterns based on an analysis of the last components of the explanations. It could, however, significantly reduce the overall number of parsing strategies required for the dictionary as a whole.

4.4 Other information needed

Reference has already been made to the usefulness of grammatical information in distinguishing different types of explanations. This information is directly available for each headword in the dictionary. What would also be of great help is some simple part-of-speech tagging of other words in the explanation text. The explanation of the word *liqueur*:

> a **liqueur** is a strong, sweet alcoholic drink, often drunk after a meal.

includes the adjectives *strong*, *sweet* and *alcoholic* as discriminators of the superordinate *drink*, and this is a relatively frequent feature of explanations of this type. If this part of the explanation could be tagged, the distinction between the two elements of the definiens would be easier to make. This type of basic tagging is not expected to represent a significant problem.

It is important to remember that this information would then be used simply to refine the strategy for parsing the explanation into its functional components as part of a dictionary database. The objective of this work is not to carry out a full syntactic parse of the explanation as a piece of text, but only to identify the functional components needed to extract the necessary information.

5. Applications

5.1 Navigating the dictionary database for natural language processing

One of the main advantages of analysing Cobuild dictionary explanations automatically lies in the potential for using the dictionary as a database. Databases which hold information on words and phrases already exist of course, but these are designed to be accessed and interpreted by human beings, not by computers. Once the explanations are amenable to automatic parsing of the type described in this paper, the computer itself will have direct access to all the information which is contained, explicitly or implicitly, in the explanation text. For example, the dictionary can in principle provide enough information to allow the disambiguation of words in natural language. Provided the definiens is analysed in sufficient detail, the computer would be able to recognise lexical associations between words and phrases in a given text. The difference between this process and the use of information already coded into a dictionary relating to synonyms, antonyms, and so on is fundamental. The parsable dictionary will allow lexical relations and other features of words which are implied by the explanation to be identified, even though they have not been explicitly considered by the lexicographer, and even though they may not be obvious to native users of the language on a conscious level.

A parsable dictionary therefore provides access to all the information contained in the explanations about the meanings of words, and with this information one of the major problems of natural language processing, the disambiguation of words in context, may become much more tractable. Where a word has several senses, Cobuild dictionaries provide sufficient contextual clues to identify each sense in a given stretch of language. The automatic analysis of explanations can make this information directly available to the computer.

5.2 The automatic thesaurus

Apart from its obvious use in natural language processing, a parsable dictionary text also has many applications in lexicography. One of the more important applications is the automatic generation of a thesaurus. If superordinates, synonyms, co-hyponyms, antonyms and so on can be identified on the basis of a functional analysis of the components of each explanation, a draft text for a printed thesaurus can be generated automatically. It may need some manual refinement, but this process would certainly reduce the amount of human effort required. Moreover, a thesaurus generated on this basis should be more comprehensive than any manually produced version, and could simultaneously be used in computer-readable form.

5.3 Dictionary refinement

The need to rewrite some of the dictionary explanations to make them more amenable to automatic parsing has already been discussed, but this rewriting would be purely for the benefit of the parser. A related result of this analysis is the scope that it gives for checking dictionary explanations to ensure that they are also in the best possible form for the human reader. Alterations made to the explanation structures purely to improve computer efficiency or save programming time are unlikely to be reflected in future editions of the dictionary, but where the analysis highlights inconsistencies in the wording or format of explanations which could confuse the human reader (and particularly the learner), improvements should result. The analysis of explanation types should make both the identification and the correction of possible problems much more straightforward and comprehensive than would be the case with any manual check.

Once the dictionary is fully accessible to the computer as a result of the parsing process, a further form of check could be carried out. The explanations normally define one word or phrase in terms of a more general one, the superordinate or its equivalent, and the potential applications of this feature in the production of a thesaurus have already been discussed. If the dictionary is to be effective, the links between explanations should be complete, and all words used in the explanations should be defined in ways that allow the user to decode the explanation that uses them. Once the dictionary is set up as a navigable database, these links can be explored automatically, and any gaps or inconsistencies highlighted and remedied in future editions.

5.4 Dictionary translation

One of the projects currently being developed by Cobuild involves the translation of the monolingual English dictionaries into forms of bilingual dictionary, currently called 'bridge bilinguals'. Bridge bilinguals are designed for learners who are not yet competent enough in English to be able to cope with monolingual explanations. The explanations in bridge bilinguals are therefore in the user's mother tongue. For example, the Brazilian Portuguese version of the *Student's Dictionary* offers the following explanation for *naive:*

> Uma pessoa descrita como **naive** acredita que as coisas são muito menos difíceis do que realmente são.

The process of training teams in different countries to translate Cobuild explanations into a variety of languages has been made much easier and more consistent by access to a systematic analysis of explanation patterns. This analysis had already been performed as a first step towards the formulation of parsing strategies. By studying the list of explanation patterns provided by the parser, various teams of translators have been able to identify problems which are likely to recur in the course of translation and to devise consistent strategies for dealing with them.

6. Conclusion

The process of developing parsing strategies for the *Student's Dictionary* has provided, among other benefits, a far clearer understanding of the problems common to all dictionaries, as well as a better grasp of the special potential of the Cobuild explanation style for use in other computer applications. The explanations, when they are properly analysed, can be seen as not merely saying what words mean: through their specialised use of natural English language they also mean exactly what they say, and the parser should make these areas of meaning fully available to the computer for use in a variety of natural language processing and other applications.

Author's Address:
School of English, The University of Birmingham, Edgbaston, Birmingham, B15 2TT. U.K.

References

Alshawi, H. 1989. "Analysing the Dictionary Definitions". *Computational Lexicography for Natural Language Processing* ed. by B. Boguraev & T. Briscoe, 153-69. London & New York: Longman.

Hanks, P. 1987. "Definitions and Explanations". *Looking Up* ed. by J. M. Sinclair, 116-36. London & Glasgow: Collins.

Longman Dictionary of Contemporary English. 1978 (P. Procter, ed.), 1987 (D. Summers, ed.). Longman: Harlow.

Oxford Advanced Learner's Dictionary. 1989 (A. Cowie, ed.). Oxford: Oxford University Press.

Sinclair, J. M., ed. 1987. *Collins Cobuild English Language Dictionary.* London & Glasgow: Collins.

___________, ed. 1990. *Collins Cobuild Student's Dictionary*

___________. 1991. *Corpus, Concordance, Collocation.* Oxford: Oxford University Press.

Teaching, Text and Technology
A Hypermedia Environment

Margaret Allan
James Cook University of North Queensland, Australia

Abstract
This paper describes a project which explores hypermedia flexibility through the development of an audio-based hypermedia network for the study of discourse intonation by students of English as a second or foreign language (ESL/EFL). A series of studies with students in a small university language centre investigates ways in which students interact with a prototype intonation study network. A transcript of two students working with an early version of the program is discussed from a Vygotskian perspective, in terms of assisted performance and self-regulation of the study resource.[1]

0. Introduction

The focus of this paper differs from others in this section in that it describes development work with a strong pedagogic aim: to use technology to provide teachers and learners with access to information about language. In particular, I am concerned with how the results of linguistic analysis can be made available to students of English as a Second or Foreign Language (ESL/EFL). The use of computer technology is also different. Ready-made computer programs are used in this project with the purpose of developing prototype materials which exemplify a set of functions for new technology in the study of ESL/EFL. This paper describes the first stage of the ongoing research and development of an intonation study network, based on the work of David Brazil on the intonation of English discourse (Brazil 1985). The research described here will later inform a second stage which will involve developing other language study networks along similar lines. In all of this, the primary focus is on ways of facilitating interaction between teachers, learners, text and technology in relation to current knowledge about language.

The Cobuild project has demonstrated how the use of computer technology to search large corpora can produce evidence which in some cases refutes our intuitions about our actual use of language and the descriptions presented in our grammars (Sinclair 1991). The new combinations of hypermedia and networking technologies could have equally profound effects on how we teach and learn languages. Even the limited form of hypermedia technology described below indicates some fairly radical shifts in perspective when its use in language learning is considered. It makes it possible to provide different levels and forms of support for the exploration of evidence about spoken language and thereby encourage teachers and learners to adopt a data-driven approach which leads the latter to make their own investigations of language (Johns 1990).

The aim of this project is to assist the development of explicit knowledge of prosodic features of English. This is clearly an approach which subscribes to the value of language awareness work in ESL/EFL teaching. This is not to deny the importance of developing learners' abilities to use language communicatively. It is simply, in James Gee's terms, to be clear that teaching knowledge about language is one thing and developing communicative use of language is another (Gee 1989). The assumption here is that there is a place and a value for both in formal language instruction and that hypermedia technology has an important role to play in the development of explicit knowledge about language as one goal of instructed second language acquisition.

John Sinclair chairs the editorial board of *Language Awareness*, a new international journal whose title represents a broad area of concern among language educators in many fields today. The inaugural editorial expresses a commitment to the extension of "potential contact between movements in pure language research, and the evaluation and *incorporation of these into the educational process*." (The Editors 1992:2; my emphasis). The project described below was set up in that spirit.

1. The research project [2]

The current project was planned with the English Department of the University of Birmingham, with the aim of using hypermedia technology to create an index to phonological features and of exploring ways in which this technology may be used in the teaching of English to speakers of other languages. The naturalistic research paradigm (Lincoln and Guba 1985) provides the context in which case studies are being conducted on the basis of developing prototype materials for the study of intonation patterns in English discourse. The materials are for a hypermedia configuration which consists of speech recordings on compact disc (CD), linked via a CD-ROM drive to a hypertext program (Hypercard) running on a Macintosh computer. The project is concerned with exploring the interaction patterns of teachers and learners working with hypermedia tools in order to develop hypotheses about ways of structuring materials for the use of this new technology in studying language.

A series of studies, recently completed with students in a small university language centre, focused largely on practical questions about the properties of the technology and student interaction with the type of program described below. After further revision of the materials, a final case study will examine working hypotheses which have come out of the current study. The rest of this paper describes the linguistic focus, the technology and the theoretical framework. A transcript from the studies is then examined to make a first pass at an analysis of the data from a Vygotskian theoretical perspective.

2. The rationale for a focus on intonation

Following a long period of neglect of prosodic features in ESL/EFL teaching, there is a revival of interest in the teaching of pronunciation and in teaching it as a feature of communicative interaction (e.g. Pennington and Richards 1986; Morley 1991). Calls for pronunciation teaching to go beyond articulatory phonetics echo the pioneering work of Brazil, Coulthard and Johns (1980) who advocated David Brazil's interactional view of discourse intonation as a central feature of pronunciation studies. In view of the overall focus of this research, the teaching of intonation recommends itself as an area where there is a particularly strong need in the form of linguistic knowledge as well as methodological ideas.

There are some difficulties in attempting to provide this type of support. Intonation is a complex phenomenon which is not fully understood and there is no consensus on the function or functions of English intonation (Coulthard 1977). Brazil (1985) argues that the intonation system has communicative value independent of its co-occurrence with syntactic choices and over and above any attitudinal value it may have. He avoids the problem of accounting for continuous variation by seeking to account only for points of choice in the system, arguing that a decision to select one from a set of available options must have some significance. Brazil (ibid) presents a full account of the four option systems he identifies: prominence, tone, key and termination operating within the tone unit, which he calls the basic building block of speech.

In common with a number of other scholars who are currently investigating possible applications of Brazil's theory, I am taking it as the most satisfactory account available so far and attempting to explore its potential as a teaching tool and to provide a critical evaluation of the model on the basis of practical implementation (cf. Hewings 1990). At least three recent or forthcoming publications have used Brazil's model as a basis for developing teaching materials (Bradford 1988; Hewings forthcoming; Brazil forthcoming). The data for the current project comes from recordings produced for Brazil's forthcoming publication, which will consist of a printed text with accompanying audiocassettes.

3. The technology

Since the focus of this research is on spoken language, I shall confine my discussion here to audio technology, but the developments outlined below also apply, actually or potentially, to the recording of video, graphics and text. The digital recording of audio on compact disc is already well established. For example, in Britain, CD sales figures rose in six years from below one million to 50 million in 1990 (Erlichman, *The Guardian Weekly*, 1991:27). The significance of recording onto disc, as opposed to tape, is that it makes the stored media capable of rapid, random access, compared to the much slower linear access afforded by tape. The significance of digital rather than analogue recording is that audio signals are encoded in the same binary format as computer programs and therefore the two can readily be combined: the audio player is controlled via the computer screen so that the whole or parts of an audio recording are played (through a speaker or headphones) while linked sets of information appear on the screen. The textbook/audiocassette combination, currently the staple fare of ESL/EFL curricula, may soon be joined by some form of CD/computer program combination which puts text on the computer screen and makes audio data available at the click of a mouse or by touching a point on the screen.

If the computer program is a hypertext program, random access to the recorded audio data is mirrored by random access to different sections of the computer program. Hypertext provides a means of structuring information in a network of nodes so that someone working with a hypertext program can move around a database, making their own connections and choices. The term 'hypermedia' has been coined to describe the interfacing of a hypertext program with media storage systems so that the nodes in the computer program network may contain data in the form of audio or video recording, graphics and photographic images as well as text.

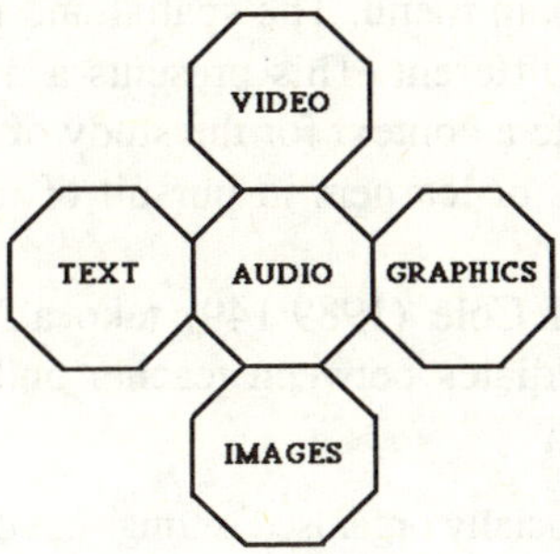

Figure 1. *A hypermedia network*

3.1 Hypermedia technology and language teaching

How significant is this technological trend for language teaching and for the design of language teaching materials? In the view of Olson and Bruner (1974:149) a new technology introduces new forms of symbolic representation and hence its own "unique pattern of skills for thinking about or dealing with the world". The characteristics of hypertext are giving rise to similar claims based on that technology's new forms of spatial and temporal array of information. Tuman (1992:61) talks of a "new spatial world" which is "non-sequential and multidimensional" and which "radically alters the connection between reader, author and text at the centre of print literacy". Much of this debate is highly speculative, as we have not yet had time to accumulate experience of using this technology in education. New technologies may be seen as potential catalysts for change in educational practice, but the reality seems to be rather that they become absorbed into prevailing institutional practice in a way which is more evolutionary than revolutionary (Mehan 1989).

In the study of language, hypermedia technology certainly has the potential to influence the way we use data such as recorded samples of speech. The design of today's print/audiocassette teaching materials assumes linear progression through examples, activities and explanation. The book is divided into units or chapters and written on the assumption that readers will start from Unit 1. Recordings are arranged on the tape in corresponding order and are supplementary to the text. With hypermedia technology the textbook is no longer the automatic starting point. Random access to any node in the network is a key characteristic of hypermedia technology and there is no obvious starting point, other than the main menu. The spatial and temporal organisation of material is thus radically different. This presents a materials designer with a very different task: to create a context for the study of a topic which allows for random access by teachers or learners in pursuit of their own interests in the topic.

Newman, Griffin and Cole (1989:149) take a Vygotskian view of the machine as a tool that mediates between teacher and student (see section 4) and they underline the need

> to be sensitive to the socially organised settings in which the machine might function in the classroom. It would not stand alone. It would be integrated into a setting in which it had a functional role.

In applying this view to the design of materials for a new technology which can perform a range of functions, I find it helpful to make an analysis of the pedagogic context which is loosely based on the conceptual framework proposed by Halliday (1978) for the context of situation, namely the field, tenor and mode of discourse. My own concern is with the field, tenor and mode of pedagogic activity. The nature of the subject and of the study activity - the field - will indicate which of the hypermedia choices available is the most suitable. In this case, the intensive study of spoken language required for the study of intonation clearly calls for an audio/text combination of the kind described above. An approach which gives teachers the role of facilitators and encourages learners to be researchers sets the tenor of the pedagogic interaction and suggests roles for the technology as a data source and as a practice instrument. The preferred mode of activity will be closely related to prevailing beliefs about methodology. For example, in a task-based approach the technology's role would be to provide data and stimuli for problem solving.

The convergence of computers and other electronic media has introduced a variety of new technological combinations, including interactive video, interactive multimedia and hypermedia. It is important to design for such technologies materials which (a) reflect unique characteristics of the particular technology they are part of, and (b) take account of the prevailing social and pedagogic context of ESL/EFL use. In addition to the factors already discussed, it is also necessary to draw on a theory of instruction which can provide a framework and a vocabulary for talking about how and why the materials work or don't work.

4. The theoretical framework: a Vygotskian theory of learning

The primary focus of this study is on the interaction between teachers, learners, spoken text and technology. Given this interactional focus, the work of those educators developing social theories of learning proved to be the most relevant to the design task. The work of Vygotsky[3] has attracted educators and psychologists who foreground social interaction as the primary influence on cognitive change. This approach has proved illuminating and productive in the work of Michael Cole and his associates for example, as well as in the work of Roland Tharp and Roger Gallimore.[4] I have found the work of these two groups particularly helpful in constructing a theoretical framework for the development of hypermedia materials and for the investigation of the ways in which students use these materials. This section provides a brief outline of the Vygotskian approach to learning and discusses its main concepts.

4.1 The zone of proximal development (ZPD)

In Vygotsky's view, learning to solve problems takes place through interaction with others (interpersonal activity) which leads later to the adoption of similar strategies in solving problems alone (intrapersonal activity). Such interactions take place within 'zones of proximal development'. These metaphorical spaces are central to Vygotsky's conceptualisation of the learning process. We all, adults as well as children, have different zones of proximal development for the wide variety of skills and knowledge we acquire and lose during our lives. In any area of learning there is a stage of development at which learners can perform, with expert assistance, tasks which they are not yet able to complete successfully on their own. The zone of proximal development represents the distance between this potential to perform a task independently and a learner's actual ability to do so. Using this concept, Tharp & Gallimore (1988:31) define teaching as "assisting performance through the ZPD". This conceptualisation highlights the need for teachers to be able to diagnose learners' capabilities in any subject area, to be prepared to adopt different strategies in providing support to different learners and to have resources which facilitate shifting goals - this last is the focus of my research with hypermedia technology.

An important implication of this spatial concept is that there is movement within it: learners may move in and out of four stages which represent shifting ratios of social and self-control. Stage 1 represents an early stage of learning where performance is assisted by *others.* In a formal instructional context, overall direction typically comes from the teacher. However, more capable peers are another source of support. I am particularly interested in the forms of support that could be built into a hypermedia program.

Stage 2 is where performance is assisted by the *self.* At this stage, learners are still in need of some support, but they are capable of judging for themselves what form of assistance they need and of making their own selections from the resources available to them. One implication of language learners becoming self-directed is that they are then in a position to decide for themselves whether it is appropriate for them to have the objective of reaching the next stage of automatic control. People have many different reasons for learning other languages and complete, automatic mastery of all aspects of a second language is often not a priority. Thus, from a Vygotskian perspective, general objectives can be expressed in terms of movement between stages within the zone of proximal development and learners can be expected to select their own objectives once they reach the second stage.

Stage 3 is one of *automatic control.* At this stage of confident control learners have no need of assistance and therefore, by definition, are no longer in the zone of proximal development. However, they have not necessarily left it for good.

Stage 4, the stage in which *regression* occurs, is a reminder that positions in the zone of proximal development are not fixed. It is quite possible to regress from automatic control to a point where some form of assistance is again required. Vygotsky postulated a recurrent cycle from self to other direction and back, thus differing radically from the Piagetian concept of fixed developmental stages.

4.2 Activity settings

Learning in this model is the result of assisted performance in pursuit of goals which are achieved through problem-solving tasks. Performance cannot be assisted in a vacuum and an activity setting provides the context for the performance of these tasks, with tools such as microcomputers providing one way of allowing learners to work without the constant supervision of teachers. Tharp & Gallimore (1988) describe classrooms organised around multiple learning centres for most of the school day, giving the teacher the opportunity to interact with small groups of children who come in rotation to the learning centre s/he manages.

4.3 The significance of a social theory of teaching and learning

As Newman et al point out, a social concept of cognitive change places the emphasis on whole tasks, in contrast to the more traditional structuring of tasks into predefined components, because "the breakdown into components is achieved in the social interaction rather than through a temporal sequence" (1989:153). The ZPD concept gives us a guiding social interactional principle which is to assist learners to take control of their own learning. This implies that they will then make their own decisions about how to approach tasks and which resources to use. This requires flexibility in the sequencing of tasks and moves us firmly away from the notion that there is one best way to approach a topic. Flexibility is a key requirement of the pedagogy here and it is also a key characteristic of hypermedia technology.

An example of the need for flexibility is the fact that not all learners interpret tasks in the same way. In relation to this project, I am already finding that there is a fuzzy border between performance of and knowledge about language. Tasks which are devised to promote the development of knowledge about language may become for particular learners a means of improving their performance in the language. This is what happens in the example presented in section 6 below.

5. Application to the present study

What does 'assisted performance' mean in terms of the present study? In this case I associate performance with the linguist's skill in perceiving, analysing and interpreting features of the intonation system in terms of their communicative value in discourse. Primary objectives concern the ability to recognise prosodic features. But, as I have already pointed out, one implication of reaching the stage of self-regulation of learning is that learners will pursue their own goals with regard to the level and type of performance they seek to achieve. Some L2 users of the program will undoubtedly wish to achieve productive control of the system, for others this will not be appropriate, while L1 speakers, for example teachers, will already have automatic productive and receptive control but will most probably have no conscious awareness of how speakers use intonation. Another factor which has to be taken into account is that intonation patterns in speech can be very difficult to hear. So different users will need different levels of assistance at different times, because they will be at different stages in their zones of proximal development.

The aim then is to develop a multifunctional hypermedia tool so that learners and their teachers can select the level of support they need at any one time to help them develop perceptual or analytic skills in relation to recorded speech. As a designer of hypermedia curriculum materials, I am creating contexts for study rather than linear sequences of instruction. Figure 2 represents an attempt to visualise the context as a network of functions. Note that all nodes are potentially connected to all others, although this is not apparent in a two-dimensional diagram.

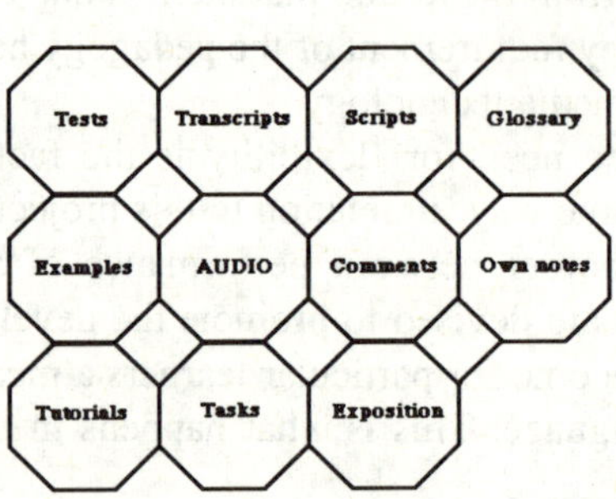

Figure 2. *A hypermedia intonation study network*

The network centres on the audio data. It is possible to replay, from within any node, any of the twelve tracks or segments of them, down to the level of individual tone units. As this is a prototype version produced without programming assistance, I have not incorporated all of the above elements in the program at this stage.

This program aims to support learners who may be at different points on a continuum between other-regulated forms of assistance and self-regulated forms of assistance. Bearing in mind that learners may choose to use the program's resources in their own way, there are some forms of assistance which are clearly quite strongly directive and which seek to provide maximum support: for example, the guided, sequenced study incorporated in a tutorial; the option to focus on one feature at a time; the availability of pre-selected and graded examples. At the other end of the continuum, the menu of choices provides for self-directed learners who are familiar with the options available and confident about using these resources to achieve their own goals in English pronunciation.

Guide to the network nodes:

Scripts	transcripts of each spoken text, with links to the glossary for selected words	**Glossary**	definitions relevant to the context in which they occur
Transcripts	intonation transcripts showing a) individual features b) full analysis	**Comment**	on the interactional significance of selected intonation choices
Examples	replay of individual tone units exemplifying each feature of the system; grouped for the level of difficulty	**Exposition**	of the linguistic description, feature by feature, with definitions of terms and links to examples
Tasks	problem-solving activities which set goals for users of the network	**Tutorials**	sequenced presentation of the system, with links to other nodes; a means of orienting new users to the network
Tests	for users to check their own abilities to perceive intonation features	**Own notes**	for users to record their own comments, and construct sequences for others to follow
Audio	12 tracks of speech recording - a mix of acted scenes and improvisation		

What is interesting here is that, even when they choose to have maximum guidance, as in the tutorial option, learners still have autonomy over replay of the recorded sound, something they do not have in a whole class, teacher-regulated setting. The transcript of two students working with the program, discussed in section 6 below, provides evidence of the use learners may make of such a function when it is placed under their control.

6. Working in a hypermedia environment

The transcript below provides an example of how two learners actually worked with an early version of the program. It comes from the final session of a series of three sessions which were designed to introduce a small group of young adult EFL learners in a university language centre to Brazil's model, via the technology described above. This recording is of two female students, one Japanese (H) and one Thai (G), who had chosen to work together in this session (their third). The task set was to prepare to make their own recording of all or part of one of the recorded texts; there was a choice of three at that point. Figure 3 shows the screen they were working with - one of a set of nine in an exercise focusing on rising and falling tones.

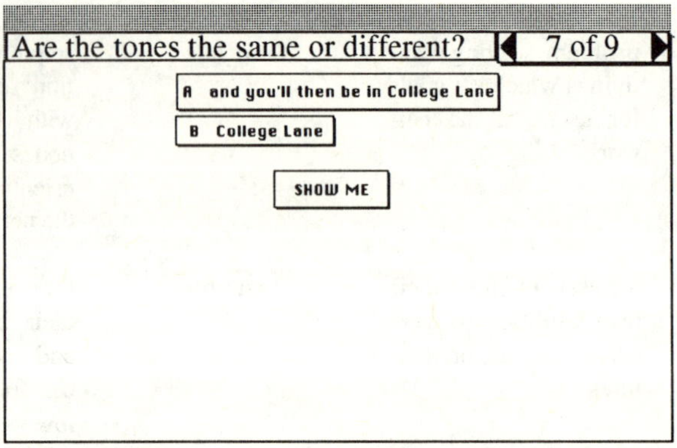

Figure 3. *A screen from a 'test yourself' set, with the SHOW ME option selected*

There was no instruction to repeat, but these two students, as almost all the others in the pilot study, chose to repeat as they listened. The following transcript[5] is taken from an audio recording of the session.

(Mandy has telephoned David to get directions to his house.)
David: // p 'n you'll THEN be in COLLege lane //
Mandy: // p COLLege lane //

(Replay from the disc is indicated in bold: RPD= replay David; RPM= replay Mandy; G and H utterances are numbered to show each attempt to produce a tone unit.)

RPD1:	// **p 'n you'll THEN be in COLLege lane** //
G1:	// p and YOU'LL then be in COLLege lane //
H1:	... and you ...
G2:	// p and YOU'LL then be in COLLege lane //
H2:	// p and YOU'LL then be in COLLege lane //
G3:	// p and YOU'LL then be in COLLege lane //
RPD2:	// **p 'n you'll THEN be in COLLege lane** //
G4:	// p COLLege lane //
H3:	... and you // p and YOU'LL then be in COLLege lane //
G5:	// p and YOU'LL then be in COLLege lane //
RPD3:	// **p 'n you'll THEN be in COLLege lane** //
G6:	// p COLLege //
H4:	// p and YOU'LL then be in COLLege lane //
G7:	// r+ COLLege LANE //
RPM1:	// **p COLLege lane** //
H5:	// p COLLege lane //
G8:	// p COLLege lane //
H6:	// p and YOU'LL then be in COLLege lane //
RPD4:	// **p 'n you'll THEN be in COLLege lane** //
H7:	//p and YOU'LL then be in COLLege lane//
H8:	// p and YOU'LL then be in COLLege lane //
G9:	// r and YOU'LL then be in COLLege lane //
G10:	// r+ LANE //
G11:	// r COLLege lane //
RPM2:	// **p COLLege lane** //
H9:	// p COLLege lane //
H10:	// p and YOU'LL then be in COLLege lane //
RPD5:	// **p 'n you'll THEN be in COLLege lane** //
G12:	// p and you'll THEN //
RPD6:	// **p 'n you'll THEN be in COLLege lane** //
G13:	// p and you'll THEN be // p in COLLege lane //
RPD7:	// **p 'n you'll THEN be in COLLege lane** //
H11:	// p and YOU'LL then be in COLLege lane //
RPD8:	// **p 'n you'll THEN be in COLLege lane** //
G14:	... and you'll ...
H12:	... and you'll the ...
G15:	... and you'll ...

H13:	... you'll
RPD9:	**// p 'n you'll THEN be in COLLege lane //**
H14:	// p and YOU'LL then be in COLLege lane //
G16:	Again
RPD10:	**// p 'n you'll THEN be in COLLege lane //**
G17:	... you THEN ...
H15:	... and YOU then ...
G18:	// p and YOU'LL then be in COLLege lane //
G19:	// ... and YOU'LL then be in coll ... //
G20:	// p and you THEN //
G21:	// p and YOU //
H16:	... you you ...
G22:	// p and you'll THEN //
G23:	// p and YOU'LL then //
RPD11:	**// p 'n you'll THEN be in COLLege lane //**
G24:	// p and you'll THEN // do again
G25:	// p and THEN //
H17:	... and YOU then be ...
RPD12:	**// p 'n you'll THEN be in COLLege lane //**
G26:	// p and THEN //
H18:	// p and THEN //
G27:	// p and you'll THEN // not // p and YOU then //
G28:	// p and you'll THEN //
H19:	... and YOU'LL ... THEN be in COLLege lane //
RPD13:	**// p 'n you'll THEN be in COLLege lane //**
G29:	// p and you'll THEN be in college lane // (LAUGHTER)
H20:	// p and THEN be in college lane //
G30:	// p and you'll THEN be // (LAUGHTER)
G31:	// p and you'll THEN be //
G32:	// p and you'll THEN be //

At this point they move on to the next screen.

6.1. An example of working in the zone of proximal development

In terms of the theoretical framework described in section 4 above, this transcript is an example of two students working in their ZPD for the perception of prominence and tone in English discourse. From their performance in this extract, it appears that they are both capable of hearing and accurately reproducing an English speaker's patterns of prominence and tone, but they have not yet achieved automatic control. In this particular instance, both students reproduced the tone accurately from the first repeat but there was a discrepancy in the onset (first prominent) syllable they produced compared to the speaker: David produced "and you'll THEN", while G and H produced "and YOU'LL then". This persisted up to the fifth replay of D's utterance, RPD5, when G makes *then* prominent in her eleventh repeat (G11). It is not until her twenty-fourth repeat that she seems to settle for that pattern and she then goes on to point it out to H (G27). H follows her lead in H20. They replay that one tone unit a total of 13 times, G repeats all or part of it 32 times, and for the last eight of those repeats she appears to have accurately perceived where the prominence was allocated. H is less certain, but is following G's lead by her last repetition (H20). G has moved forward in her ZPD and helps her companion to make a similar advance in her ability to perceive prominence in spontaneous English conversational speech. In the recording they made immediately after this session, H took the part of David and on that particular line she reverted to her earlier allocation of prominence to *you*. Without the assistance of G, she was not able to sustain her performance.

6.2. The other-to-self-regulation continuum

H and G control the computer technology to regulate the focus and pace of their learning activity. They are still other-regulated in the sense that they are directed to undertake a task. They are self-regulated in their control of the technology and in the choices they are able to make regarding their use of the program.

The most obvious feature of this session was the number of times the two students repeated one utterance. The pattern was similar for other tone units they replayed during this session, which lasted over 45 minutes. The simple logistics of a teacher/whole class activity setting preclude learner choices of this kind. Working directly with this program therefore opens up more possibilities of self-regulation, simply in terms of the choices available. However, I hypothesise that it has more significance than this. It must be part of achieving perceptual control of a second language to be able to identify one's own mishearings. Thus the possibility of being left to make those discoveries, instead of having a teacher point them out, must contribute to the achievement of self-regulation in a listening ability. I would argue that the breakthrough in perception that G makes is an example of a small advance towards automatic control of this aspect of perceiving English intonation.

Of course, not all learners are ready to handle autonomy. Many classrooms are heavily oriented to other-regulation of learning activity, and students become socialised into this mode of working. One student in my pilot study interview commented that lack of guidance was a danger for 'lazy students'. "No one control you ... you can go to another part. It doesn't in order. Then we should go one two three four five. It didn't do like that. It isn't good I think." A desire to be told what to do next seems clear evidence of other-regulation on the part of a student who has not had this kind of choice on offer before. The tutorial option may help in such cases.

6.3. An example of peer support

In an earlier sequence of trial sessions, students worked in groups of three for periods of fifteen minutes at a time. Of eight students interviewed after that trial, five said that they would have preferred to work alone, while two found group work helpful. One of those two reasoned that, in mixed nationality groupings, they make different mistakes and can help each other. In the trials from which this transcript comes, students had the choice of working on their own or with others by their own arrangement. G and H chose to work together on this occasion, although both had worked alone at other times. My reading of this session suggests that by working together these two students encouraged each other to persist in their efforts to perceive and reproduce tone units accurately. G, who turned out to be the more capable listener in this instance, provided the support H needed to perform beyond her own ability at that stage. The implications are that, given appropriate access to machines, tasks associated with the program should involve a mix of individual and group work in order to allow for individual reflection and peer-supported performance when students are undertaking private rehearsal of a task (Willis and Willis 1989).

6.4. Students' own goals and perceptions

Designing for random access is very different from designing for linear access. It is pointless to plan a step-by-step approach to a subject when the essence of the technology is that users can choose their own paths through the information. These two students chose the text that several had found the most difficult because it is an improvised dialogue with all the usual features of spontaneous interaction - overlapping turns, unfinished turns, fast delivery and so on (Hammond 1990). They then chose to spend time on the exercise that presented two tone units side by side and asked them to judge whether the tones were the same or different. In the process of working through this, they chose to repeat each tone unit. And in the extract quoted above, they are in fact concentrating on prominence rather than tone.

The two students undoubtedly had their own goals in relation to listening and speaking. They were about to sit an oral language test for university entrance, which both had previously failed. This presumably contributed to the fact that they continued well into their lunch break, although there was no pressure on them to do so - another example of behaviour associated with self-regulation.

H and G took evident pleasure in listening to the playback of the recording they went on to make and seemed to feel that they had achieved something. This seems to be what G is expressing in her comment about this particular session: "The last time I did it I really like it ... We enjoyed it very much. I think that is perfectly. We used for a long time and then we put the tape on (and recorded ourselves) and then we listened to it. I didn't believe it - oh, is good. Pronunciation is better."[6]

7. Conclusion

John Sinclair (1992) signals the profound effect that the advent of computers will have on the study of language through the application of analytic software to large text corpora. He also looks forward to the time when personal computers will be capable of running this software and therefore of making both the tools and the information they gather more readily available to the general public. The main concern of the project outlined in this paper is to include ESL/EFL teachers and learners among those who are able to use such linguistic resources for their own, diverse purposes. The intonation study network is seen as one, specialised tool, fulfilling a particular set of language awareness functions. It will eventually form part of a kit of networked pedagogic tools designed to support the study of lexico-grammatical as well as prosodic features of spoken language: a language study network which links directly into the new agendas John Sinclair has set through his pioneering work in corpus linguistics.

We have here a historical convergence: technology which could give us more direct access to the results of current linguistic research; new knowledge about real language emerging from corpus linguistics; a social theory of pedagogy and cognition which shifts the perception of learning away from Piagetian psychology and makes it more accessible to observation by foregrounding its interpersonal character. These possibilities present a challenge to our ability to reconceptualise pedagogic practice. I have described a first attempt to bring together particular theoretical, technological and empirical perspectives in the conviction that it is through pedagogic practice and application rather than grand theory that breakthroughs will be made.

Author's Address:
James Cook University of North Queensland, School of Education, Townsville Qld 4811, Australia.

Notes

1. I owe a considerable debt to my colleague Allan Luke whose detailed comments on several drafts of this paper provided much-needed support in my journey towards this final version. I also thank Joanna Channell and Martin Hewings for their helpful comments. Any remaining errors are my own.
2. The groundwork for this project was laid during a study leave attachment to the Department of English, University of Birmingham. Dr David Brazil and Richard Cauldwell had central roles in the development of the project outline and Richard Cauldwell went on to produce a prototype research tool using Voyager software to provide access to intonation transcriptions of spoken data recorded on CD. We were greatly helped in our initial struggles with programming by Tom Davies, who was generous with his time and impressive in his ability to make Hypercard do what we wanted it to. I also wish to acknowledge, for their contributions in kind, Cambridge University Press who granted permission to use audio recordings produced for Brazil (forthcoming), Nimbus Records who transferred these to CD and New Media Productions who oversaw this operation and purchased the Voyager software for us.
3. The seminal work of the Russian theorist Lev Semenovich Vygotsky (1896-1934) has become known in English-speaking countries largely through the efforts of North American psychologists and educators who have sought to explore and develop his interdisciplinary perspective on the human sciences. Good introductions to this work are: Kozulin (1990); Moll (1990); Newman et al (1989); Tharp & Gallimore (1988); Wertsch (1985a; 1985b). See Vygotsky (1978; 1986) for translations of his major work. Sinha (1989) is an excellent review of current interpretations of Vygotsky's method and theory. Applications to second language acquisition are discussed by Foley (1990), Frawley & Lantolf (1985), John-Steiner (1985), Lantolf & Ahmed (1989), Lantolf & Frawley (1985) and Roy (1989).
4. For detailed accounts of these works, see Newman et al (1989) and Tharp & Gallimore (1988).
5. Transcription conventions:

tone unit boundaries	//
incomplete tone unit	...
capitals, underlined	tonic syllable (carries tone movement)
capitals, not underlined	onset syllable (prominent but no tone movement)
tones:	p = fall; r = fall-rise; r+ = rise

6. Interviewer clarifies that she didn't believe it was her voice.

References

Bradford, B. 1988. *Intonation in Context*. Cambridge: Cambridge University Press.

Brazil, D. 1985, 1992. *The Communicative Value of Intonation in English*. Discourse Analysis Monograph no. 8. Birmingham: University of Birmingham, English Language Research.

______. Forthcoming. *Advanced English Pronunciation*. Cambridge: Cambridge University Press.

______ M. Coulthard & C. Johns. 1980. *Discourse Intonation and Language Teaching*. London: Longman.

Brown, G. 1977, 1990. *Listening to Spoken English*. London: Longman.

Coulthard, M. 1977, 1985. *An Introduction to Discourse Analysis*. London: Longman.

The Editors. 1992. "Language Awareness: Wat is Dat?". *Language Awareness* 1,1. 1-3.

Erlichman, J. 1991. "Sudden Death of the Long-playing Record." *The Guardian Weekly*. 17.3.1991.

Foley, J. A. 1990. "Task-based Approaches to Language Learning from the Learner's Point of View." *Language and Education* 4,2. 81-102.

Frawley, W. & J. P. Lantolf. 1985. "Second Language Discourse: A Vygotskian Perspective." *Applied Linguistics* 6,1. 19-44.

Gee, J. P. 1990. *Social Linguistics and Literacies: Ideology in Discourses. (= Critical Perspectives on Literacy and Education.)* London: The Falmer Press.

Goody, J. 1977. *The Domestication of the Savage Mind*. Cambridge: Cambridge University Press.

Halliday, M.A.K. 1978. *Language as Social Semiotic*. London: Edward Arnold.

Hammond, J. 1990. "Is Learning to Read and Write the Same as Learning to Speak?" *Literacy for a Changing World* ed. by F. Christie, 26-53. Hawthorne: Australian Council for Educational Research.

Hewings, M. 1990. "Introduction". *Papers in Discourse Intonation* ed. by M. Hewings, 1-4. Discourse Analysis Monograph 16. Birmingham: University of Birmingham, English Language Research.

______. Forthcoming. *Pronunciation Tasks: A Course for Pre-intermediate Students*. Cambridge: Cambridge University Press.

John-Steiner, V. 1985. "The Road to Competence in an Alien Land: a Vygotskian Perspective on Bilingualism". *Culture, Communication and Cognition: Vygotskian Perspectives* ed. by J. Wertsch, 348-71. Cambridge: Cambridge University Press.

Johns, T. 1991. "Should you be Persuaded: Two Examples of Data-driven Learning". *ELR Journal (New Series)* 4. 1-16.

Kozulin, A. 1990. *Vygotsky's Psychology. A Biography of Ideas*. Hemel Hempstead: Harvester Wheatsheaf.

Lantolf, J. P. & M. K. Ahmed, 1989. "Psycholinguistic Perspectives on Interlanguage Variation: A Vygotskian Analysis." *Variation in Second Language Acquisition* (Volume 2) ed. by S. Gass, C. Madden, D. Preston & L. Selinker, 93-108. Clevedon: Multilingual Matters.

Lantolf, J. P. & W. Frawley. 1985. "On Communication Strategies: A Functional Perspective". *Rassegna Italiana di Lingua Applicata* 17,2-3. 143-57.

Lincoln, Y. S. & E. G. Guba. 1985. *Naturalistic Inquiry*. Beverly Hills: Sage Publications.

Mehan, H. 1989. "Microcomputers in Classrooms: Educational Technology or Social Practice?" *Anthropology and Education Quarterly* 20. 4-22.

Moll, L. C, ed. 1990. *Vygotsky and Education. Instructional Implications and Applications of Sociohistorical Psychology.* Cambridge: Cambridge University Press.

Morley, J. 1991. "The Pronunciation Component in Teaching English to Speakers of Other Languages." *TESOL Quarterly* 25,3. 481-520.

Newman, D., P. Griffin & M. Cole. 1989. *The Construction Zone: Working for Cognitive Change in School (= Learning in Doing: Social, Cognitive, and Computational Perspectives.)* Cambridge: Cambridge University Press.

Olson, D. R. & J. S. Bruner. 1974. "Learning through Experience and Learning through Media." *Media and Symbols: The Forms of Expression, Communication and Education* ed. by D. R. Olson 125-50. Chicago: University of Chicago Press.

Pennington, M. C. & J. C. Richards. 1986. "Pronunciation Revisited." *TESOL Quarterly* 20,2. 207-25.

Roy, A. M. 1989. "Developing Second Language Literacy: A Vygotskian Perspective". *Journal of Teaching Writing* 8,1. 91-98.

Sinclair, J. 1991. *Corpus, Concordance, Collocation. (= Describing English Language.)* Oxford: Oxford University Press.

______. 1992. "The Automatic Analysis of Corpora". *Directions in Corpus Linguistics. Proceedings of Nobel Symposium 82, Stockholm 4-8 August 1991* ed. by J. Svartvik, 379-97. Berlin and New York: Mouton de Gruyter.

Sinha, Chris. 1989. "Reading Vygotsky." *History of the Human Sciences* 2,3. 309-31.

Tharp, R. G. & R. Gallimore. 1988. *Rousing Minds to Life: Teaching, Learning, and Schooling in Social Context.* Cambridge: Cambridge University Press.

Tuman, M. 1992. *Word Perfect: Literacy in the Computer Age (= Critical Perspectives on Literacy and Education.* London: Falmer Press.

Vygotsky, L.S. 1978. *Mind in Society: The Development of Higher Psychological Processes* ed. by M. Cole, V. John-Steiner, S. Scribner, & E. Souberman. Cambridge, Mass.: Harvard University Press.

______. 1986. *Thought and Language* ed. by A. Kozulin. Cambridge, Mass.: The M.I.T. Press.

Wertsch, J., ed. 1985a. *Culture, Communication and Cognition: Vygotskian Perspectives.* Cambridge: Cambridge University Press.

______. 1985b. *Vygotsky and the Social Formation of Mind.* Cambridge, Mass.: Harvard University Press.

Willis, D. & J. Willis. 1987. "Varied Activities for Variable Language". *ELT Journal* 41,1. 12-18.

Index